Archaeobotanical studies of past plant cultivation in northern Europe

Advances in Archaeobotany

Volume 5

Website

www.archaeobotany.eu

ISSN

2405-7452

Contact the Publisher

Barkhuis
Kooiweg 38
9761 GL Eelde
the Netherlands
info@barkhuis.nl
www.barkhuis.nl

Archaeobotanical studies of past plant cultivation in northern Europe

EDITED BY

Santeri Vanhanen & Per Lagerås

Barkhuis Publishing
Groningen 2020

Book design: Roelf Barkhuis
Frontcover desgin: Santeri Vanhanen
Frontcover image: Santeri Vanhanen & Per Lagerås
Copy editor: Suzanne Needs-Howarth

Copy editing was subsidized by The Archaeologists, National Historical Museums, Sweden; Arkeologisk Museum, University of Stavanger, Norway; and Department of Archaeological Science and Conservation, Moesgaard Museum, Denmark.

ISBN 9789493194113

Table of contents

Introduction to
Archaeobotanical studies of past plant cultivation in northern Europe

Santeri Vanhanen[1] & Per Lagerås[2]

Author addresses and e-mail:
[1] The Archaeologists, National Historical Museums, Sweden, santeri.vanhanen@arkeologerna.com
[2] The Archaeologists, National Historical Museums, Sweden, per.lageras@arkeologerna.com

Agriculture has a long and successful history, and today it is practised in most parts of the world. It has been, and still is, the economic basis for all populous societies, and its long-term history is tightly linked to social development and to cultural traditions and diversity.

The study of agricultural history is a fundamental part of historical and archaeological research. First, it provides important contributions to the study of social organisation, cultural contacts, food traditions, social stratification, technological development, settlement geography, migration, colonisation, abandonment and population development. Second, it combines perspectives from both the humanities and the natural sciences by studying past interaction between humans and their environment. Third, it reveals ecological aspects of past societies and their impact on plants, animals and natural resources, and of their dependency on climate. It also contributes with historical perspectives on present-day environmental concerns, including those about the future.

The history of agriculture stretches back long before the earliest written records, and the study of it thus depends on archaeological investigations, including different empirical and analytical methods. Most important methods, or subdisciplines, in this regard are zooarchaeology (the study of animal bones and teeth) and archaeobotany (the study of plant remains), providing insights into past animal husbandry and plant cultivation, respectively. The source materials for these methods are routinely collected from archaeological sites during excavation, but also from peat and soil profiles and other natural historical archives.

This book presents archaeobotanical studies on past plant cultivation in northern Europe. The studies cover the Nordic countries and adjacent parts of the Baltic countries and Russia, and they span more than 5 000 years of agricultural history, from the Neolithic to the Middle Ages. They highlight and discuss many different aspects of early agriculture, from the first introduction of cultivation, to crop choices, expansions and declines, climatic adaptation, and vegetable gardening. Together, they testify to the great importance of agriculture, even at these northerly latitudes. They also reveal a great agricultural diversity, both in time and in space.

The diversity may, to a large degree, be explained by contrasting environmental and climatic conditions. From south to north, the area covered by this book ranges from temperate deciduous woodland (today to a large degree replaced by agricultural landscapes), via coniferous taiga, to arctic shrub-tundra. Mean temperatures typically decrease from south to north, but they are also affected by elevation, which is most evident in the mountainous areas of Norway and northern Sweden. Another important factor influencing vegetation and land use is the range from oceanic to continental climatic conditions. Typical oceanic conditions are found along the North Atlantic coast of Norway and in Iceland, where summers are relatively cool but long, due to mild temperatures in spring and autumn. Typical continental conditions, with warm summers and cold winters, are found in eastern Finland, the Baltic countries and Russia.

In addition, the area shows a diversity of bedrock, ranging from old granites and gneisses, which dominate much of Finland and Sweden, to younger sedimentary rocks, such as sandstones, marls, limestones, which dominate Denmark and parts of southern Sweden, Norway and the Baltic countries. The mountain range of Sweden and Norway shows a complex mix of folded and transformed bedrock, whereas Iceland and other Atlantic islands are characterised by young volcanic bedrock.

The most common type of Quaternary deposit is till deposited by the Weichselian ice sheet. Till on old gneiss and granite is usually sandy and poor in nutrients (but rich in stones and boulders), whereas till on sedimentary bedrock may be clayey, calcareous and much more fertile. Glaciofluvial deposits of sand and gravel are widespread, whereas clay sediment is restricted to some lowland plain regions.

This climatic and environmental diversity has provided very different preconditions for agriculture, which is reflected in a great variation in both the onset and the development of plant cultivation within this wide

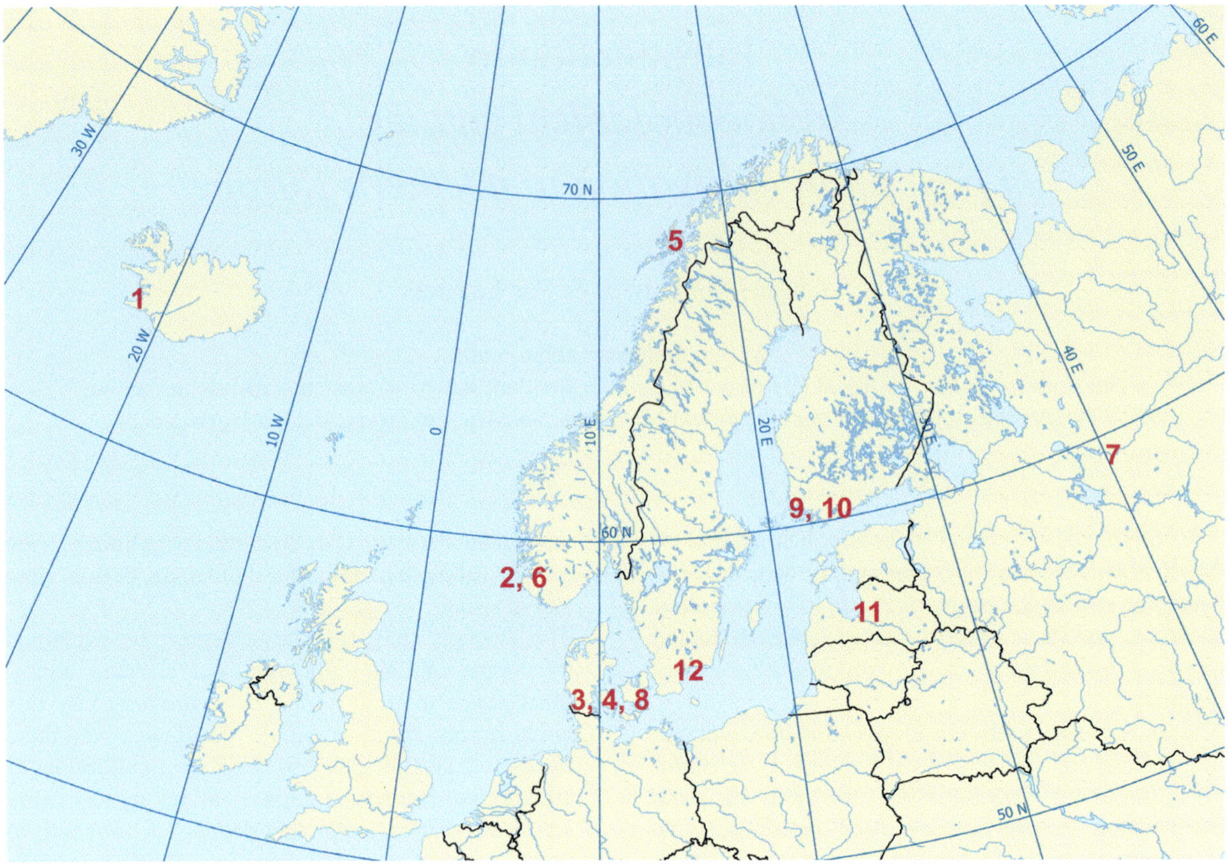

Figure 1. Location of geographical areas discussed in the articles. 1. Dawn Elise Mooney and Lísabet Guðmundsdóttir; 2 Eli-Christine Soltvedt; 3 Marianne Høyem Andreasen; 4 Peter Mose Jensen et al. 5 Christin Eldegard Jensen; 6 Sara Westling and Christin Eldegard Jensen; 7 Wiebke Kirleis et al.; 8 Kirstine Krath Helweg; 9 Santeri Vanhanen; 10 Terttu Lempiäinen et al.; 11 Mindaugas Grikpėdis and Giedre Motuzaite Matuzeviciute; 12 Per Lagerås and Mikael Larsson. (map: Santeri Vanhanen)

region. Cultivation has been conducted continuously for six thousand years in Denmark and southern Sweden but has a shorter history farther north and east. In many areas, the establishment of agriculture appears to have been a slow process, and harsh climate has resulted in frequent retreats.

Archaeobotany is an expanding field in northern Europe. Today, the study of plant remains has become an integral part of archaeological investigation, and plant remains are retrieved from numerous excavations. Archaeobotany in northern Europe includes a diversity of scientific approaches and publication traditions. Archaeobotanical investigations can be included in archaeological reports or published in local or international books or journals. Most of the reports and a large proportion of the publications are written in local languages. This plenitude of publication channels makes it difficult for a northern European, let alone an international scholar, to track the development of ancient cultivation in this northern region. Our aim with this book is to fill this gap and to showcase archaeobotany dealing with past plant cultivation in northern Europe.

The book consists of twelve chapters, which cover a wide geographic area, from Iceland in the west to the north-western Russian taiga zone in the east and from Denmark and Lithuania in the south to northern Norway in the north (Fig. 1). The chapters also cover a long time span, from the Neolithic to the Middle Ages.

Dawn Elise Mooney and Lísabet Guðmundsdóttir discuss a large find of barley grain from a Viking Age settlement in south-western Iceland, proposing that barley, and probably also flax, were cultivated locally in Iceland during the Viking Age and were not only imported from Norway.

Eli-Christine Soltvedt discusses the earliest cultivation in the region of Rogaland, Norway. Cultivation was introduced during the late 3rd millennium BC, increased during the Late Neolithic, but decreased thereafter, during the Late Bronze Age.

Marianne Høyem Andreasen reviews finds of naked wheats in Denmark from the Neolithic onwards. Based on the widespread distribution but often small numbers of the finds, she concludes that the finds indicate local cultivation of a crop suited for special purposes.

Peter Mose Jensen, Livija Ivanovaitė and Anja Vegeberg Jensen discuss finds of charred cereal grains from Denmark dated to the Late Neolithic and the Bronze Age. They identified macroremains from an unusually large storage find dated to the Bronze Age, which included the earliest large concentration of hulled barley in Denmark. The authors show that different cereal crops were stored in different containers.

Christin Eldegard Jensen discusses new finds related to the earliest agriculture close to the Arctic Circle in northern Norway. She shows that the earliest agricultural expansion in that region was during the Late Bronze Age, followed by alternating phases of abandonment and expansion.

Sara Westling and Christin Eldegard Jensen presents a large find of rye from Norway, dated to the 6th century AD. They discuss the find in relation to the 6th-century crisis, and whether people tried to adapt to a cooler climate by introducing a new crop.

Wiebke Kirleis, Magdalena Wieckowska-Lüth, Henny Piezonka, Nadezhda Nedomolkina, Sebastian Lorenz, Vanessa Elberfeld and Jens Schneeweiß present new data on plant gathering and cultivation in the north-west Russian taiga zone. They discuss the importance of yellow water lily seeds as a prehistoric food supply and show that the earliest evidence of cultivation in the region dates from the Middle Ages.

Kirstine Krath Helweg reviews current knowledge regarding medieval garden cultivation in Denmark and southern Sweden. She shows how both macrofossil studies and written sources can make important contributions within this field.

Santeri Vanhanen revisits a large collection of archaeobotanical material from sites in south-western Finland dated to the early first millennium AD. He found several crops, dominated by barley, as well as remains of hemp, which indicates contacts between Finland and Continental Europe.

Terttu Lempiäinen, Maija Helamaa, Heli Lehto, Ulla Moilanen, Markku Oinonen, Sakari Salomaa and Kari Uotila discuss cultivation in south-western Finland from the Late Bronze Age to the Middle Ages and present evidence of early barley and naked-wheat cultivation in the region.

Mindaugas Grikpėdis and Giedre Motuzaite Matuzeviciute review old and new evidence of cultivated plants in the Baltic countries from the Neolithic to the 14th century AD, creating a new basis for future research.

Per Lagerås and Mikael Larsson present Iron Age finds of emmer and spelt in Sweden and show that at least emmer was grown locally in some places, even long after the general decline of these crops. They discuss whether the presence of these ancient crops during the Iron Age could indicate influences from the east.

Barley cultivation in Viking Age Iceland in light of evidence from Lækjargata 10–12, Reykjavík

Dawn Elise Mooney[1], Lísabet Guðmundsdóttir[2]

Author addresses:
[1] Arkeologisk Museum, University of Stavanger,
Peder Klows gate 31A, PB 8600, Forus. 4036 Stavanger, Norway
[2] Fornleifastofnun Íslands, Bárugötu 3, 101 Reykjavík, Iceland

E-mail: dawn.e.mooney@uis.no

Abstract

Recent excavations of a Viking Age settlement at Lækjargata, in the centre of modern-day Reykjavík, produced one of the largest assemblages of charred botanical remains found so far in Iceland. Iceland lies at the northern limits of productive arable agriculture today, however this was not always the case in the past, and there has been much debate surrounding the history of cultivation on the island. This paper addresses the question of whether the cereal remains found at early Icelandic sites are likely to have been locally grown or imported. Often only small quantities of charred plant macrofossils are recovered from Icelandic archaeological sites, and therefore discussions surrounding agriculture have frequently relied on evidence from palynological studies, landscape surveys, and historical documentary sources. In contrast, the assemblage from Lækjargata included large quantities of barley grain, along with associated weeds, edible wild plants, and a single seed of flax, which is thought to have been rare in Icelandic cultivation systems. Comparison of this assemblage with others from Iceland and the North Atlantic region suggests that the grain was cultivated close to the settlement, and prepared for storage on-site. These findings add significant weight to a growing body of evidence for the early cultivation of barley in south-western Iceland, and suggest that some Settlement-era Icelandic communities produced a significant proportion of the cereals they consumed, rather than relying on imports. The field would benefit from more comprehensive archaeobotanical sampling and analysis, in order to determine the extent to which the trends identified here can be applied elsewhere on the island.

Introduction

Iceland lies on the northerly limits of modern arable agriculture (Reykdal et al. 2014), and its rugged landscape, short growing season, and poor climate must have rendered cereal cultivation a challenging pursuit throughout its history. Due to the island's late colonisation, in the 9th century AD, its history of cultivation is shorter than that of southern Scandinavia (Rowley-Conwy and Legge 2015), but the subject still provokes debate. There has been a tendency to assume that large-scale grain import was necessary in the past. Zori et al. (2013: 154) assert that 'high-status Icelanders were unable to grow adequate amounts of grain to support their needs' (see also Gelsinger 1981), and there is historical documentary evidence to support the practice of importing cereals or cereal products. A letter from Pope Celestine III to the Archbishop of Norway, for example, approves the export of flour to Iceland in the late 12th century AD (Sigurdsson 1857–76: 291–295). Grain imports are certainly common in modern Iceland: in 2014, Icelandic farmers produced just under 14,000 tonnes of barley grain, while imports of cereal grain to the country stood at 79,472 tonnes (Martin et al. 2016: 11).

There is, however, substantial evidence that cereal cultivation, especially of *Hordeum vulgare* L. (barley), did take place in Iceland during the Viking Age and early medieval period. Yet most of the archaeobotanical studies conducted on Icelandic material have focused on small assemblages. The lack of larger-scale analyses is often attributed to the poor preservation of fragile charred plant macrofossils in soils subject to extreme temperature fluctuations (Zutter 1999; Martin 2014). Therefore, much of the available evidence for cultivation in Iceland derives from historical documentary sources, place names, studies of ancient fields, and palynological investigations (Guðmundsson et al. 2004). The following text presents the results of an archaeobotanical analysis from the site of Lækjargata 10–12, in the centre of

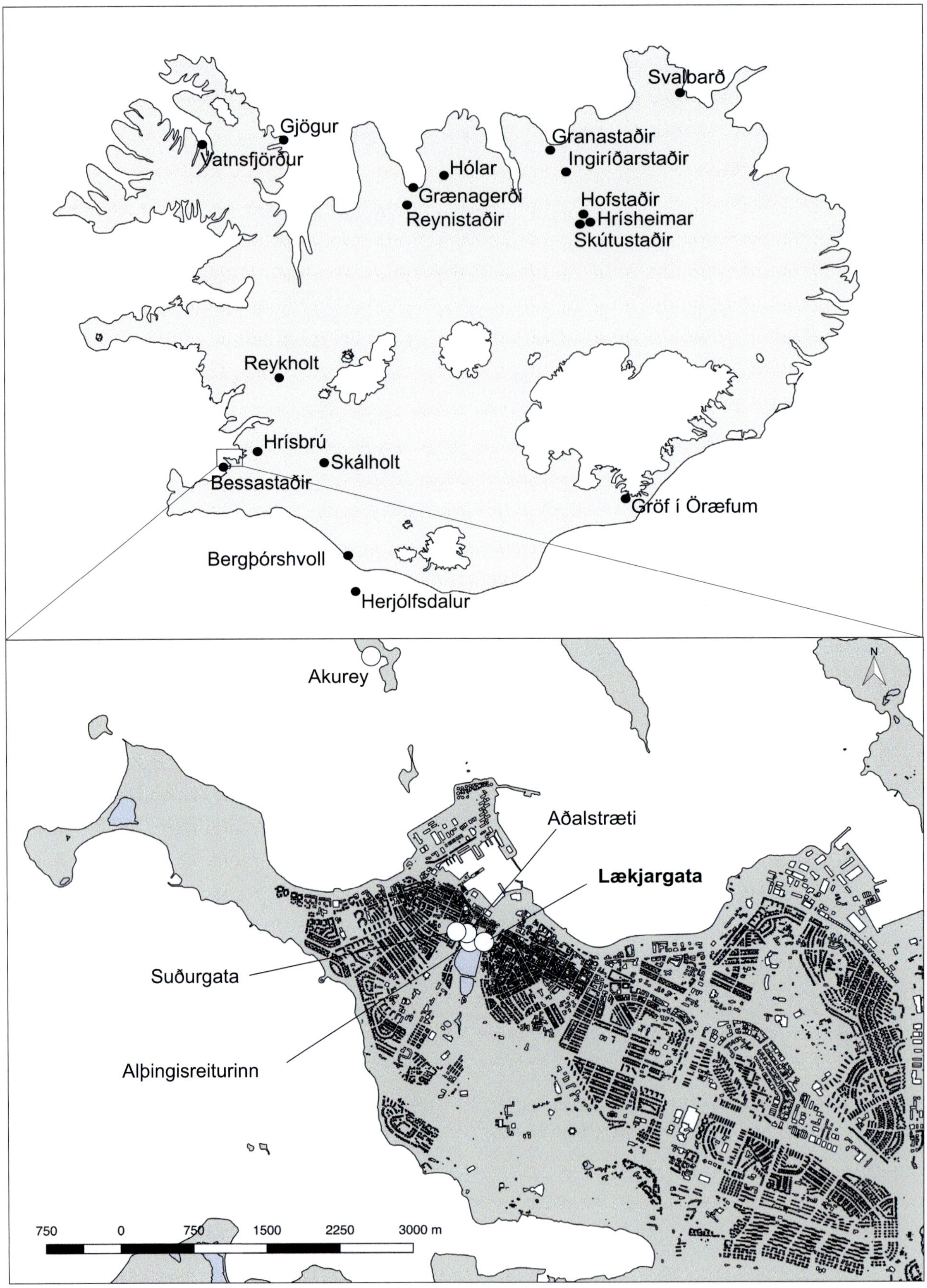

Figure 1. Map showing the location of sites mentioned in the text within Iceland, with an inset of the Reykjavík area showing the location of Lækjargata and nearby sites. (map: Dawn Elise Mooney and Lísabet Guðmundsdóttir)

modern-day Reykjavík (Fig. 1). The plant macrofossil assemblage from this site was unusually large, and provides an opportunity to examine direct evidence of arable agriculture in the 9th–11th century AD in comparison with studies from contemporary sites. The site will be described below, but first it is necessary to provide a brief history of arable agriculture in Iceland and a background to the discussions of cereal cultivation on the island.

The archaeology of cereal cultivation in Iceland

Proxy records and historical sources indicate that the climate of Iceland during the Settlement (c. 870–930 AD) and Commonwealth (930–1262 AD) periods was suitable for cereal cultivation, but that conditions worsened through the medieval period leading up to the onset of the Little Ice Age (Patterson et al. 2010; Vésteinsson et al. 2002). It is likely that this climatic deterioration played a role in the cessation of cereal cultivation, along with increased trade with Europe after Iceland came under the rule of the Norwegian crown in AD 1262 (Karlsson 2000; Zori et al. 2013).

In order to cultivate cereal crops, the settlers would have had to bring seed grain with them. The settlers of Iceland originally came from Scandinavia and the British Isles (Helgason et al. 2000), particularly western Norway (Karlsson 2000). Barley has been one of the most important cereal crops in Scandinavia since the Neolithic (Robinson 1993), and from the Roman Iron Age onwards *Hordeum vulgare* var. *vulgare* (hulled barley) was the main cereal crop in Norway (Prøsch-Danielsen and Soltvedt 2011; Bjørnstad 2012). This dominance is possibly due to changes in manuring practices (Viklund 1998), or simply due to the fact that hulled barley is very tolerant to cool, damp conditions, in terms of both growth and storage (Prøsch-Danielsen and Soltvedt 2011). It is hulled barley which is therefore most likely to have been brought to Iceland as seed grain by the Norse settlers.

There is some debate as to the original extent of cereal cultivation in Iceland. Perdikaris (1999) questions the reliability of barley as a staple food, and Zori et al. (2013) suggest that locally grown barley would only have been available to high-status groups, who would have used the grain to brew beer. However, Zori et al.'s (2013) hypothesis is not supported by the archaeobotanical evidence: brewing involves allowing the grains to sprout during malting, and none of the grains found in the published Icelandic archaeobotanical assemblages surveyed for this study have been identified as sprouted. This lack of evidence does not, of course, rule out the use of barley in brewing, but further evidence is needed to support this hypothesis.

It is probable that barley was not available in large quantities at all early Icelandic farms, but this may be more related to the local environment than to social status. For example, excavations at Granastaðir in northern Iceland produced no botanical or archaeological evidence of cereal processing (Einarsson 1994; Bending 2007), and barley was also rare at the northern site of Hofstaðir (Guðmundsson 2009). In comparison, relatively large quantities of barley have been found at sites in the southern and western parts of the island, such as Reykholt (Guðmundsson et al. 2012), Hrísbrú (Martin 2014; Zori et al. 2013), Gröf (Friðriksson 1959), Bergþórshvoll (Friðriksson 1960), Suðurgata (Nordahl 1988), and Herjólfsdalur (Hermanns-Auðardóttir 1989). This does not appear to be an artificial divide: the area around Lake Mývatn in north-eastern Iceland has been the subject of intensive archaeological investigation, yet cereal remains are found in much smaller quantities in this region than in the south and west. Nonetheless, barley grains have been found in early deposits at sites in the northern region of Skagafjörður (Trigg et al. 2009; Bolender and Trigg 2013; Catlin et al. 2018).

There is place name evidence for cereal cultivation in Iceland: the element 'akur' means 'arable field', and is relatively common. The island of Akurey is the site of one of four supposed ancient barley fields that were studied by Garðar Guðmundsson and colleagues (2004). All of these fields were identified as anthropogenic features where cereal cultivation, along with tilling and probable manuring, took place (Guðmundsson et al. 2004). Possible arable fields are also known from northern Iceland, for example at Ingiríðarstaðir in Þegjandadalur (Hreiðarsdóttir and Roberts 2009).

Palynologists often use occurrences of *Hordeum*-type pollen in Iceland to infer barley cultivation; however, this is problematic as the pollen of the native grass *Elymus arenarius* L. also falls into this category (Erlendsson et al. 2014). Nevertheless, *Hordeum*-type pollen identified around farms away from the coastal habitat of *E. arenarius* seem likely to indicate cereal cultivation and are commonly encountered in palynological studies in south-western Iceland (e.g. Dugmore et al. 2005; Einarsson 1962, 1963; Erlendsson et al. 2006; Hallsdóttir 1987; Vickers et al. 2011). At the site of Hrísbrú, *Hordeum*-type pollen is only present from the mid-9th to early 13th century AD, after which changes in the environmental record have been interpreted as representing a change in land use towards more grazing and hay production (Erlendsson et al. 2014). The same pattern is suggested by pollen studies from Reykholt, where historical sources also indicate the cessation of cereal cultivation by the 13th century AD (Sveinbjarnardóttir 2012).

Many of the sites mentioned here are large, high-status farms, and this may not be coincidental. However, it should be noted that very little archaeobotanical analysis has been conducted on material from low-status settlements in Iceland, so there is also a bias in the available data towards larger farms. Nonetheless, cereal cultivation is a labour-intensive

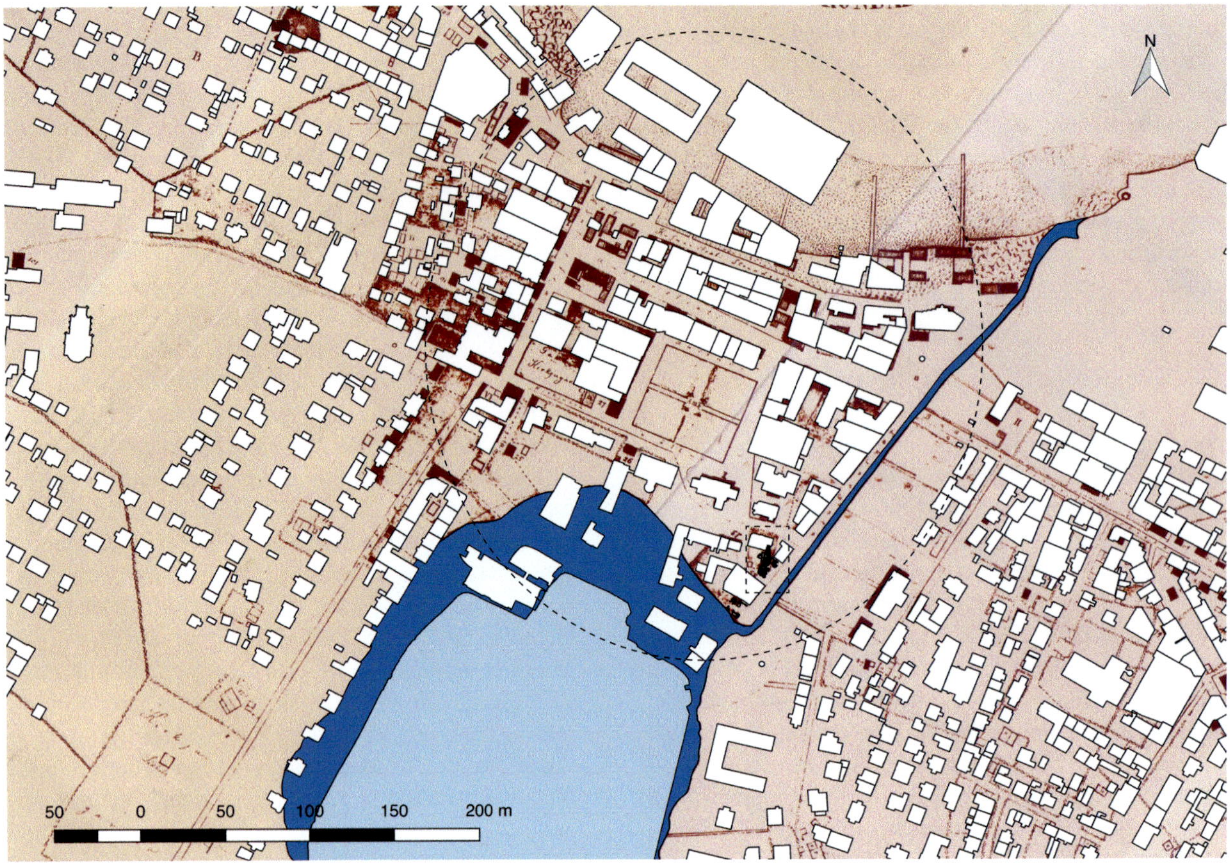

Figure 2. Map of Reykjavík in which modern structures (white) are overlain on an 1876 map by Sveinn Sveinsson and Benedikt Gröndal (Sveinsson and Gröndal 1876). The modern extent of Tjörnin is shown in light blue, and the water level in 1876 is indicated in dark blue. The Lækjargata longhouse is located within the black square, and the area of Kvosin lies approximately within the black circle. (map: Dawn Elise Mooney and Lísabet Guðmundsdóttir)

activity, requiring the seasonal mobilisation of a large workforce, which was perhaps less feasible at isolated farmsteads. The archaeological structures and associated finds of barley grain at Lækjargata 10–12 suggest that the area covered by modern-day central Reykjavík may have been a focal point for cereal cultivation and processing during the Viking Age.

Excavations at Lækjargata 10–12

Excavations at Lækjargata 10–12 were carried out in summer 2015, in advance of construction work. While documentary sources established the presence of 18th–20th century buildings at the site (Líndal 1988; Þrastardóttir and Gunnarsdóttir 2005), the excavations revealed earlier structures, including a Viking Age longhouse and associated outbuildings. These were heavily truncated by modern construction, but their discovery challenges the conventional view of the history of settlement in Reykjavík, in which the main settlement was thought to be close to the northwest corner of the lake Tjörnin (Vésteinsson 2006).
Reykjavík appears to have been the site of one of the earliest settlements after the colonisation of Iceland (Þorláksson 2006). This is supported not only by historical documentary sources (Íslenzk Fornrit 1968), but also by archaeological investigations that indicate settlement in the area before the fall of the so-called settlement tephra (*landnámslag*) around AD 870–880 (Roberts 2001; Schmid et al. 2017). The main settlement occupies a low-lying area (Kvosin) bordered by hills to the east and west, with the lake Tjörnin to the south and the sea around 300 m to the north (Fig. 2), in which four longhouses have been excavated (Einarsson 1995; Nordahl 1988; Roberts 2001). About 20 m east of these longhouses, adjacent to the lake, was a large activity area with smithies, middens, corn-drying ovens, a woodworking area, and a wooden trackway (Garðarsdóttir 2010). This concentration of structures could have supported at least four households, but further work is needed to determine which structures were contemporary. The Lækjargata excavations established the presence of a fifth household in the area. Unlike previously excavated dwellings, the newly found structures were located on the eastern side of Kvosin, next to a stream running from Tjörnin to the sea. The farmstead was established in the first decades after the initial settlement of Iceland and appears to have been in use for at least 150 years. The area was abandoned after the farm burned to the ground in the late 11th or early

Figure 3. Overview photo of the Lækjargata longhouse during its excavation. North is at the top of the photograph. The walls of the longhouse can be seen running north–south, more clearly visible on the less-truncated western edge. The central hearth is at the top of the photo, truncated by the basement of a house and surrounded by peat ash floor layers. South of this are the living area, the weaving room, and the byre. Photograph: Fornleifastofnun Íslands.

12th century AD (Guðmundsdóttir 2017). Figure 3 shows the Lækjargata longhouse (Area A) under excavation in summer 2015.

Three phases of occupation were identified at Lækjargata. Radiocarbon results indicate that the oldest phase, Phase C3, dates to the 9th century, or shortly

Table 1. Radiocarbon dates on barley grains from the longhouse and Structure 2 at Lækjargata. Calibrated with OxCal v4.3.3. (Bronk Ramsey 2017) using IntCal 13 atmospheric curve (Reimer et al. 2013). SUERC: Scottish Universities Environmental Research Centre.

Context number	Context type	Sample number	Sample type	Lab identifier	Radiocarbon age BP (1 sigma level)	$\delta^{13}C$ values	Probability range (95.4% probability)
39833	Longhouse floor – Area A, Phase C3	123	Barley (*Hordeum vulgare*)	SUERC-73980	1164±30	−25 ‰ assumed	772–905 calAD (75.6%) 915–967 calAD (19.8%)
40120	Longhouse floor – Area A, Phase C2	242	Barley (*Hordeum vulgare*)	SUERC-73973	1086±30	−23.2 ‰	937–1016 calAD (64.4%) 894–932 calAD (31.0%)
263	Structure 2 – Area B hearth fill, Phase C3–C2	429	Barley (*Hordeum vulgare*)	SUERC-73982	1185±34	−24.3 ‰	722–740 calAD (2.9%) 767–902 calAD (84.5%) 920–967 calAD (8.0%)

after the settlement tephra fell (Sigurgeirsson 2015) (Table 1). At this point in time, the longhouse was at least 22 m long and 4.5–5 m wide and comprised three activity areas: a byre to the south, an activity area with a wooden bench and thick floors in the centre, and an unusually long stone-lined fireplace (5.2 m) to the north. During Phase C2, the longhouse remained a similar size. It was still divided into three areas; however, the byre was smaller, and the northern part of this area became a separate room, probably used for weaving. A floor layer from this phase was dated to the mid-10th to early 11th century AD (Table 1). By Phase C1, the byre was no longer in use but the rest of the longhouse retained its earlier structure. A collapsed turf wall from this phase showed evidence of the KR 920 tephra (from the eruption of Katla in AD 920) in the turves, indicating that part of the wall was constructed after this date (Sigurgeirsson 2015).

South of the longhouse were the remains of three contemporary structures (Area B). Radiocarbon results indicate that they were in use during the 10th to 11th century, during the same time period as the longhouse (Fig. 4). These were heavily truncated, but included several hearths or ovens. These are thought to have been linked to cereal processing, due to the presence of large quantities of barley grain along with a quern stone. One of the ovens, hearth [237] in Structure 2, was stone-lined and measured 2 m long, 0.7 m wide, and 0.4 m deep. It had three fills which produced both charred grain and slag fragments, suggesting that the feature was used both for drying grain and for metalworking (Guðmundsdóttir and Davíðsdóttir 2017).

Materials and methods

The excavations employed single-context excavation and recording. Charred plant macrofossil remains from 83 samples were analysed, from all floor deposits and hearth fills. The standard sample size was 10 l, or 100% of the material from contexts smaller than 10 l in volume. All floor layers were sampled on a 1 m grid system. The samples ranged in volume from 0.5 l to 28 l, and the total volume of soil processed was 752.6 l. The samples were processed by water flotation. The light fraction of the samples was retained on a 500µm mesh and air dried. Material smaller than 500µm was discarded. The samples were then sorted and plant remains retrieved for analysis.

Identifications of plant macrofossils were made by comparing observed characteristics with modern reference material, along with published and digital sources (Anderberg 1994; Beijerinck 1947; Berggren 1969, 1981; Cappers et al. 2006; Jacomet 2006; Neef et al. 2012). Nomenclature used in the text, diagrams, and tables follows Mossberg and Stenberg (2007), and English and Icelandic common names follow Kristinsson (2001).

Results

The charred plant macrofossils recovered from the site are presented in Appendix 1, divided into two tables presenting results from the longhouse (Area A, Table A1) and the outbuildings (Area B, Table A2). Full analysis of the botanical assemblage can be found in the technical report (Mooney 2017). Charred plant macrofossils were present across all areas of the site but were most abundant in floors and hearths in the living area of the longhouse and the outbuilding Structure 2. Preservation of the charred plant macrofossil assemblage was generally moderate to good, in that diagnostic features were often preserved despite the fact that many seeds and grains showed low levels of abrasion. Cereal grains were often fragmented and pitted, and unidentifiable Cerealia grains and fragments made up almost 18% of the total grain assemblage (see Table 2).

Most cereal grains were identified as *H. vulgare* or *H. vulgare* var. *vulgare*, with a small component of *Avena* sp. (oats) also noted. A single, poorly-preserved seed of *Linum usitatissimum* L. (flax) was recorded from Phase C3. A wide range of wild plants were also not-

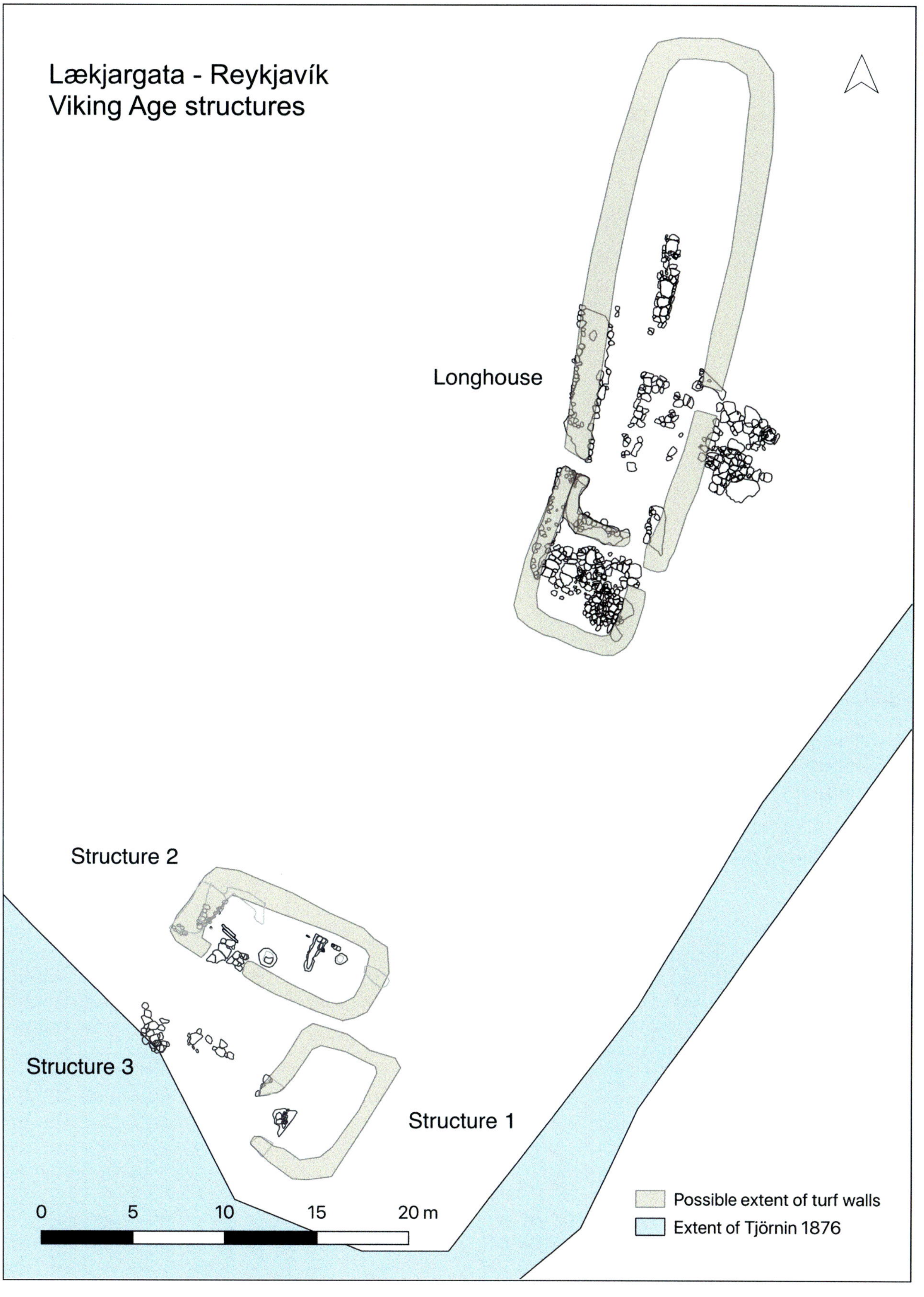

Figure 4. Plan showing the longhouse and contemporary structures at Lækjargata 10–12, with the position of the lake and stream from 1876 (see Fig. 2). (map: Lísabet Guðmundsdóttir)

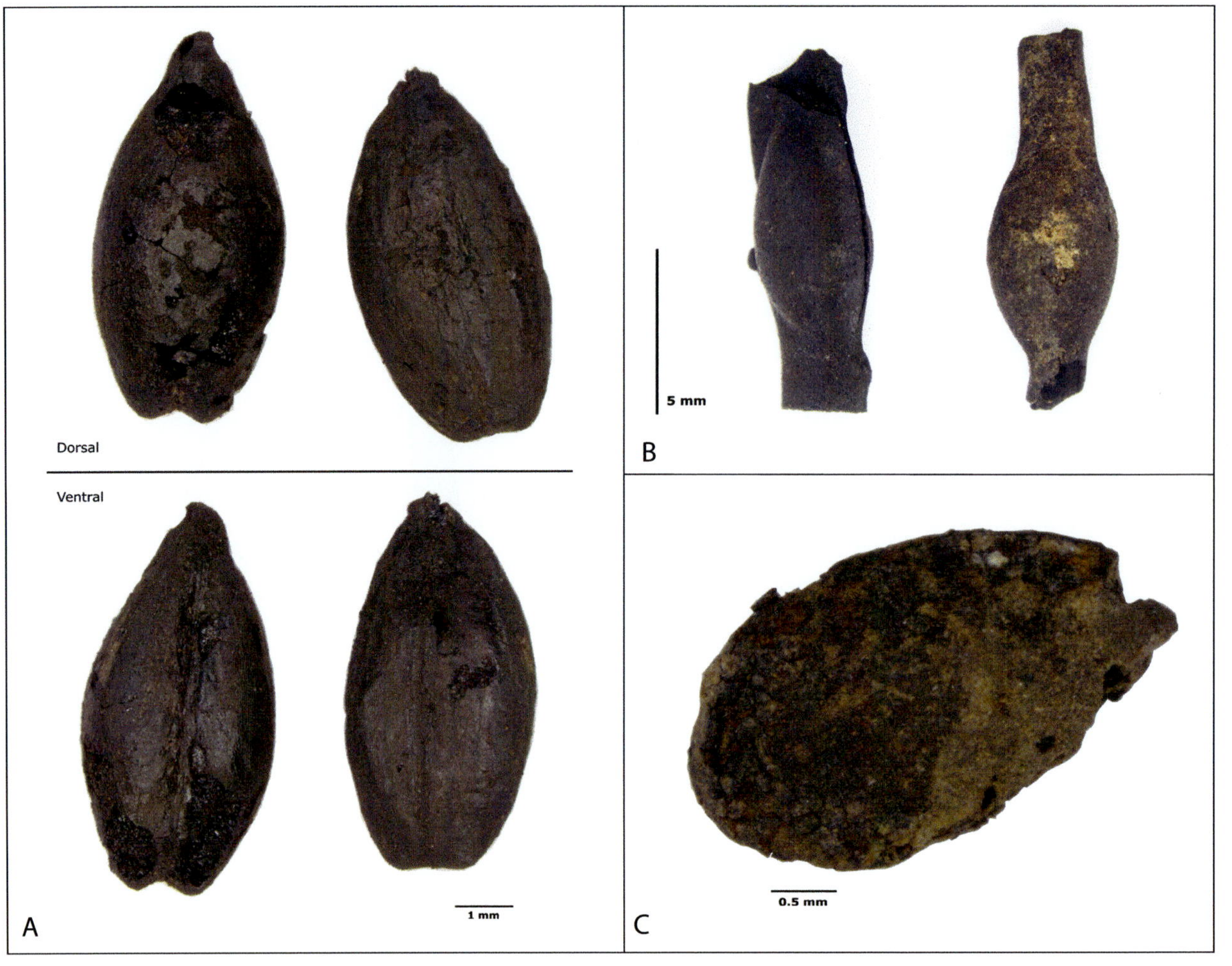

Figure 5. Examples of charred plant macrofossil remains from Lækjargata 10–12. A: *Hordeum vulgare* var. *vulgare* from the living area of the longhouse; B: *Fucus*-type seaweed cf. *Ascophyllum nodosum* from outside the longhouse; C: *Linum usitatissimum* from the byre area of the longhouse. (figure: Dawn Elise Mooney)

ed, most of which were either segetal weeds or taxa that would have been found growing in and around the homefield of the farmstead. Heathland plants such as *Empetrum nigrum* L. and *Vaccinium uliginosum* L., and aquatic taxa including *Menyanthes trifoliata* L. and *Potamogeton* sp., were likely components of habitats in the close vicinity of the site. Fragments of charred *Fucus*-type seaweed, mostly comparable to *Ascophyllum nodosum* (L.) Le Jolis, were found across the site, with large quantities present in one of the outbuildings (Structure 2) and in a deposit outside of the west side of the longhouse. A single fragment of charred cf. *Prunus* sp. stone found in the outbuilding Structure 1 may represent the consumption of imported fruit, however the fragment was too small to conclusively identify. Examples of some of the charred plant macrofossils found at Lækjargata are shown in Figure 5.

Discussion

While the quantity of plant remains recovered varied across the site, the species composition was relatively constant. There was no variation over time in the cereals recorded: barley, mostly of the hulled six-row variety, is the only major food crop identified. Flax was also noted, but only as a single seed. The variety of wild flora is consistent across the samples and suggests the presence of arable fields, manured homefield areas, damp grassland, and wetland and aquatic environments. The exploitation of heathland, probably for both the collection of berries as food and the collection of brushwood and peat for fuel, is evidenced by *E. nigrum* seeds and leaves throughout the samples. *Fucus*-type seaweed also appears to have been exploited throughout the occupation of the site.

High concentrations of charred plant remains in the floors are likely to represent the spreading of ash to reduce moisture and odours. The concentrations of grain around hearth [237], combined with the morphology of the feature, suggest that it could have functioned as a grain-drying oven. Furthermore, while cereal grains were common in samples from this structure, they were rare in samples from other structures in the same area. Samples from the weaving room were particularly low in charred plant remains. This reflects the fact that no burning occurred

Table 2. Summary of finds of cereal grains from Lækjargata 10-12

	Total number of fragments	% of cereal assemblage
Cerealia	35	1.99
Cerealia fragments	278	15.78
Avena sp.	7	0.40
Hordeum vulgare	189	10.73
Hordeum vulgare var. *vulgare*	1,253	71.11

Table 3. Symmetry of *Hordeum vulgare* var. *vulgare* grains from Lækjargata 10–12

	Total number of fragments	% of hulled barley assemblage
Symmetrical	335	27
Asymmetrical	597	47
Undetermined symmetry	321	26

in situ in this area, and may also indicate that spreading of ash on the floors occurred to a lesser extent in this room than in other areas of the longhouse. It is most likely that this reflects the specific function of the room, and that it was separated by panelled walls from the main living space in the longhouse.

The two large concentrations of charred seaweed fragments, representing material which must have been brought to the site from the seashore, may represent specific activity areas. Both deposits consisted almost exclusively of charred seaweed, and may be related to activities such as salt production, textile processing or metalworking. The large quantities of charred seaweed at the site are unlikely to be related to cultivation (Mooney 2017, 2018). Occasional finds of *Fucus*-type seaweed fragments throughout both areas probably represent the spreading of burnt waste on floors, or the accidental tracking in of charred material.

Cereal remains

The grain assemblage from the site is unusually large compared to other Icelandic sites (e.g. Hermanns-Auðardóttir 1989; Mooney 2008, 2011; Guðmundsson 2009; Trigg et al. 2009; Martin 2010, 2014; Bold 2011; Guðmundsson et al. 2012; Bolender and Trigg 2013). For example, at the Viking Age site of Hrísbrú, where barley grains are described as ubiquitous (Martin 2014: 203), only 44 grains were reported. The most well-known large find of Medieval charred barley in Iceland, from Gröf í Öræfum, consisted of around 500 grains (Friðriksson 1959). In contrast, the assemblage from Lækjargata comprised a total of 1,762 caryopses and caryopsis fragments, along with occasional examples of chaff. Taxonomic identifications of the cereals are summarised in Table 2.

The 189 *H. vulgare* grains from Lækjargata for which the variety could not be determined almost certainly represent poorly-preserved examples of *H. vulgare* var. *vulgare*, and this is also likely to be the case for the 313 grains and grain fragments which could only be identified as Cerealia. The *H. vulgare* var. *vulgare* grains from the site were also analysed to determine their symmetry. Of a total of 1,253 *H. vulgare* var. *vulgare* grains identified at Lækjargata, 321 were of undetermined symmetry (see Table 3). Of the remainder, 335 were symmetrical and 597 were asymmetrical, equivalent to a ratio of 1:1.78. It is therefore probable that the assemblage derived from the cultivation of six-row hulled barley (cf. Jacomet 2006).

Barley is by far the most common cereal found in archaeobotanical assemblages in Iceland and is generally thought to be six-row hulled barley (Friðriksson 1959, 1960; Guðmundsson et al. 2012; Guðmundsson and Hillman 2012; Guðmundsson 2009; Trigg et al. 2009; Bold 2011). However, at sites with small quantities of grain, finds are identified as *H. vulgare* var. *vulgare* (Martin 2010) or simply as *H. vulgare* (Hermanns-Auðardóttir 1989; Guðmundsson 2001; Martin 2003; Mooney 2008, 2011; Martin 2014). Oats, which were also identified at Lækjargata, have been found at Hofstaðir (Guðmundsson 2009) and Hrísheimar (Bold 2011), but only in very small quantities. Larger quantities of probable oat grains have been found at the site of Grænagerði in Skagafjörður, however research on this material is ongoing (Catlin et al. 2018). Table 2 shows that the overall occurrence of oats at Lækjargata is only 0.4% of the total assemblage, which would suggest that these occurrences are incidental rather than resulting from cultivation. Oats often grow as a weed in fields of barley, and are known to be a common component of barley seed grain (Zohary et al. 2012; Guðmundsson 2009). Despite this, it has been proposed that oats may have been cultivated alongside barley in the area around Reykholt in the period AD 1000–1200 (Sveinbjarnardóttir et al. 2007). However, recent work (Joly et al. 2007; Erlendsson 2012) has shown ambiguity in the differentiation of *Hordeum*-type pollen from those of other cereals and large grasses, so

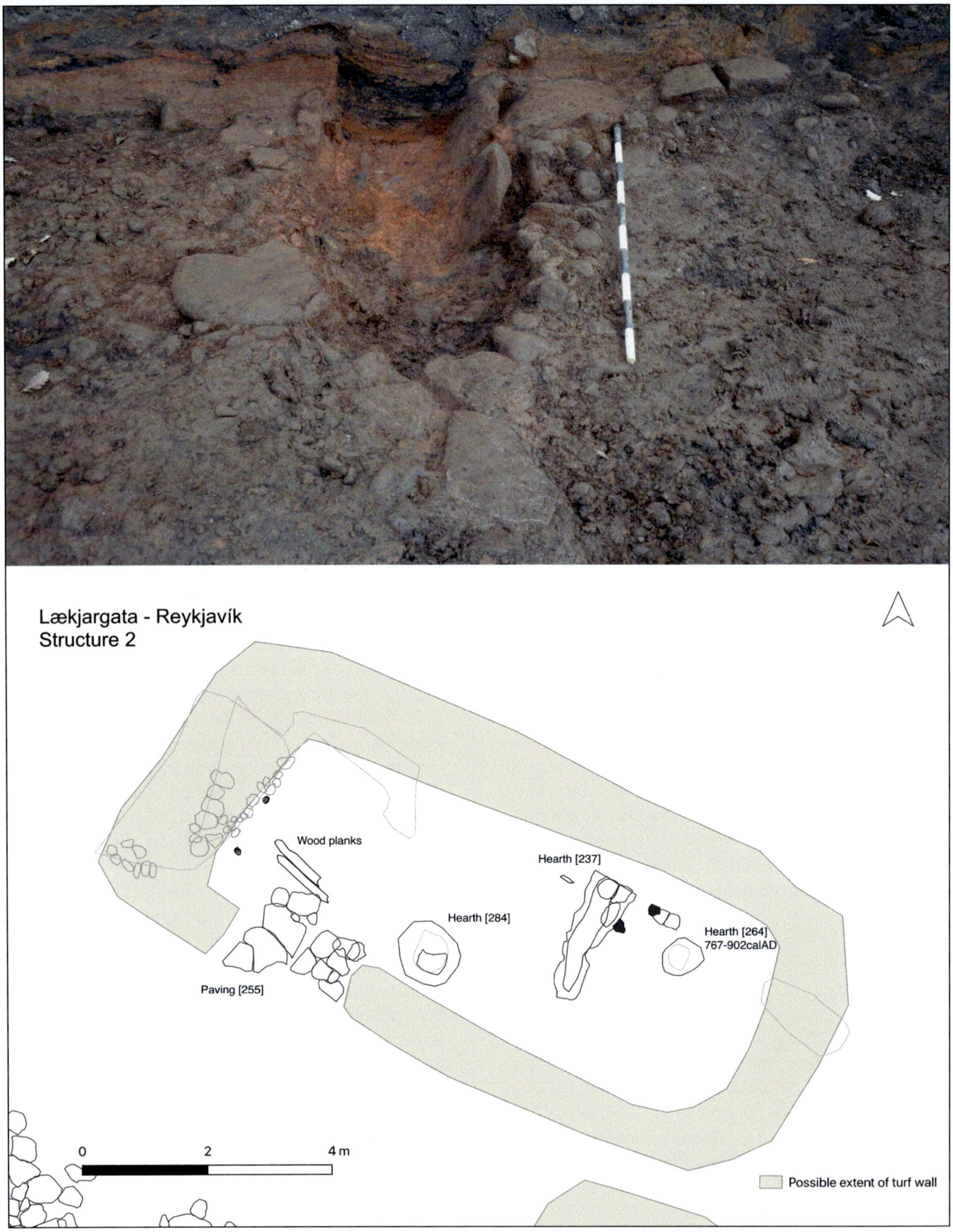

Figure 6. Hearth [237] under excavation (top) and mapped in plan view within Structure 2 (bottom) (figure: Lísabet Guðmundsdóttir and Fornleifastofnun Íslands).

the distinction between oat and barley pollen here may not be valid.

While the evidence for cultivation of oats in Iceland is sparse, the evidence for cultivation of barley in Iceland is significant. However, the presence of barley does not indicate that the crop was grown locally – it may have been imported, either from abroad or from elsewhere in Iceland. Factors such as the quantity of grain found, the wild seed assemblage, and the presence of chaff can be useful in identifying local cultivation. Some of the strongest evidence of Icelandic cereal cultivation comes from the site of Gröf (Friðriksson 1959), where a large assemblage of barley was found in the remains of a drying kiln. The assemblage contained not only grain, but also substantial amounts of chaff, which would not have been present in cleaned grain imported to the site. Further evidence is found in a 10th-12th century hearth at Reykholt (Guðmundsson et al. 2012; Guðmundsson and Hillman 2012), where an assemblage of charred grain with very frequent weed seeds is interpreted as waste from crop processing.

Weed flora

The segetal weeds found at Lækjargata include *Galeopsis bifida/speciosa*, *Spergula arvensis,* and *Stellaria media*. Guðmundsson et al. (2012: 115) have described these species as 'archetypical weeds of crops' that are widely found in Icelandic assemblages. Examples of *Galeopsis* spp. have been found at Reykholt (Guðmundsson et al. 2012), Bessastaðir (Guðmundsson 2009), and Bergþórshvoll (Friðriksson 1960). *S. arvensis* is also common and has previously been noted at Bergþórshvoll (Friðriksson 1960), Vatnsfjörður (Mooney 2008), Hofstaðir (Guðmundsson 2009), Bessastaðir (Guðmundsson 2009), and Reykholt (Guðmundsson et al. 2012), while *S. media* is almost ubiquitous.

Neither *S. arvensis* nor *G. tetrahit/bifida/speciosa* are thought to be native to Iceland. *S. arvensis* is naturalised in Iceland and is now common in the south-western region (Kristinsson 2010), but was originally introduced (Guðmundsson 2009). It is a common arable weed across Europe and may have been introduced as a component of seed grain. *G. tetrahit* is described in archaeological literature as native to Iceland (Guðmundsson 2009; Guðmundsson et al. 2012), but this is in disagreement with botanical sources that state that *Galeopsis* spp. were introduced within the past 300 years (Kristinsson 2010; Wasowicz et al. 2013). These taxa could represent a weed seed component in poorly-cleaned imported grain. However, if so, a variety of other seeds of plants alien to Iceland should also be present, and this is not the case. None of the comparative assemblages cited in this text contain other segetal weeds that are not widely found in Iceland. Given this, it seems more likely that *Galeopsis* spp. were initially accidentally introduced to Iceland with seed grain, and thereafter became common weeds of arable fields. The cessation of cereal cultivation in Iceland by the 15th century, combined with climatic deterioration during the Little Ice Age, may have led to the temporary extinction of these taxa in Iceland, followed by later re-introduction.

The low species diversity of the weed flora from Lækjargata, and the strong similarity of the grain assemblage to contemporary assemblages, indicates that the barley found at the site was probably grown locally, although the low quantities of chaff recovered suggest that cereal processing occurred elsewhere. The lack of other cereal taxa such as wheat and rye is also an argument for local cultivation – these taxa would likely have been present in imported grain. This hypothesis is also supported by evidence of corn-drying at the site. The high quantity of burnt grain found at the site suggests that either corn-drying or malt roasting for brewing took place at Lækjargata. As only a single sprouted grain was observed in the barley assemblage, the latter appears unlikely. The presence of seeds of *Menyanthes trifoliata*, which can be used as a substitute for hops in flavouring beer (Mabberley 2008: 539) could be indicative of brewing, but could also be a component of the local aquatic flora. It is more probable that corn drying was the source of much of the charred grain at the site. The stone-lined hearth [237] (Fig. 6) is the most likely location for this, as it resembles other corn-drying structures recorded in Iceland, at Gröf í Öræfum (Gestsson 1959), Herjólfsdalur (Hermanns-Auðardóttir 1989), and Alþingisreiturinn (Garðarsdóttir 2010).

Conclusion

The abundance of grain from Lækjargata, along with a limited weed flora, suggests the local cultivation of barley from the earliest period of occupation. The presence of a single seed of flax is insufficient evidence to suggest flax cultivation, although limited evidence of this does exist elsewhere in Iceland in the form of place name evidence and a single pollen grain from Skálholt (Einarsson 1962: 459). Further research is needed to address this issue (cf. Mooney 2017). The likelihood of local cultivation of barley is also attested by the presence of a possible corn-drying oven at Lækjargata. The site of cultivation is unclear, although it has been suggested that the south-facing slopes at Landakot, c. 500 m west of Lækjargata, are a likely location (Líndal 1988). However, this still does not amount to conclusive proof that the Lækjargata barley was grown locally, or even within Iceland. Corn-drying ovens could have been used to dry grain which had become damp during importation by sea from Scandinavia or Britain. The seed grain used by the earliest settlers, for example, was certainly imported to Iceland. The weeds identified are all also native to northern Europe, and the palynological evidence of cereal cultivation could relate to the growth of native large grasses (c.f. Joly et al. 2007). However, although the analysis of material from Lækjargata has raised further questions, it has also made a significant contribution to the consolidation of knowledge regarding the history of cultivation in Iceland. New applications of stable isotope analysis of plant remains could be used to address the uncertainty surrounding to what extent barley was cultivated in Iceland. Carbon ($\delta^{13}C$) and nitrogen ($\delta^{15}N$) isotope analysis can be used to assess the growing conditions of plants (c.f. Aguilera et al. 2008) and, if applied to Icelandic material, could potentially help identify imported grain. Furthermore, due to the island's distinct volcanic geology, strontium isotope ($^{87}Sr/^{86}Sr$) analysis could also provide an opportunity to conclusively identify locally-grown grain (cf. Heier et al. 2008).

Some scholars (e.g. Zutter 1992, 1999; Ross and Zutter 2007; Martin 2014) have stated that very little archaeobotanical research has been conducted in Iceland. Martin (2014: 194) goes so far as to say that "most Icelandic archaeologists see little value in macrobotanical data". This is demonstrably untrue. It is true that, with a few notable early exceptions (e.g. Friðriksson 1959, 1960; Hermanns-Auðardóttir 1989; Zutter 1992), archaeobotanical analysis only began to be systematically employed in Iceland in the 2000s, but since then archaeobotany has contributed significantly to the interpretation of a wide variety of sites. Some of these studies are yet to be published but are accessible online in grey literature or dissertations (e.g. Vatnsfjörður, Hrísheimar, Skútustaðir, Aðalstræti), while many others (e.g. Reykholt, Hofstaðir, Hrísbrú, Svalbarð, Gjögur, Reynistaðir) are the subject of easily accessible publications.

However, it is also true that archaeobotany in Iceland is an expanding field, and many more studies are in progress or unpublished even as grey literature at the time of writing. Even more material exists in the form of samples which have been taken during excavations, but are yet to be analysed. For this reason, it is essential that archaeobotanists working on Icelandic material strive not only to disseminate their results through articles, conference presentations, and interim technical reports, but also to communicate and discuss their findings with one another and with other archaeologists as part of the research process. It is only through increased cooperation with each other, and through interdisciplinary investigations, that we will gain a holistic understanding of Iceland's complex land-use history.

References

Anderberg A-L (1994) Atlas of seeds and small fruits from northwest-European plant species: Part 4. Resedaceae – Umbellifereae. Swedish Museum of Natural History, Stockholm

Aguilera M, Araus JL, Voltas J, Rodríguez-Ariza MO, Molina F, Rovira N, Buxó R, Ferrio JP (2008) Stable carbon and nitrogen isotopes and quality traits of fossil cereal grains provide clues on sustainability at the beginnings of Mediterranean agriculture. Rapid Commun Mass Spectrom 22:1653–1663

Beijerinck W (1947) Zadenatlas der Nederlandsche flora. Wageningen University, Wageningen

Bending J (2007) The economy of the Norse settlement of the North Atlantic islands and its environmental impact: An archaeobotanical assessment. Dissertation, University of Sheffield

Berggren G (1969) Atlas of seeds and small fruits of northwest-European plant species: Part 2. Cyperaceae. Swedish Natural Science Research Council, Stockholm

Berggren G (1981) Atlas of seeds and small fruits of northwest-European plant species: Part 3. Salicaceae–Cruciferae. Swedish Natural Science Research Council, Stockholm

Bjørnstad Å (2012) Vårt daglege brød: Kornets kulturhistorie. Vidarforlaget AS, Oslo

Bold R (2011) Norse utilisation of archaeobotanical resources within the Mývatnssveit locale, northern Iceland. Dissertation, University of Durham

Bolender DJ, Trigg HB (2013) Barley ubiquity in Langholt Skagafjörður. Conference paper, North Atlantic Biocultural Organisation meeting, 12–13 July 2013, Akureyri, Iceland

Bronk Ramsay C (2017) Methods for summarizing radiocarbon datasets. Radiocarbon 59(2):1809–1833

Cappers RTJ, Bekker RM, Jans JEA (2006) Digitale zadenatlas van Nederland. Barkhuis, Groningen

Catlin K, Steinberg J, Bolender D (2018) Fornbýli landscape and archaeological survey on Hegranes (FLASH) interim report 2017. Fiske Center for Archaeological Research, University of Massachusetts, Boston

Dugmore AJ, Church MJ, Buckland PC, Edwards KJ, Lawson I, McGovern TH, Panagiotakopulu E, Simpson IA, Skidmore P, Sveinbjarnardóttir G (2005) The Norse *Landnám* on the North Atlantic islands: An environmental impact assessment. Polar Rec 41:21–37

Einarsson BF (1994) The settlement of Iceland: A critical approach: Granastaðir and the ecological heritage. Department of Archaeology, Gothenburg University, Gothenburg

Einarsson BF (1995) Fornleifarannsókn í Aðalstræti 12 1993. Skýrslur Árbæjarsafns 49. Árbæjarsafn, Reykjavík

Einarsson Þ (1962) Vitnisburður frjógreiningar um gróður, veðurfar og landnám á Íslandi. Saga 3:442–469

Einarsson Þ (1963) Pollen analytical studies in the vegetation and climate history of Iceland in Late and Post-Glacial times. In: Löve Á, Löve D (eds) North Atlantic biota and their history. Pergamon, Oxford, pp 355–365

Erlendsson E (2012) Pollen analysis on samples from context [577]. In: Sveinbjarnardóttir G (ed) Reykholt: Archaeological investigations at a high status farm in western Iceland. Snorrastofa/ The National Museum of Iceland, Reykjavík, pp 247–249

Erlendsson E, Edwards KJ, Gísladóttir G (2014) Landscape change, land use and occupation patterns inferred from two palaeoenvironmental datasets from the Mosfell valley, SW Iceland. In: Zori D, Byock J (eds) Viking archaeology in Iceland: Mosfell Archaeological Project. Brepols, Turnhout, pp 181–192

Erlendsson E, Edwards KJ, Lawson I, Vésteinsson O (2006) Can there be a correspondence between Icelandic palynological and settlement evidence? In: Arneborg J, Grønnow B (eds) Dynamics of northern societies. Nationalmuseet, Copenhagen, pp 347–354

Friðriksson S (1959) Korn frá Gröf í Öræfum. Árbók hins Íslenzka Fornleifafélags 1959. Hið Íslenzka Fornleifafélag, Reykjavík, pp 88–91

Friðriksson S (1960) Jurtaleifar frá Bergþórshvoli á söguöld. Árbók hins Íslenzka Fornleifafélags 1960. Hið Íslenzka Fornleifafélag, Reykjavík, pp 64–75

Garðarsdóttir V (2010) Fornleifauppgröftur á Alþingisreitnum 2008–2010. Bindi I, Reykjavík

Gelsinger BE (1981) Icelandic enterprise: Commerce and economy in the Middle Ages. University of South Carolina Press, Columbia

Gestsson G (1959) Gröf í Öræfum. Árbók hins Íslenzka Fornleifafélags 1959. Hið Íslenzka Fornleifafélag, Reykjavík, pp 5–87

Guðmundsdóttir L (2017) Fornleifarannsókn á Lóð Lækjargötu 12. Svæði A. Framvinduskýrsla. Fornleifastofnun Íslands, Reykjavík

Guðmundsdóttir L, Davíðsdóttir LL (2017) Fornleifarannsókn á lóð Lækjargötu 12. 2. bindi – svæði B. Fornleifastofnun Íslands, Reykjavík

Guðmundsson G (2001) The environmental evidence. In: Roberts HM (ed) Fornleifarannsókn á lóðunum/archaeological excavations at Aðalstræti 14–18, 2001: A preliminary report/framvinduskýrslur. Fornleifastofnun Íslands, Reykjavík, pp 83–88

Guðmundsson G (2009) The plant remains. In: Lucas G (ed) Hofstaðir: Excavations of a Viking Age feasting hall in northeastern Iceland. Fornleifastofnun Íslands, Reykjavík, pp 353–367

Guðmundsson G, Hillman G (2012) Charred remains of grains and seeds from hearth [99]. In: Sveinbjarnardóttir G (ed) Reykholt: Archaeological investigations at a high status farm in western Iceland. Snorrastofa/The National Museum of Iceland, Reykjavík, pp 242–243

Guðmundsson G, Snæsdóttir M, Simpson I, Hallsdóttir M, Sigurgeirsson MÁ, Árnason K (2004) Fornir Akrar á Íslandi: Meintar minjar um kornrækt á fyrri öldum. Árbók hins Íslenzka Fornleifafélags 2002–2003. Hið Íslenzka Fornleifafélag, Reykjavík, pp 78–106

Guðmundsson G, Sveinbjarnardóttir G, Hillman G (2012) Charred remains of grains and seeds from the high-status farm site of Reykholt in Iceland. Environ Archaeol 17(2):111–117

Hallsdóttir M (1987) Pollen analytical studies of human influence on vegetation in relation to the Landnám tephra layer in southwest Iceland. Lund University, Lund

Heier A, Evans J, Montgomery J (2008) The potential of carbonised grain to preserve biogenic 87Sr/86Sr signatures within the burial environment. Archaeometry 51(2):277–291

Helgason A, Sigurðardóttir S, Nicholson J, Sykes B, Hill EW, Bradley DG, Bosnes V, Gulcher JR, Ward R, Stefánsson K (2000) Estimating Scandinavian and Gaelic ancestry in the male settlers of Iceland. Am J Hum Genet 67(3):697–717

Hermanns-Auðardóttir M (1989) Islands Tidiga Bosättning. Umeå Universitet Arkeologiska institutionen, Umeå

Hreiðarsdóttir EÓ, Roberts H (2009) Þögnin rofin: Fysrtu niðurstöður rannsókna á eyðibyggð á Þegjandadal. In: Ingolfssón B (ed) Árbók Þingeyinga 2009. Menningarmiðstöð Þingeyinga, Húsavík, pp 5–24

Íslenzk Fornrít (1968) Volume I. Hið Íslenzka Fornritafélag, Reykjavík

Jacomet S (2006) Identification of cereal remains from archaeological sites. Basel University, Basel

Joly C, Barillé L, Barreau M, Mancheron A, Visset L (2007) Grain and annulus diameter as criteria for distinguishing pollen grains of cereals from wild grasses. Rev Palaeobot Palynol 146:221–233.

Líndal P (1988) Reykjavík. Sögustaður við Sund. 1.-3. Örn og Örlygur, Reykjavík

Karlsson G (2000) Iceland's 1100 years: The history of a marginal society. Mál og Menning, Reykjavík

Kristinsson H (ed) (2001) Íslensk plöntuheiti. Náttúrufræðistofnun Íslands Akureyrarsetur, Akureyri

Kristinsson H (2010) Flowering plants and ferns of iceland, 3rd edn. Mál og Menning, Reykjavík

Mabberley DJ (2008) Mabberley's plant-book: A portable dictionary of plants, their classifications and uses, 3rd edn. Cambridge University Press, Cambridge

Martin P, Reykdal O, Halland H (2016) Current cereal growing situation in five northern regions and the potential for using local cereals in food and drink. Agronomy Institute UHI/Matís/NIBIO. http://www.nibio.no/prosjekter/northern-cereals--new-markets-for-a-changing-environment/_/attachment/ download/7aa902f1-da83-4ac5-bcd0-3eff93813a54/Cereal%20situation%20&%20markets%20final%20report%20NPA%20V2.pdf Accessed 20 September 2017

Martin SL (2003) Macrofossil plant remains 2003. In: Hellqvist M (ed) Fossilfynd på Hólar 2003: Rapport över fund av fossila insekter, osteologisk analys av ben, makrofossila växtdelar, metallurgiska analyser och från Hólar i Hjaltadalur, norra Island, 2003. Hólarannsóknin, Hólar, pp 30–36

Martin SL (2010) Viðauki IV: Jurtagreining. In: Garðarsdóttir V (ed) Alþingisreiturinn 2008–2010. Bindi II, Reykjavík, pp 641–643

Martin, SL (2014) The recovery and analysis of macrobotanical remains from Hrísbrú. In: Zori D, Byock J (ed) Viking archaeology in Iceland: Mosfell Archaeological Project. Brepols, Turnhout, pp 193–206

Mooney DE (2008) Assessment of botanical remains in Structures 1 and 3 in the Viking Age area. In: Milek K (ed) Vatnsfjörður 2007: Framvinduskyrslur/Interim Report. Fornleifastofnun Íslands Reykjavík, pp 111–116

Mooney DE (2011) Botanical remains from a sample column through the farm mound. In: Milek K (ed) Vatnsfjörður 2010: Framvinduskýrslur/Interim Report. Fornleifastofnun Íslands, Reykjavík, pp 155–172

Mooney DE (2017) Analysis of macrobotanical remains from excavations at Lækjargata 10–12, Reykjavík, Iceland (LÆK 2015-10). Arkeologisk Museum, Universitetet i Stavanger, Stavanger

Mooney DE (2018) Charred Fucus-type seaweed in the North Atlantic: a survey of finds and potential uses. Environmental Archaeology. DOI:10.1080/14614103.2018.1558805

Mossberg B, Stenberg L (2007) Gyldendals Store Nordiske flora. Gyldendal Norsk Forlag AS, Oslo

Neef R, Cappers RTJ, Bekker RM, Boulous L, Dinies M, Ertuğ ZF, Keller N, Lahitte M, Meulenbeld GJ, Zhu YP (2012) Digital atlas of economic plants in archaeology. Barkhuis, Groningen

Nordahl E (1988) Reykjavík from the archaeological point of view. Societas Archaeologica Uppsaliensis, Uppsala

Patterson WP, Dietrich KA, Holmden C, Andrews JT (2010) Two millennia of North Atlantic seasonality and implications for Norse colonies. PNAS 107(12):5306–5310

Perdikaris S (1999) From chiefly provisioning to commercial fishery: Long-term economic change in arctic Norway. World Archaeol 30(3):388–402

Prøsch-Danielsen L, Soltvedt E-C (2011) From saddle to rotary hand querns in south-western Norway and the corresponding crop plant assemblages. Acta Archaeol 82:129–162

Reimer PJ, Bard E, Bayliss A et al (2013) IntCal13 and Marine13 radiocarbon age calibration curves 0–50,000 Years cal BP. Radiocarbon 55(4):1869–1887

Reykdal O, Kristjánsdóttir ÞA, Hermansson J, Martin P, Dalmannsdóttir S, Djurhuus R, Kavanagh V, Frederiksen A (2014) Status of cereal cultivation in the North Atlantic region. Matís, Reykjavík

Roberts HM (2001) Fornleifarannsókn á lóðunum/Archaeological excavation at Aðalstræti 14–18, 2001: A preliminary report/ Framvinduskýrslur. Fornleifastofnun Íslands, Reykjavík

Robinson DE (1993) Dyrkede planter fra Danmarks forhistorie. Arkeologiske Udgravninger i Danmark 1993:22–39

Ross JM, Zutter C (2007) Comparing Norse animal husbandry practices: Palaeoethnobotanical analyses from Iceland and Greenland. Arctic Anthropol 44(1):62–85

Rowley-Conwy P, Legge T (2015) Subsistence practices in western and northern Europe. In: Fowler C, Harding J, Hofmann D (ed) The Oxford handbook of Neolithic Europe. Oxford University Press, Oxford, pp 429–446

Schmid ME, Dugmore AJ, Vésteinsson O, Newton AJ (2017) Tephra isochrons and chronologies of colonisation. Quat Geochronol 40:56–66

Sigurdsson J (ed) (1857–76) Diplomatarium Islandicum 834–1264. Volume 1. S.L. Möller, Copenhagen

Sigurgeirsson MÁ (2015) Fornleifarannsókn í Lækjargötu 2015. Gjóskulagarannsókn. Fornleifastofnun Íslands, Reykjavík

Sveinbjarnardóttir G (2012) Reykholt: Archaeological investigations at a high status farm in western Iceland. Snorrastofa/The National Museum of Iceland, Reykjavík

Sveinbjarnardóttir G, Erlendsson E, Vickers K, McGovern TH, Milek KB, Edwards KJ, Simpson IA, Cook G (2007) The palaeoecology of a high status Icelandic farm. Environ Archaeol 12(2):187–206

Sveinsson S, Gröndal B (1876) Reykjavík: Mælt hefir S. Sveinsson 1876. B. Gröndal málaði. Reykjavík

Trigg HB, Bolender DJ, Johnson KM, Patalano MD, Steinberg JM (2009) Note on barley found in the lowest levels of the farm mound midden at Reynistaður, Skagafjörður, Iceland. Archaeol Isl 7:64–72

van Wyk B-E, Wink M (2004) Medicinal plants of the world. Timber Press, Portland

Vésteinsson O (2006) Bæjarstæðið. In: Sverrisdóttir B (ed) Reykjavík 871±2. Landnámssýningin/The Settlement exhibition. Minjasafn Reykjavíkur, Reykjavík, pp 94–96

Vésteinsson O, McGovern TH, Keller C (2002) Enduring impacts: Social and environmental aspects of Viking Age settlement in Iceland and Greenland. Archaeol Isl 2:98–136

Vickers K, Erlendsson E, Church MJ, Edwards KJ, Bending J (2011) 1000 Years of environmental change and human impact at Stóra-Mörk, southern Iceland: A multiproxy study of a dynamic and vulnerable landscape. Holocene 21:979–995

Viklund K (1998) Cereals, weeds and crop processing in Iron Age Sweden: Methodological and interpretational aspects of archaeobotanical evidence. Umeå University Department of Arcchaeology, Umeå

Wasowicz P, Przedpelska-Wasowicz EM, Kristinsson H (2013) Alien vascular plants in Iceland: Diversity, spatial patterns, temporal trends, and the impact of climate change. Flora 208:648–673

Zohary D, Hopf M, Weiss E (2012) Domestication of plants in the Old World, 4th edn. Oxford University Press, Oxford

Zori D, Byock J, Erlendsson E, Martin S, Wake T, Edwards KJ (2013) Feasting in Viking Age Iceland: Sustaining a chiefly political economy in a marginal environment. Antiquity 87:150–165

Zutter C (1992) Icelandic PLANT AND LAND-USE PATTerns: Archaeobotanical analysis of the Svalbarð midden (6706-60), northeastern Iceland. In: Morris C, Rackham D (ed) Norse and later settlement and subsistence in the North Atlantic. University of Glasgow Press, Glasgow, pp 139–148

Zutter C (1999) Congruence or concordance in archaeobotany: Assessing micro- and macro-botanical data sets from Icelandic middens. J Archaeol Sci 26:833–844

Þorláksson H (2006) Forystuafl í Reykjavík á 10. Öld. In: Sverrisdóttir B (ed) Reykjavík 871±2. Landnámssýningin/The Settlement exhibition. Minjasafn Reykjavíkur, Reykjavík, pp 44–45

Þrastardóttir DK, Gunnarsdóttir GG (2005) Húsakönnun Vonarstræti – Templarasund – Kirkjutorg – Skólabrú – Lækjargata. Minjasafn Reykjavíkur, Reykjavík

Appendix 1

The charred plant macrofossils recovered from the site, divided into two tables presenting results from the longhouse (Area A, Table A1) and the outbuildings (Area B, Table A2).

Table A1. Summary of identifications of charred plant material from the longhouse at Lækjargata 10–12 (Area A). Unless otherwise stated, totals represent both whole and partial macrofossil remains. Calculations of identifications as a percentage of the Area A assemblage do not include charred seaweed fragments. Ubiquity scores represent the percentage of samples in which a taxon was identified.

		Phase C3	Phase C2/C3	Phase C2	Phase C1/C2	Phase C1									
Taxonomic identification	Element	Byre	Living area	Outside	Living area	Living area	Weaving room	Living area	Pit [52]	Hearth [53]	Living area	Weaving room	Total	% Area A assemblage	Ubiquity
Cerealia	Caryopsis	-	12	-	-	-	-	-	2	5	-	-	19	1.3	16.3
Cerealia	Caryopsis fragment	14	39	-	-	1	-	-	15	21	35	-	125	8.3	55.8
Hordeum vulgare	Caryopsis	34	34	-	1	3	-	-	3	17	40	1	133	8.8	55.8
Hordeum vulgare var. *vulgare*	Caryopsis	120	144	2	2	-	1	1	13	212	304	3	802	53.3	74.4
Linum usitatissimum	Seed/fruit	1	-	-	-	-	-	-	-	-	-	-	1	<1	2.3
Empetrum nigrum	Seed/fruit	-	1	-	-	-	-	-	-	8	2	-	11	<1	6.9
Rubus saxatilis	Seed/fruit	-	-	-	-	-	-	-	-	-	1	-	1	<1	2.3
Poaceae	Seed/fruit	1	-	-	-	-	-	-	-	-	7	-	8	<1	11.6
Poa sp.	Seed/fruit	1	4	-	2	-	-	-	-	2	-	-	9	<1	9.3
Poaceae large cf. *Bromus/Elymus*	Seed/fruit	-	1	-	-	-	-	-	-	-	7	-	8	<1	4.7
Bromus/Festuca	Seed/fruit	-	1	-	-	-	-	-	-	-	-	-	1	<1	2.3
Carex distigmaticae	Seed/fruit	2	5	-	-	1	-	1	-	-	4	1	14	<1	23.3
Carex tristigmaticae	Seed/fruit	6	1	-	-	-	-	-	-	-	26	-	33	2.2	19.9
Carex elata/acuta/dioica-type	Seed/fruit	9	14	-	-	-	-	-	-	2	9	-	11	<1	18.6
Galeopsis bifida/speciosa	Seed/fruit	-	-	-	-	-	-	-	-	-	2	-	2	<1	2.3
Potentilla sp.	Seed/fruit	1	-	-	-	-	-	-	-	-	-	-	1	<1	2.3
Persicaria maculosa	Seed/fruit	-	-	-	-	-	-	-	-	-	1	-	1	<1	2.3
Spergula arvensis	Seed/fruit	-	4	-	-	-	-	-	-	2	1	-	7	<1	9.3
Alchemilla sp.	Seed/fruit	-	-	-	-	-	-	-	1	-	-	-	1	<1	2.3
Atriplex sp.	Seed/fruit	-	2	-	-	-	-	-	-	2	7	-	11	<1	6.9
Lamiaceae	Seed/fruit	-	-	-	-	-	-	-	-	1	-	-	1	<1	2.3
Brassicaceae	Seed/fruit	-	-	-	-	-	-	-	-	-	1	-	1	<1	2.3
cf. *Medicago* sp.	Seed/fruit	-	-	-	-	-	-	-	-	-	2	-	2	<1	2.3
Potamogeton sp.	Seed/fruit	-	-	-	-	-	-	-	-	-	2	-	2	<1	2.3
Vaccinium sp.	Seed/fruit	1	1	-	-	-	-	-	-	-	-	-	2	<1	4.7
cf. *Vicia/Lathyrus*	Seed/fruit	1	-	-	-	-	-	-	-	-	-	-	1	<1	2.3
Stellaria media	Seed/fruit	44	51	-	6	1	2	-	4	56	134	-	298	19.8	65.1
Fucus-type seaweed	Thallus fragment	19	34	3164	-	7	2	2	-	2	4	-	3234	-	30.2

Table A2. Summary of identifications of charred plant material from the outbuildings at Lækjargata 10–12 (Area B). Unless otherwise stated, totals represent both whole and partial macrofossil remains. Calculations of identifications as a percentage of the Area B assemblage do not include charred seaweed fragments. Ubiquity scores represent the percentage of samples in which a taxon was identified.

Taxonomic identification	Element	Structure 1	Structure 2	Structure 3	Total	% Area B assemblage	Ubiquity
Cerealia	Caryopsis	3	13	-	16	1.3	20
Cerealia	Caryopsis fragment	4	149	-	153	12.6	47.5
Avena sp.	Caryopsis	-	7	-	7	<1	5
Hordeum vulgare	Caryopsis	2	55	-	57	4.7	30
Hordeum vulgare var. *vulgare*	Caryopsis	10	439	-	449	36.9	52.5
Hordeum vulgare	Glume base	-	2	-	2	<1	2.5
Empetrum nigrum	Seed/fruit	1	2	-	3	<1	7.5
Empetrum nigrum leaves	Leaves	2	29	-	31	2.6	17.5
cf. *Prunus* sp.	Nut/stone fragment	1	-	-	1	<1	2.5
Poaceae	Seed/fruit	8	16	-	24	1.9	22.5
Poaceae small cf. *Alopecurus*	Seed/fruit	-	16	-	16	1.3	17.5
Poaceae large cf. *Bromus/Elymus*	Seed/fruit	1	-	2	3	<1	5
Poa sp.	Seed/fruit	5	48	-	53	4.4	35
Carex distigmaticae	Seed/fruit	-	8	-	8	<1	10
Carex tristigmaticae	Seed/fruit	3	96	2	101	8.3	35
Carex elata/acuta/dioica-type	Seed/fruit	-	12	-	12	<1	15
Luzula multiflora	Seed/fruit	-	6	-	6	<1	2.5
Potentilla anserina	Seed/fruit	-	5	-	5	<1	5
Ranunculus acris	Seed/fruit	-	1	-	1	<1	2.5
Galeopsis bifida/speciosa	Seed/fruit	-	7	-	7	<1	10
Polygonum aviculare	Seed/fruit	-	43	-	43	3.5	5
Rumex sp.	Seed/fruit	1	-	-	1	<1	2.5
Rumex acetosella	Seed/fruit	-	5	-	5	<1	2.5
Spergula arvensis	Seed/fruit	-	3	-	3	<1	7.5
Atriplex sp.	Seed/fruit	-	2	-	2	<1	5
Atriplex glabriuscula	Seed/fruit	-	6	-	6	<1	5
Carduus/Cirsium	Seed/fruit	-	1	-	1	<1	2.5
Lamiaceae	Seed/fruit	1	-	-	1	<1	2.5
Mentha sp.	Seed/fruit	1	-	-	1	<1	2.5
Menyanthes trifoliata	Seed/fruit	1	26	-	27	2.2	20
Caryophyllaceae	Seed/fruit	-	14	-	14	1.2	7.5
Stellaria media	Seed/fruit	5	151	-	156	12.8	55
Fucus-type seaweed	Thallus fragment	21	845	-	866	-	17.5

Cereal cultivation in south-western Norway: Boom and bust in the Late Neolithic and Early Bronze Age

Eli-Christine Soltvedt

Author address:
Arkeologisk Museum, University of Stavanger, Norway

E-mail:eli.c.soltvedt@uis.no

Abstract

Macrofossils recovered from settlement structures and features have contributed to an understanding of how agrarian life arose and developed during the first period of cereal cultivation in Norway. The earliest dated carbonised cereals from Rogaland, c. 2300 cal BC, were discovered in a pit that could not be linked to any contemporaneous buildings. Relatively high densities of naked barley (*Hordeum vulgare* var. *nudum*) and emmer (*Triticum dicoccum*) are a consistent element in the earliest longhouses that appear in Rogaland, about 200 years later. There is an uneven distribution of cereals across Rogaland, with houses located in Sola and Klepp found to contain large amounts of cereals. The so-called 'Neolithic package' was complete in south-western Norway by 2100-2000 cal BC. Numbers of direct radiocarbon dates for cereals show an increase towards the end of the Late Neolithic (1950 cal BC) and then decrease in the Early Bronze Age (1650 cal BC). A decrease has also been documented in the number of two-aisled longhouses from around 1500 cal BC. There is a link between large amounts of charred cereals and the two-aisled houses in Rogaland. When three-aisled houses appear, the concentrations of cereals found within these buildings are smaller than in the previous two-aisled houses. Cereal cultivation fluctuates through the period from 2200-1400 BC. But although there is a distinct decrease in the Early Bronze Age, cereal cultivation did not cease at this time.

Introduction

This chapter presents an overview of the plant species grown and used in south-western Norway during the Late Neolithic and Early Bronze Age and discusses the findings in a northern European context. Carbonised cereals are reported from post-built longhouses, hearths and prehistoric field layers elsewhere in southern Norway from c. 2400 BC onwards (Bårdseth and Sandvik 2010; Glørstad 2003; Hjelle and Solem 2008; Soltvedt 2000; Soltvedt and Henningsmoen 2016). In the south-western part of Norway, two domesticated plant species become established in the Late Neolithic: emmer (*Triticum dicoccum*) and naked barley (*Hordeum vulgare* var. *nudum*) (Soltvedt 2000). In contrast, in southern Sweden, Denmark, northern Germany, Ireland and Britain, cereal cultivation is first recorded in the Early Neolithic (Høyem Andreasen 2009; Kirleis and Fischer 2014; McClatchie et al. 2014; Stevens and Fuller 2012; Sørensen and Karg 2014).
Domesticated cereals are present at well-known settlements in the Middle East, such as Tell Aswad, Abu Hureyra and Chayöhü, as early as 11 000-12 000 years ago (Zohary et al. 2012). Agriculture became established along the major river valleys in Germany from 5500-4500 BC, and cultivation of cereals and other crops was a crucial element in the economy of the Linear Pottery culture (Zohary et al. 2012). Early domesticated plants, such as flax (*Linum usitassimum*), pea (*Pisum sativum*), lentil (*Lens culinaris*), vetch (beans) (*Vicia* spp.), wheat (*Triticum* spp.) and barley (*Hordeum* spp.), were cultivated. As agriculture spread northward and westward in Europe, several species were lost from the suite of crop plants. This was due to the plants' demands for a tolerable climate and a lack of local food-preparation traditions (Bakels 2014). After domestication in the Middle East, approximately 7000-8000 years elapsed before emmer and barley were cultivated in Norway. The transfer of knowledge and skills, and the development of cereal crops genetically adapted to a colder and more humid climate, a shorter growing season and a longer day length in summer, were major challenges for early farmers in Norway.
No charred cereals have thus far been recorded from the Early Neolithic and first part of the Middle Neolithic (4000-2600 BC) in Norway. Despite this, a modest number of 'agro-Neolithic' artefacts have been found, and limited deforestation can be observed

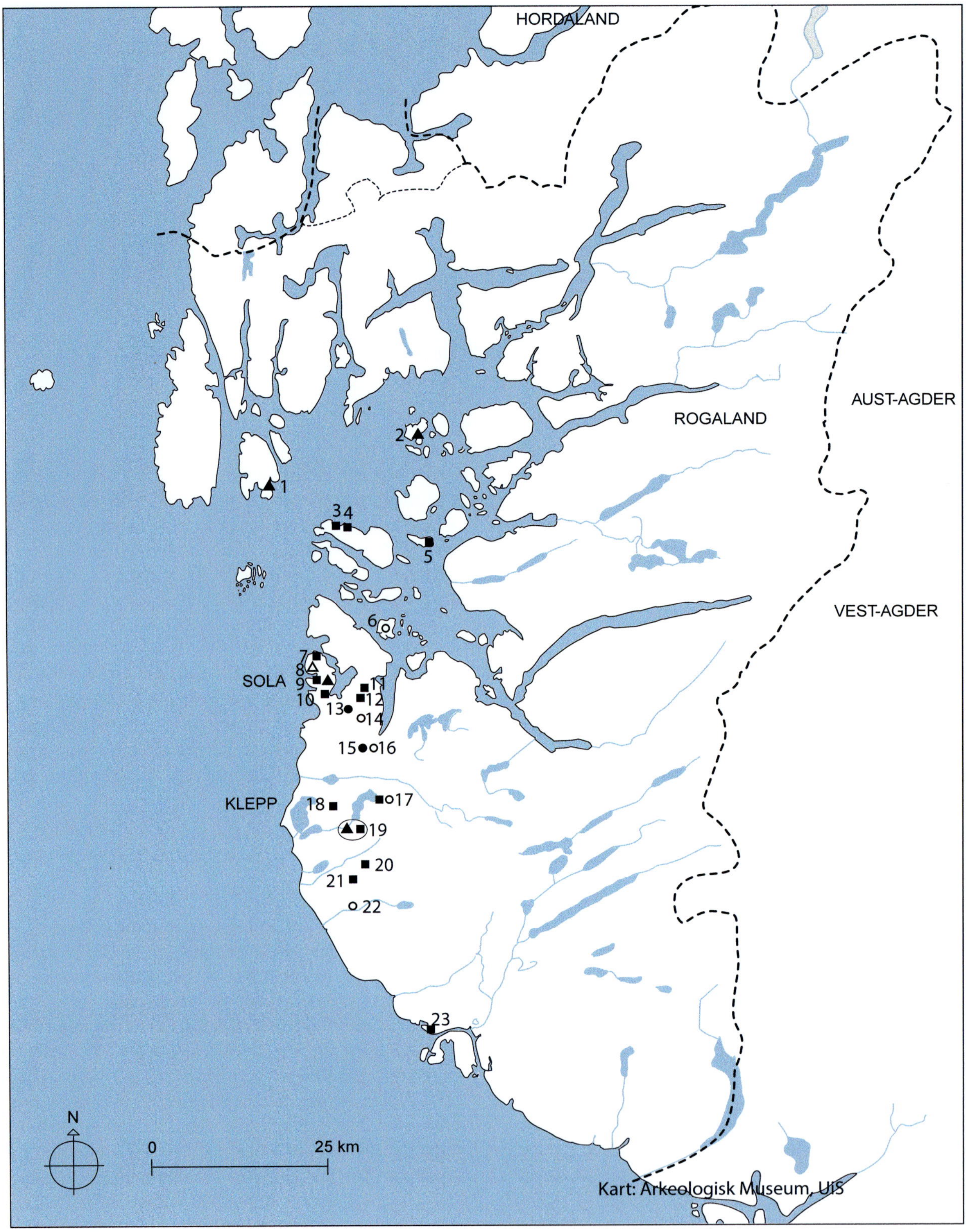

Figure 1. Map of the county of Rogaland, Norway, showing analysed sites and contexts.

- ■ Two-aisled house
- ● Unassociated posthole
- ▲ Prehistoric field layer
- ○ Pits, cultural layer
- ∆ Ritual place

1 Bokn; 2 Eik; 3 Voll; 4 Sørbø; 5 Ystebø; 6 Austbø; 7 Jåsund; 8 Myklebust; 9 Tjora; 10 Sømme; 11 Jåttå; 12 Røyneberg; 13 Skadberg; 14 Soma; 15 Sørbøtunet; 16 Sørbø Hove; 17 Frøyland; 18 Kleppestemmen; 19 Kvåle; 20 Bjorhaug; 21 Kvia; 22 Tjemslandsmarka; 23 Hellvik.

in the pollen diagrams for this period in Rogaland (Høgestøl and Prøsch-Danielsen 2006; Simonsen and Prøsch-Danielsen 2005). The latter is interpreted as reflecting foraging activity for fodder from broad-leaved trees. Bones of domesticated animals from the same period have been found in the rockshelter site of Stangelandshelleren, in Rogaland, Norway (Høgestøl and Prøsch-Danielsen 2006). Cereal pollen records from southern Norway imply small-scale cereal cultivation prior to the Late Neolithic (2300-1800 BC) (Hjelle et al. 2006; Høgestøl and Prøsch-Danielsen 2006; Prøsch-Danielsen 1996; Soltvedt and Henningsmoen 2016). In the late Middle Neolithic and Late Neolithic (2600-1800 BC), the number of artefacts related to the Single Grave and Battle Axe cultures increases in Rogaland, and the deforestation process clearly continued (Høgestøl and Prøsch-Danielsen 2006).

Charred cereals have been found inside two-aisled houses, demonstrating the existence of an agrarian society in Rogaland from 2200 cal BC (Soltvedt 2000, 2005). Agriculture-related artefacts are also present in Rogaland in the Late Neolithic and Early Bronze Age. Pollen evidence reveals an opening up of the forest, leading towards the development of heathland or grassland and permanent fields (Høgestøl and Prøsch-Danielsen 2006). The breakthrough in cereal cultivation evident in the Late Neolithic in western Norway can be compared to the abrupt transition to agriculture seen in Ireland and the British Isles around 3750 BC (Early Neolithic) (McClatchie et al. 2014; Stevens and Fuller 2012; Whitehouse and Kirleis 2014).

New archaeobotanical data have become available due to strategic and extensive sampling initiatives in Rogaland over the past 20 years (Hjelle et al. 2016). Most of the data have only been published in grey literature, and the cultivation of domesticated plants has therefore not been fully considered in the debate on neolithisation in Norway.

The aim of this paper is to elucidate cereal production in Rogaland, southern Norway, in the Late Neolithic and Early Bronze Age, through examining the uneven distribution of cereal remains, both in chronological and geographical terms, and the artefacts recovered from the sites of the time. The paper further gain insight into agriculture and economy and how it changed over time.

Materials, methods and sites

Archaeobotanical data from Late Neolithic and Early Bronze Age sites (2300-1500 BC) in Rogaland are presented below. The sites were investigated in connection with rescue archaeology undertaken prior to the construction of roads, buildings and water pipelines, undertaken by both the public and the private sector. The Museum of Archaeology at the University of Stavanger was the administrator of these archaeological projects. The interpretations and conclusions presented in this paper are based on charred cereal grains, weed seeds, seeds of collected wild plants and hazelnut shell fragments recovered from features related to 23 sites. A total of 22 two-aisled houses were recognised at 14 of these sites. The background for the interpretation made in this paper are the analyses of samples from approximately 500 features. The sample strategy has changed over time. Initially, few features were sampled and analysed for palaeobotanical remains and, in some cases, radiocarbon dated. Later on, nearly100% of the features were sampled and analysed. The postholes for the roof supports were prioritised, and features in wall rows and the interior were sampled from every part of the house. A total of 111 AMS ^{14}C-dates have been obtained for cereals from the Late Neolithic and Early Bronze Age. The ^{14}C-dated cereals derive from selected features at the 23 sites (Fig. 1). At Bokn site 1, Eik site 2, Skadberg site 14, Sørbøtunet site 15 and Hove Sørbø site 16, traces of settlement activity from the Late Neolithic and Early Bronze Age were either weak or absent, but cereals were found and ^{14}C dated (Table 1).

Mechanical stripping of the topsoil was the primary field method used in investigating the settlement areas (Løken et al. 1996). Samples were processed in a machine designed to separate organic from minerogenic material by water flotation (Bakkevig et al. 2002), and the organic remains were retained on a sieve with a mesh aperture of 0.5 mm. Sample volumes before processing were 0.3-5.5 litres.

The plant nomenclature follows Lid and Lid (2005) and Streeter et al. (2010) for scientific names. Table 2 shows the prehistoric periods and abbreviations used in the text. All ^{14}C dates were calibrated using the OxCal v 4.3.2 programme at two δ (Stuiver et al. 1998). Calibrated dates are given in years BC and AD.

Results

Charred cereal grains were recovered from house features in the two-aisled houses excavated and sampled in Rogaland (Table 1). Cereal grains dated to the Late Neolithic and Early Bronze Age were recovered from a further nine sites; they derived from a ritual place, pits and field layers, in locations where two-aisled houses were not found in the vicinity. Charred cereals dated to the Late Neolithic were also recovered from two Pre-Roman Iron Age houses, demonstrating agricultural activity at the sites in the Late Neolithic (sites 13 and 15, Fig. 1, Table 1).

Cereal grains from the two-aisled houses have been ^{14}C dated, and the results fall within the period 2200-1450 BC (Fig. 2). ^{14}C-dated cereal grains from the house areas at Kvia and Voll and from houses 10 and 13 at Tjora suggest that these are some of the earliest two-aisled houses in Rogaland, while the house areas

Table 1. Information about sites in Rogaland, from which charred cereals from the Late Neolithic and Early Bronze Age have been recovered. For abbreviations, see Table 2.

Site name and site number. Refers to map in figure 1		Number two-aisled houses	Cereal finds contexts for Late Neolithic and Early Bronze Age	Late Neolithic and Early Bronze Age artefacts	Later periods	Cereals, later date, recovered from the site	References
Bokn	1		Field layers		Three-aisled houses	EBA and VP	Reiersen et al., in progress
Eik	2		Field layers		Field layers	LBA and MP	Jensen, Fredh, & Ahlquist, 2015
Voll	3	1	Two-aisled house	Bifacial flint arrowheads	Culture layer	EBA	Høgestøl, 1995; Soltvedt, 2000
Sørbø	4	1	Two-aisled house	Flint scraper and flint sickle fragments	Field layer	PRIA	Høgestøl, 1995; Soltvedt, 2000
Ystebø	5	1	Two-aisled house	Loom weight	Settlement features RIA	RIA	Soltvedt, 2000
Austbø	6		Hearth and culture layers	Bifacial flint arrowheads, pottery fragments	Three-aisled houses EBA, EIA	EBA	Meling, 2015; Meling, 2016
Jåsund	7	1	Two-aisled house	Pottery fragments	Field layer, homestead RIA	RIA	Fyllingen, 2012; Soltvedt & Jensen, 2012
Myklebust	8		Pit, ritual place	Pottery fragments, probably from small cups	Three-aisled houses BA, EIA	EBA, RIA, MiP	Dahl, 2014; Gil, 2012; Sandvik, 2018
Tjora	9	7	Two-aisled house, field layer	Bifacial flint arrowheads, spoon-shaped scrapers, flint sickle, thick-butted axe, pottery fragments	Three-aisled houses PRIA, RIA, burials LN, BA, PRIA, RIA, MiP, LIA, field layers	EBA, LBA, PRIA, RIA, MiP, VP	Fyllingen & Armstrong, 2012; Soltvedt & Jensen, 2011
Sømme	10	1	Two-aisled house	Pottery fragments, flint scrapers	Three-aisled houses, BA-PRIA, MeP-MP	LBA, PRIA, MeP, VP, MP	Meling, 2017; Meling et al., 2020
Jåttå	11	1	Two-aisled house	Flint scraper	Settlement features	EBA, LBA, MeP	Soltvedt, 2005
Røyneberg	12	1	Two-aisled house	Flint borer	Three-aisled house, LBA	LBA	
Skadberg	13		Posthole		Three-aisled houses IA, burials VA	LBA-ERIA	Bjørlo, 2011; Soltvedt, 2011
Soma	14		Clearance and culture layers	Bifacial flint tools	Fairy circles	RIA	Holst, 1997
Sørbøtunet	15		Posthole		Three-aisled houses, IA	LBA, MeP, MiP	Sandvik, 1999
Hove Sørbø	16		Culture layers and pit		Approximately 100 houses, BA, IA	LBA, IA, MP	Bjørdal 2014; in progress; Westling in progress
Frøyland	17	2	Two-aisled house	Bifacial tools fragments, pottery fragments	Burials, IA, VA		Bjørdal, 2009
Kleppestemmen	18	1	Two-aisled house		Three-aisled house, EBA		Bakkevig, 2008; Berge, 2007
Kvåle	19	2	Two-aisled house	Bifacial flint arrowheads, flint dagger shaft, pottery fragments	Three-aisled house, EBA, MiP, field layers, cairns, settlement features	LBA, PRIA, MiP, VP, MP	Soltvedt, Løken, Prøsch-Danielsen, Børsheim, & Oma, 2007
Bjorhaug	20	1	Two-aisled house		Three aisled houses IA, field layers RIA	PRIA, RIA, MeP	Fredh & Westling, 2014; Fyllingen, 2015
Kvia	21	1	Two-aisled house	Pottery fragments, probably from small cups	Burial mound, field layers	PRIA	Bjørlo & Denham, 2011; Westling & Overland, 2012
Tjemslandsmarka	22		Burial cairn, bottom layer	Pottery fragments	Cairns, settlement feature	EBA, PRIA	Bortheim & Dahl, 2014; Fredh & Soltvedt, 2014
Hellvik	23	1	Two-aisled house				Sandvik & Jensen, 2009

Table 2. Norwegian prehistoric periods and abbreviations used in this chapter.

Abbreviation	Periods	Uncalibrated years BP	Calibrated years BC/AD
EN	Early Neolithic	5200-4700 BP	4000-3300 BC
MN	Middle Neolithic	4700-3900 BP	3300-2400 BC
LN	Late Neolithic	3900-3500 BP	2400-1800 BC
EBA	Early Bronze Age	3500-3000 BP	1800-1200 BC
LBA	Late Bronze Age	3000-2440 BP	1200-500 BC
PRIA	Pre-Roman Iron Age	2400-2010 BP	500 -1 BC
RIA	Roman Iron Age	2010-1680 BP	AD 1-400
MiP	Migration Period	1680-1500 BP	AD 400-570
LIA	Late Iron Age	1500-100 BP	AD 570-1050
MeP	Merovingian Period	1500-1210 BP	AD 570-800
VP	Viking Period	1210-1000 BP	AD 800-1050
MP	Medieval Period	1000-	AD 1050-1536

at Sørbø, Ystebø, Sømme, Bjorhaug and Hellvik, and house 7 at Tjora, have some of the latest two-aisled houses within the prehistoric period. Small amounts of cereal grains were recovered from these 'late' houses (Table 3, Fig. 3a). The dates obtained from the Jåsund (site 7, Fig. 1), Jåttå (site 11, Fig. 1) and Kvia (site 21, Fig. 1) house remains differ substantially within the period. The earliest dated cereal grains are from Frøyland, in Rogaland (3875±40, 2468-2211 cal BC, TUa-6803), and were recovered from a posthole located within the outline of a two-aisled house (site 17: Frøyland house 1) that is c. 400 years later than the cereal grain. The second- and third-earliest cereal grains derive from a pit (Austbø, site 6) and a ritual place (Myklebust, site 8) with no relation to early houses. The probability curve for 111 ^{14}C-dated cereal grains shows a relatively strong representation of cereals between 1950-1650 BC in Rogaland (Fig. 2).

Samples were taken and analysed from 100% (Kvåle), 19% (Voll) and 39% (Frøyland) of the features belonging to the houses. At Kvåle house 1 (site 19, Fig. 1), Voll (site 3, Fig. 1) and Frøyland (site 17, Fig. 1), 100% of the analysed features contained cereals (Table 3, Fig. 3b). In only two of the 22 houses were cereal grains present in less than 70% of the sampled features (Table 3, Fig. 3b). Although cereal grains were found in a high proportion of features, the actual numbers of cereals in these features were generally low (Table 3, Fig. 3a). About 20% of the features containing cereals had less than one cereal grain per litre of soil. Samples of less than 1 litre contained little if any archaeobotanical material. Only one feature contained more than 1000 cereal grains, and 35 features contained more than 100 cereal grains.

Naked barley was dominant in the sampled features, while emmer grains were less common, corresponding to about one third of the amount of naked barley (Table 3, Fig. 4). Hulled barley grains are also represented, but only in small numbers. Most of the emmer grains came from Tjora house 15 at site 9, Frøyland house 2 at site 17 and Kvåle houses 1 and 2 at site 19 (tab 3). The emmer found in Tjora house 15 and Kvåle house 2 came from a single feature in each case (Soltvedt and Jensen 2011; Soltvedt et al. 2007).

The largest numbers of cereal grains were found at Kleppestemmen, Frøyland and Kvåle (Fig. 3a), which are all located in the Klepp district (Fig. 1). At both Frøyland and Kvåle, two two-aisled houses had been built on the same spot, but they could be distinguished because they were oriented at a slightly different angle (Fig. 5). Nevertheless, we acknowledge the possibility of contamination between the houses. Many samples from each house were analysed (Table 3). A trend in the results is recognisable. The latest two-aisled houses at both sites were the longest and broadest; they were occupied between 1900 and 1700 BC. It is possible to construct a relative chronology for the five houses based on the ^{14}C-dates: The earliest is Frøyland house 1, followed by Kvåle house 2, Kleppestemmen and Frøyland house 2. Kvåle house 1 as the last in the series (c. 1780-1620 BC). It has the greatest concentration of both naked barley and emmer grains.

Seven to eight two-aisled houses from the Late Neolithic and Early Bronze Age were recorded at the Tjora site, and some of these buildings had been in use simultaneously (Fyllingen and Armstrong 2012). Field layers containing charred cereals dated to the Late Neolithic (c. 2000-1825 BC) provide evidence of cereal cultivation at the site. The distribution of ard marks, but also the charred, ^{14}C-dated cereals from field layers, indicate a field size of at least 800 m^2.

The samples contained relatively few seeds of common weeds (4%) (Fig. 6). Seeds of pale persicaria (*Persicaria lapathifolium*) and redshank (*Persicaria maculosa*) were most numerous, followed by those of corn spurry (*Spergula arvensis*) and common chickweed (*Stellaria media*) (Fig. 7). Collected wild plants were also represented (Fig. 8), among which fragments of hazelnut shell dominated. Hazelnut (*Corylus avellana*), raspberry (*Rubus idaeus*) and crowberry (*Empetrum nigrum*) could have been used as food. Berries of juniper (*Juniperus communis*) and bearberry (*Arctostaphylos uva-ursi*) could have been used

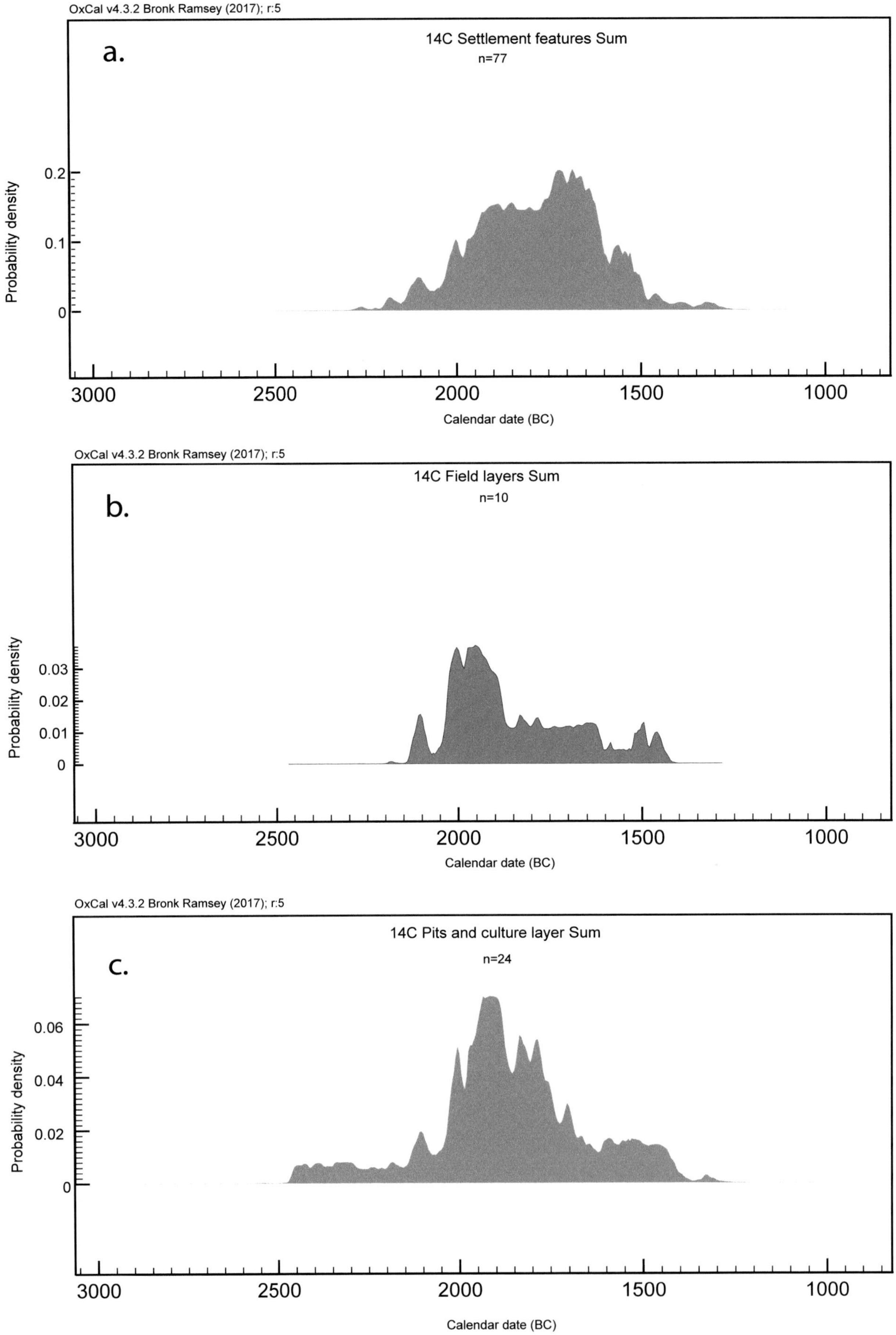

Figure 2. Summed probability distribution of radiocarbon-dated charred cereals from Rogaland, in the period 2500–1250 BC. a charred cereals from houses; b charred cereals from field layers; c charred cereals from pits, cultural layers and settlement features.

Table 3. Charred cereal grains found in features related to two-aisled houses in Rogaland.

Site number (fig. 1, tab. 1) : Site name house number	Number of features with cereals / Number of features sampled	Litres analysed from features containing cereals / total litres analysed from house	Number of cereal grains	Features with more than 1000 cereal grains	Features with more than 100 cereal grains	Cereal grains per litre soil	Minimum cereal grains per litre	Maximum cereal grains per litre	Hordeum vulgare var. vulgare, hulled barley	Hordeum vulgare var. nudum, naked barley	Hordeum vulgare sp, barley	cf. Triticum aestivum, bread wheat	Triticum dicoccum, emmer	Triticum sp, wheat	Avena, oat	Cereal unidentified	Cereal fragments unidentified
19:Kvåle house 1	47/47	242	5296	1	13	21.9	0.5	117	105	1004	352	0	681	221	0	715	6654
19:Kvåle house 2	20/23	99.5/105	3075.5		4	30.9	1	35	51	790	116	0	296	48	0	481	3582
3:Voll	9/9	38	144			3.8	0.4	15	0	71	0	0	0	5	0	68	
5:Ystebø. Talgje	27/44	77.8/102	145			1.9	0.3	6.7		82	3.0	0	0	5	0	25	92
11:Jåttå	39/49	265/309	346.3			1.3	0.1	11.7	2	97	13.0	0	0	17	0	85	397
12:Røyneberg	12/13	73/75	76			1	0.1	1.7		11	5.0	0	0	17	0	17	78
9:Tjora house 7	7/8	28/33	31			1.1	0.7	1.8	2	19	0	0	2	0	0	0	24
9:Tjora house 9	14/17	59/65	100			1.7	0.2	6.5	0	42	4	0	0	7	0	26	63
9:Tjora house 10	10/15	33/47	48			1.4	0.1	7.5	1.0	27.0	0	0	1	1	0	6	35
9:Tjora house 11	12/16	88/103	109			1.2	0.1	3.6	1	24.0	14.0	0	9	2	0	31	84
9:Tjora house 13	10/12	29/36	23			0.8	0.1	1.4	0	8	4	0	0	1	0	7	11
9:Tjora house 15	6/8	21/25	481		1	23.1	0.3	255	0	11	1	0	246	73	0	10	420
9:Tjora house 12	14/33	36/62	30			0.8	0.3	2.1	0	13	0	0	0	0	0	12	17
7:Jåsund	58/79	151/192	528		1	3.5	0.1	55	0	232	14	0	12	8	0	178	252
21: Kvia	25/28	86/90	1181		3	13.7	0.4	68	2	80	401	3	15	9	0	417	763
17:Frøyland house 1	12/12	68	305		1	4.4	0.1	9	0	101	42	0	5	0	0	105	158
17:Frøyland house 2	14/16	100/105	2902		7	29.0	0.2	90	0	1387	10	0	193	0	0	620	2075
4:Sørbø	8/10	12.4/13.2	207		1	16.7	0.5	10	0	49	9	0	6	1	0	142	0
23:Hellvik	11/27	32/75	41			1.3	0.3	3.3	6	0	13	0	0	3	1	18	
18:Kleppestemmen	27/35	112/127	1685		4	15.0	0.3	38	0	1064	88	0	41	0	0	149	1028
20:Bjorhaug	11/17	20/32	65			3.3	0.1	154	0	3	10	0	1	2	0	34	44
10:Sømme	14/15	44.1/44.6	127			2.9	0.2	5	0	48	0	0	6	0	0	49	73

as medicine and as spices. Remains of weeds and collected wild plants were present at low concentrations in this material, relative to the cereal grains (Fig. 6). Charred seeds of sedges (Cyperaceae) are generally damaged, poorly preserved, and therefore difficult to identify to species. These seeds are constantly present in the features from prehistoric houses, and it has been argued that artefacts made of basketry probably were produced from collected sedges (Soltvedt and Henningsmoen 2016).

House remains, artefacts, clearance cairns, burials and field layers from later periods have been recorded at the sites presented in this paper (Table 1). Because of the re-use of the sites, there will always be a risk of contamination, which can be reduced by employing ^{14}C-dating, knowledge about crop-species and knowledge about the patterns of different features in prehistoric houses. The data presented in this paper show trends in cereal cultivation in Late Neolithic and Bronze Age Rogaland. Cereals from the later Bronze Age, Iron Age, Viking Age or Medieval period have been ^{14}C dated from 85% of the sites. Few artefacts were found within the two-aisled houses in Rogaland (Table 1). Pottery fragments were the most abundant. At the Myklebust and Kvia (sites 8 and 21, fig 1), it is suggested that some of these potsherds derived from small cups (Bjørlo and Denham, 2011; Gil, 2012). It is difficult to determine the functions of the various vessel types, but ethnographical studies referred to by Haaland (2006) suggest that small cups were used for serving and drinking. Flint sickles were found at only two of the sites (Tjora site 9 and Sørbø site 4, fig 1). A few arrowheads and spoon-shaped scrapers were recovered too at Voll, Austbø, Tjora and Kvåle (Table 1), the arrowheads indicating that the economy was also based on hunting. The Late Neolithic and Early Bronze Age sites from which charred cereals have been recovered are more numerous than typical hunting and fishing sites in Rogaland, which is partially a result of where rescue excavations have been carried out. At these typical hunting and fishing sites, a few tools and some ^{14}C-dates reveal activity in the Late Neolithic and Early Bronze Age (Gjerland 1990; Solberg 2014).

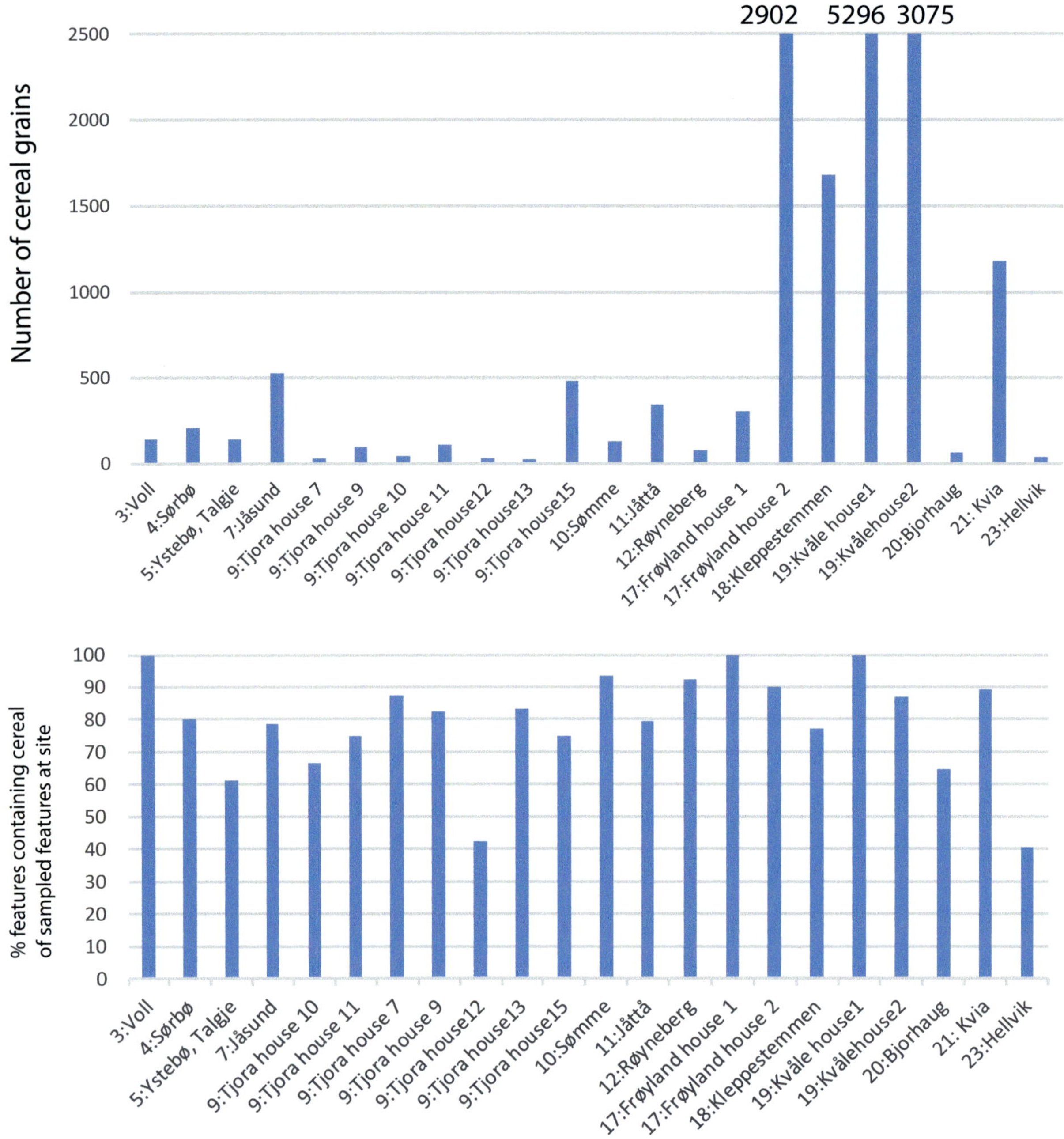

Figure 3. Occurrence of cereal grains in two-aisled houses in Rogaland. a total number of cereal grains; b percentage of features containing cereal grains.

Discussion

To date, there is evidence from 23 sites demonstrating cereal cultivation in Rogaland during the Late Neolithic and Early Bronze Age. The areas of Klepp and Sola are special with regard to cereal cultivation (Fig. 1). In the former, the sites of Kleppestemmen, Frøyland and Kvåle (sites 18, 17 and 19) have a high proportion of features containing cereals, which occur in relatively large amounts (Fig. 3). At Kvåle and Frøyland, two houses were built after each other on the same spot, suggesting that people lived here for a long time (Fig. 5). The latest of these two-aisled houses (1780-1620 BC) are the longest and broadest, most probably due to an expansion of the household. At both Kvåle (Soltvedt et al., 2007) and Kleppestemmen (Bakkevig 2008; Berge 2007), a three-aisled house from the Early Bronze Age (c. 1650-1450 BC) was encountered, further emphasising the extended stability and importance of the area. The Sola area, where sites Jåsund 7, Myklebust 8, Tjora 9 and Sømme 10 are located (Fig. 1), appears to be special because of the number of two-aisled houses (at least 10), all of which contained charred cereals. Charred cereals were also found in field layers at the Tjora site (site 9, Fig. 1), confirming cereal cultivation in the Late Neolithic (Soltvedt and Jensen 2011).

At Sola, the earliest ^{14}C-dated cereals (3775±40 BP,

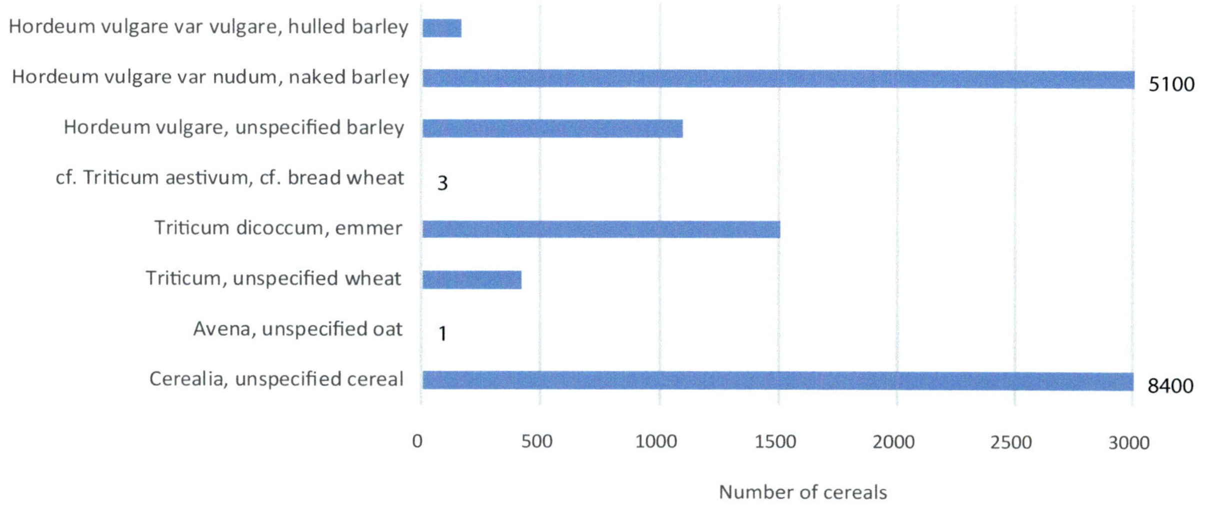

Figure 4. Number of cereal grains per species in two-aisled houses in Rogaland.

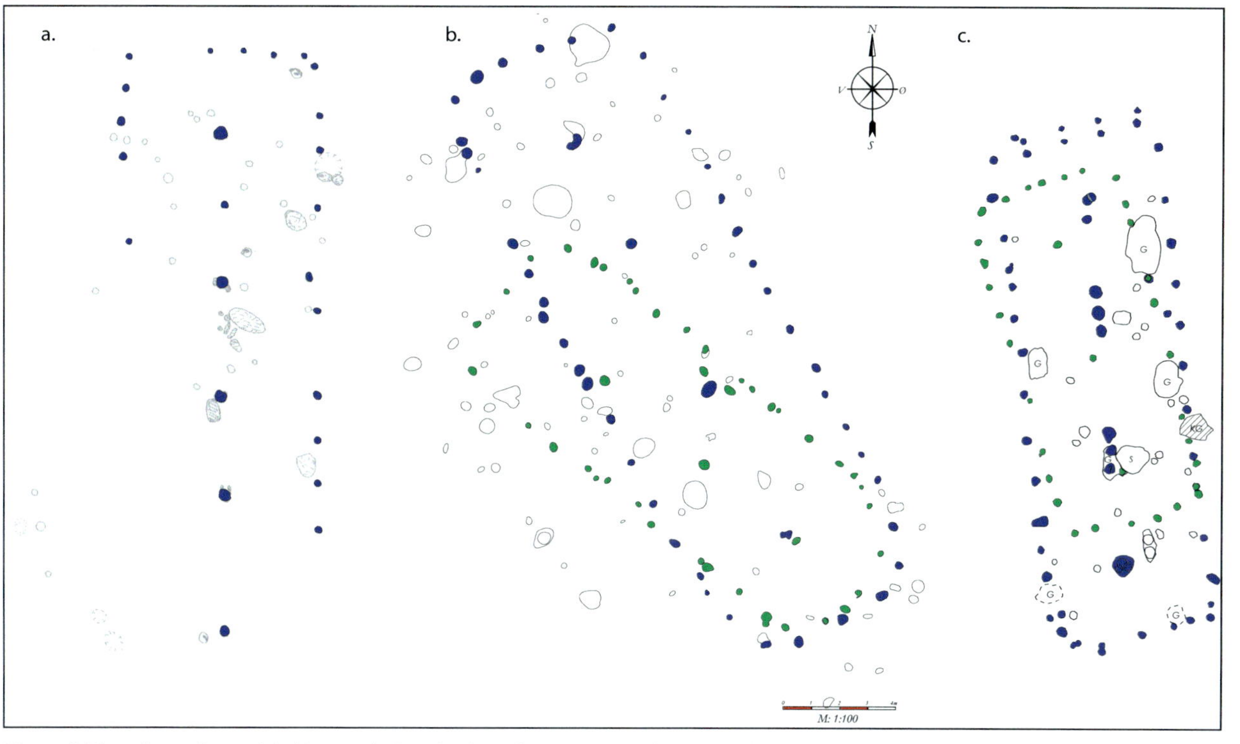

Figure 5. Plan views of two-aisled houses in Rogaland. a Kleppestemmen site no. 18; b Kvåle site no. 19 (houses 1 and 2); c Frøyland site no. 17 (houses 1 and 2). Green marks the earliest houses, blue the latest.

cal 2338-2039 BC, TRa-4046) are from the ritual place of Alvasteinen (elf stone) at Myklebust (site 8, Fig. 1). A broad spectrum of plants and plant products have been found within the framework of ritual contexts (Hansson and Heiss 2014). Elsewhere in the world, in Egyptian contexts (tombs), a number of corn-mummies depicting animals and humans have been encountered, and cereals have been found placed or scattered around the deceased (Centrone 2005). It is assumed that this is an expression of the idea of renewal and resurrection for the deceased. The presence of charred cereals at the ritual place at Myklebust (site 8) suggests that cereals may have been of ritual importance even before the appearance of the first two-aisled houses.

The concentrations of cereals in the Rogaland two-aisled houses are low, but not exceptionally so when compared with those of Neolithic sites elsewhere in Europe (Linear Pottery culture sites) and the British Isles (Bogaard and Jones 2007; McClatchie et al. 2014). Concentrations of less than one cereal grain per litre occur in 20% of the features containing cereals in Rogaland, while at British Neolithic sites more than 60% of the features have concentrations lower than this (Bogaard and Jones 2007).

In Rogaland, naked barley was the dominant crop species in the Late Neolithic and Early Bronze Age. A range of cereal species are reported from Late

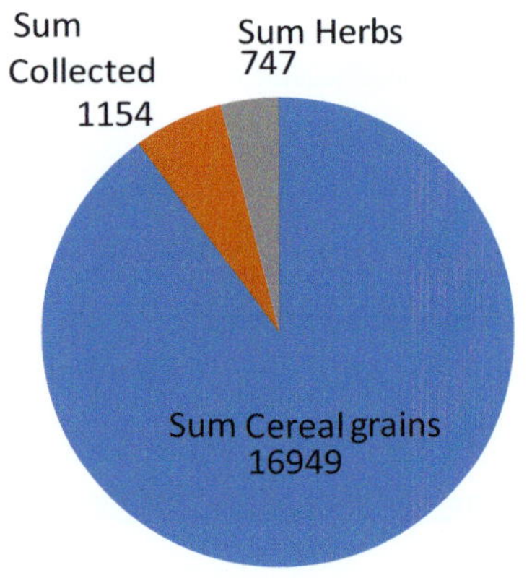

Figure 6. Sum of items per plant category in two-aisled houses in Rogaland.

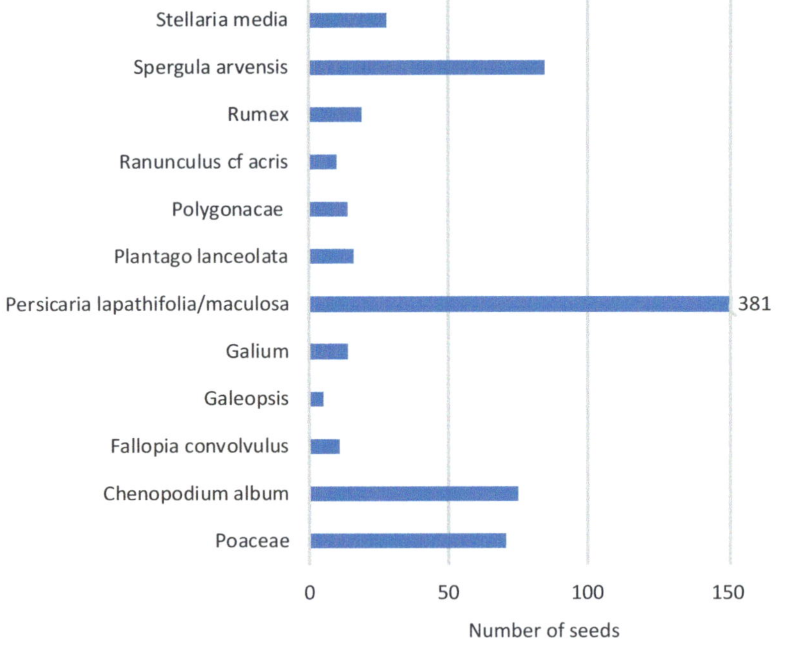

Figure 7. Number of items of weed species in two-aisled houses in Rogaland.

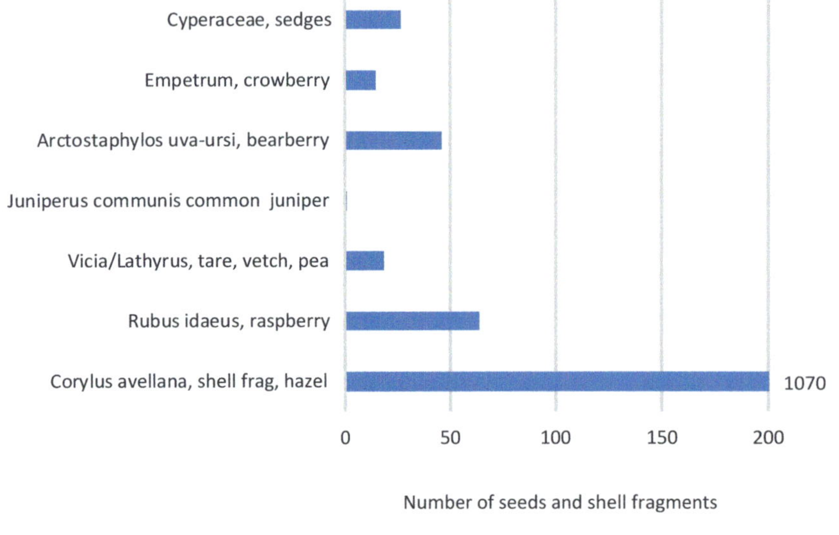

Figure 8. Number of items of collected plants in two-aisled houses in Rogaland.

Neolithic sites in Jutland, Denmark, but again the dominant crop was naked barley (Høyem Andreasen 2009; Robinson 2003). Naked barley has also been reported from northern Sweden (Viklund 2011), and a review of Neolithic plant remains from Scotland (Bishop et al. 2009) shows that barley was fairly common along the Atlantic coast of Scotland in the Late Neolithic. At some sites in Scotland, emmer is represented by one third of the amount represented by barley. McClatchie et al. (2014) discuss the presence of the various cereal species in Neolithic Ireland, England and Scotland. In England, as in Scotland, barley became more important in the Late Neolithic. It should be emphasised, when comparing plant use in the British Isles with that in Norway, that the Late Neolithic in Britain extended from 3000 to 2500 BC, while in Norway it lasted from 2300 to 1800 BC. According to Stevens and Fuller (2012) there is a lack of reported cereals from the Late Neolithic and especially from the Early Bronze Age. In the Middle Bronze Age (ca. 1700-1200 BC), there was a re-introduction of cereal cultivation, together with a fully fledged agricultural society. Barley, and especially naked barley, must have been the preferred crop in the outer zone of north-western Europe in the period 3000-1700 BC. A wet climate during this period may have been one of the reasons for this choice of cereal type.

In the Late Neolithic, pollen evidence reveals an opening up of the woodland leading towards heathland or grassland and permanent fields (Høgestøl and Prøsch-Danielsen 2006). According to these authors, animal husbandry was also part of the economy of the first cereal cultivators in Rogaland. In the two core districts of Klepp and Sola, the available resources (landscape, soils, etc.) were appropriate for marine and freshwater fishing, crop cultivation, animal husbandry and hunting. Even nowadays, the districts of Klepp and Sola have the highest number of acres for cereal cultivation in Rogaland (data for 2017; Landbruksdirektoratet.no 2017). The focus on cereal cultivation evident in Rogaland in the Late Neolithic is not so prominent in other areas of Norway. In south-eastern Norway and in Trøndelag, there are extensive areas suitable for cereal production (Fig. 9) (Moen 1999). Flint sickles are strongly associated with cereal cultivation in the Late Neolithic and Early Bronze Age, and 103 sickles from Rogaland are recorded in the Musit database, a national digital asset regarding archaeological collection, (Musit database, 2018). There is also an abundance of sickles from south-eastern Norway, but the number is lower in Trøndelag. Artefacts found as stray finds in western Norway show a clear southern influence (Solberg 1993).

Solberg (1993) reviewed stray finds of flint artefacts related to the Late Neolithic and Early Bronze Age in western Norway, in the counties of Rogaland, Hordaland and Sogn and Fjordane. The greatest numbers of flint daggers and flint sickles are found in the districts of Klepp and Sola (Fig. 1). Solberg explains the number and distribution of artefacts in terms of a high population density and a developed social hierarchy in Klepp and Sola during the Late Neolithic and Early Bronze Age. Extensive cereal cultivation and consumption from 2100 to 1500 BC probably constituted an important factor in the formation of a stable, long-term foundation for these societies. From southern Sweden and Denmark, monumental two-aisled houses are known from the Late Neolithic (1950-1700 BC) (Brink 2013; Johannsen 2017). It has been suggested that the variety of crop species found within these houses reflects an important economic contribution to a surplus in a stable society. The settlements in Rogaland may have been influenced by the focus on cereal cultivation in these societies in southern Scandinavia.

Seeds of various weed species of *Persicaria* are most dominant in Rogaland (Fig. 7), and a similar situation prevails in Denmark (Robinson 2003). Similarly, the emphasis on cultivation of naked barley during the Late Neolithic in Rogaland may have been influenced by the focus on naked barley that has been documented during the Late Neolithic in Jutland, Denmark (Høyem Andreasen 2009; Robinson 2003). Even if the two-aisled houses were smaller in Rogaland, the distribution of Late Neolithic artefacts, the long-term occupation of sites in the Late Neolithic and the focus on cultivation of naked barley all indicate close links with societies in southern Scandinavia.

In the rescue archaeology projects discussed here, features strongly associated with houses were sampled preferentially, but cairns and graves also had a high priority. Features with no evident links with settlement structures yielded some of the oldest ^{14}C-dated cereal grains, i.e. those from Austbø site 6, Myklebust site 8 and Frøyland 17 (figs. 1 and 2c). Future archaeobotanical analyses may push the onset of cereal cultivation back a century or so in Rogaland, but the clear increase evident from 2000 BC onwards will probably still stand (Fig. 2).

As is the case for Rogaland, seed assemblages containing low numbers of weed seeds have been reported from the Neolithic in northern Germany (Kirleis et al. 2012) and Denmark (Robinson 2003). The low number of weed seeds may be caused by different factors, such as pioneering cultivation without manuring or the long chain of processing stages before the seeds end up as charred cereals in the features. Samples from British Neolithic sites also contain low numbers of weed seeds, of which a high proportion are annuals (Bogaard and Jones 2007). Field cultivation experiments cited by Bogaard and Jones (2007) suggest a proportion of 75% perennials in fields with shifting cultivation regimes. But when fields are reworked and tilled every year, perennials are disfavoured by the disturbance of the soil and do not become established. In Hordaland (Fig. 9), both shift-

Figure 9. Map of Norway showing regions mentioned in the text. The darker green marks the most important cereal-growing districts in Norway around 1990AD (from Moen 1999 fig. 99).

ing and permanent cultivation may have taken place from the very beginning of cereal cultivation, in the Late Neolithic and Early Bronze Age (Halvorsen and Hjelle 2017). The high proportion of annuals seen in Rogaland indicates that the fields were cultivated permanently, i.e. the earliest farmers maintained their fields for extended periods.

Cereal grains can be kept for many years and therefore represent a reliable and predictable food resource, if yields are acceptable. Cereals must be processed before they can be utilised, but cereal products, such as bread, porridge or beer are difficult to identify archaeologically. The cereals, as archaeobotanists find them, are products of a working process, but they still require further processing and preparation to be edible. The absence of saddle querns in and around the houses (Table 1) indicates that porridge or beer rather than some kind of bread may have been the main cereal product in the Late Neolithic and Early Bronze Age. Three saddle querns have been recorded in Rogaland that can be related to Late Neolithic and Bronze Age contexts (Prøsch-Danielsen and Soltvedt 2011), but none of the cereal remains in the same contexts have been ^{14}C dated.

The archaeobotanical record from Rogaland shows no cereal grains prior to the Late Neolithic (c. 2300 BC), something that has given rise to considerable debate, given the strong evidence of cereal cultivation in Scotland and southern Scandinavia as much as 1500 years earlier (Bishop et al. 2009; Sørensen and Karg 2014). Summed probability distribution of ^{14}C-dates on charred cereals between 2400 and 1 BC are presented in figure 10. The cereals are from excavations conducted over the past 30-40 years in Rogaland. Future excavations and further research may adjust the summed probability distribution curve in figure 10. The distribution shows an increase in finds of charred cereals, here viewed as a proxy for cereal cultivation, around 2100-1950 BC and a decrease in the Early Bronze Age, c. 1600-1500 BC in Rogaland (Fig. 2 and 10). In Ireland and the British Isles, a discernible decline in cereals is reported in the period 2800-2500 BC (McClatchie et al. 2014; Stevens and Fuller 2012). For the British mainland, Stevens and Fuller (2012) have collected 249 ^{14}C-dated cereals. These data show a modest peak around 2200-2000 BC. A distinct increase in cereals from c. 1700 BC is interpreted as a reintroduction of cereal cultivation, together with a well-developed agricultural society (Stevens and Fuller 2012).

Høyem Andreasen (2009) interprets the greater diversity of cereal species demonstrated in the Late Neolithic as reflecting a need for dependable yields and a greater focus on crop cultivation relative to animal husbandry during this period. Archaeobotanical data from northern Sweden indicate a similar trend, with the earliest cereal use or cultivation being recorded between 2600 and 2100 BC. After that time, traces of cereal use are intermittent, and they continue do be so until permanent cereal cultivation is verified in the Early Iron Age in Norrland, northern Sweden (Viklund 2011).

There are several taphonomic events that can contribute to changes in the summed probability of ^{14}C-dated cereals. Both shifting traditions of processing, cleaning, storage, milling and cooking inside the house and ritual behaviour have influenced the deposition of cereals in the past. Food represents sustenance for the body, but scholars have also highlighted its importance as an expression of our relations with other people, and the link that exists between food and ideology (Haaland 2006). Additionally, the changes evident in figure 10 probably also reflect demographic and climatic fluctuations. Analyses of ice-cores, glacier advances and permafrost changes indicate a drop in mean annual temperature around 3000 BC, an increase around 2400 BC, and another drop around 1000 BC (Lilleøren et al. 2012). Based on the distribution of artefacts from the Late Neolithic and the Early Bronze Age, Solberg (1993) interpreted a relatively high population density and a developed social hierarchy, at least in Sola and Klepp in Rogaland. The scale of cereal cultivation evident in Rogaland may reflect a society that, due to agriculture, could maintain a high population density. The prominent decrease in the number of ^{14}C dates of cereals towards the end of the Early Bronze Age (Fig.10) suggests

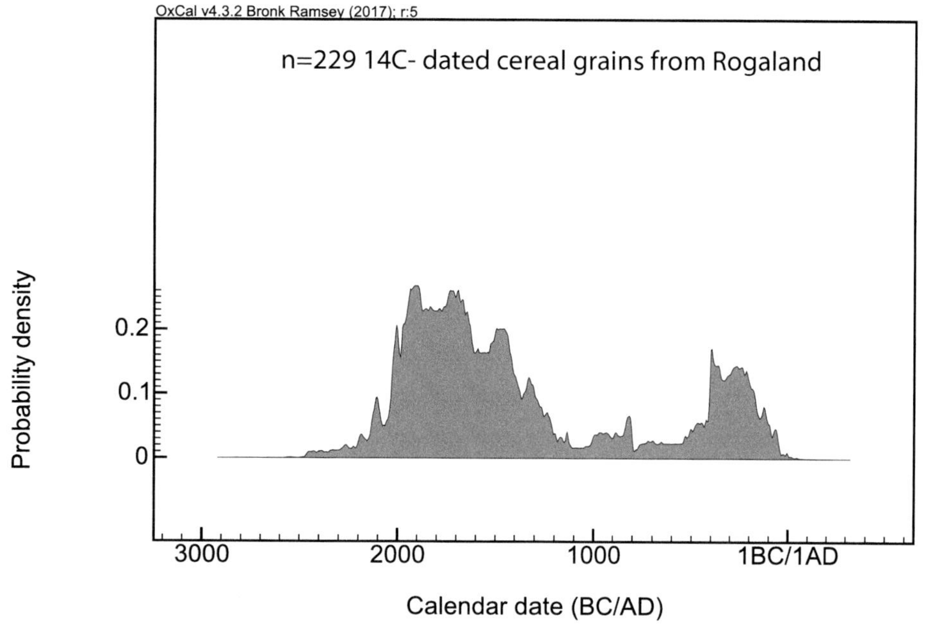

Figure 10. Summed probability of available radiocarbon dates, 2500–1 BC, of cereal grains from excavated sites in Rogaland.

that cereal cultivation was more limited during the later Bronze Age. The continuation of cereal cultivation indicates the importance for cereal products for later Bronze Age society.

Conclusions

The earliest cereals have been found in isolated postholes, pits and a ritual place. In future excavations, archaeobotanical analysis of material from features of this kind should be given high priority. The cereals recovered from the ritual place at Myklebust are from around 2250 BC. They therefore predate documented longhouses in Rogaland and indicate a focus on cereal use around 2300-2200 BC. Based on the material recovered from the 23 sites, an increase in cereals is evident between 2000 and 1950 BC, and a decrease between 1500 and 1400 BC. The most intense period of cereal cultivation according to this study was between 1900 BC and 1650 BC. The archaeobotanical results bear witness to a culture in which cereal cultivation made an important contribution to the economy. This is particularly evident in the material from Klepp and Sola, dated to between 1900 BC and 1650 BC.

At Kvia, Jåsund and Jåttå, ^{14}C-dated cereals indicate household activity from 2200 BC, and over the following 600 years this may have been either continuous or discontinuous. These two possibilities give fascinating alternatives: people living at the same place for several hundred years or successive human populations settling at the same place where people had lived before. At the Tjora, Røyneberg and Kvåle sites, the ^{14}C-dates for cereal grains bear witness to cereal use over a relatively long period during the Late Neolithic. These settlements had, to a major degree, a sedentary and stable life in which cereal production was a significant element.

Naked barley and emmer were the main cultivated species during the Late Neolithic in Rogaland, with naked barley as the dominant species, as seen in Scotland, Ireland and Denmark too. The low incidence of weed seeds is a further similarity in this respect. The coincidence in material culture between Denmark and Norway is understandable because of their geographic proximity, while the similarities with the British Isles indicate that natural circumstances, such as a wet climate and how crops and weeds react to cultivation, are also important.

Charred cereals are a persistent occurrence at archaeological sites from the Late Neolithic (2300 BC) onwards, even if the numbers and concentrations of finds are low in some periods. The presence of cereals in every two-aisled house, and the relatively high concentration of cereals in the features, suggests that cereals were a 'must-have' in the Late Neolithic. The focus on cereal cultivation is related to a wish to maintain the consumption of porridge or beer, achieve predictability in food production and a surplus in the agrarian economy. An important goal may also have been to maintain a high population density and a similar lifestyle and ideology as the elites in central areas of southern Scandinavia.

Acknowledgements

This study was funded by Museum of Archaeology, University of Stavanger, and inspired by the participation in the Agrarian network part of the project 'Forskning i Felleskap' (joint research program) and the research cluster FIA (Prehistoric Farming at the North European fringe—Interdisciplinary Aspects) at the Museum of Archaeology, University of Stavanger. I am thankful to my colleges at the museum for fruitful discussions and give special thanks to Theo Bell Gil for helping with different IT problems. I also appreciate valuable comments on the manuscript from two anonymous reviewers.

References

Bakels C (2014) The first farmers of the Northwest European Plain: Some remarks on their crops, crop cultivation and impact on the environment. J Archaeol Sci 51:94–97.

Bakkevig S (2008) Planterestanalyse av jordprøver fra arkeologiske undersøkelser på Kleppestemmen, Kleppe, gnr 1/22, Klepp kommune, Rogaland (Oppdragsrapport Vol. 2008/23). Universitetet i Stavanger Arkeologisk Museum, Stavanger

Bakkevig S, Griffin K, Prøsch-Danielsen L, Sandvik, PU, Simonsen A, Soltvedt, EC, Virnovskaia T (2002) Archaeobotany in Norway: Investigations and methodological advances at the Museum of Archaeology, Stavanger. In: Viklund K (ed) Nordic archaeobotany: NAG 2000 in Umeå. Archaeology and Environment 15. University of Umeå, Umeå, pp 23–48

Berge J (2007) Undersøking av hustomter frå slutten av yngre steinalder og eldre bronsealder: Kleppe, gnr. 1, bnr. 22 "Kleppestemmen" Klepp kommune (Oppdragsrapport Vol. 2007/5). Universitetet i Stavanger Arkeologisk museum, Stavanger

Bishop R, Church M, Rowley-Conwy P (2009) Cereals, fruits and nuts in the Scottish Neolithic. Proceedings of the Society of Antiquaries of Scotland, 139:47–103

Bjørdal E (2009) Arkeologisk utgraving av toskipa langhus frå overgangen yngre steinalder–eldre bronsealder og graver frå vikingtid og yngre jernalder på Frøyland, Frøyland, gnr. 28 / bnr. 1 mfl. (Time kommune Oppdragsrapport Vol. B 2009/4). Universitetet i Stavanger Arkeologisk Museum, Stavanger

Bjørdal E (2014) Garder og graver frå steinalder til mellomalder på Hove og Sørbø i Sandnes. Frå haug ok heiðni, 3:10–19

Bjørlo AH (2011) Arkeologisk utgraving av bosetningsspor fra eldre jernalder og graver fra yngre jernalder på Skadberg gnr. 32, bnr. 1 og 2, Sola k., Rogaland (Oppdragsrapport Vol. 2011/12). Universitetet i Stavanger Arkeologisk Museum, Stavanger

Bjørlo AH, Denham SD (2011) Arkeologisk og naturvitenskapelig undersøkelse av toskipet langhus fra eldre bronsealder og grav fra eldre jernalder på Kvia: Kvia gnr. 19, bnr. 31, Motland gnr 20, bnr. 3, Hå kommune, Rogaland (Oppdragsrapport Vol. 2011/28). Universitetet i Stavanger Arkeologisk Museum, Stavanger

Bogaard A, Jones G (2007) Neolithic farming in Britain and central Europe: Contrast or continuity? In: Whittle A, Cummings V (eds) Going over the Mesolithic–Neolithic transition in northwest Europe. British Academy, London, pp 357–376

Bortheim K, Dahl BI (2014) Arkeologisk undersøking av røysfelt fra E.BRA og bosettingsspor fra FØRROM i Tjemslandsmarka, Tjemsland Nordre gnr. 53 bnr. 1 og 68, Hå kommune, Rogaland (Oppdragsrapport Vol. 2014/2). Universitetet i Stavanger Arkeologisk Museum, Stavanger

Brink K (2013) Houses and hierarchies: Economic and social relations in the Late Neolithic and Early Bronze Age of southernmost Scandinavia. Eur J Archaeol 16(3):433–458. doi:10.1179/1461957113Y.0000000033

Bårdseth GA, Sandvik PU (2010) An interdisciplinary approach to the understanding of past settlement and farming: Two cases from SE Norway. J Archaeol Sci 37(12):3281–3293. doi: 10.1016/j.jas.2010.07.031

Centrone MC (2005) Behind the corn-mummies. In: Piquette K, Love S (eds) Current research in Egyptology 2003: Proceedings of the fourth annual symposium. Oxbow, Oxford, pp 11–28

Dahl BI (2014) Arkeologisk utgraving av hus og graver, Myklebust gnr 3, Sola kommune (Rogaland Oppdragsrapport Vol. B 2014/20). Universitetet i Stavanger Arkeologisk Museum, Stavanger

Fredh D, Soltvedt EC (2014) Naturvitenskapelige analyser på Tjemslandsmarka: Tjemsland Nordre, gnr 53, bnr 1 og 68, Hå kommune, Rogaland (Oppdragsrapport Vol. 2014/6). Universitetet i Stavanger Arkeologisk Museum, Stavanger

Fredh D, Westling S (2014) Naturvitenskapelige undersøkelser på Bjorhaug: Bjorhaug gnr 25 bnr 2, 18 og Bøhagen, gnr 26 bnr 9, 14, 130 m.fl., Hå kommune, Rogaland (Oppdragsrapport Vol. 2014/25). Universitetet i Stavanger Arkeologisk Museum, Stavanger

Fyllingen H (2012) Arkeologisk utgravning på Jåsund gnr 1, bnr 6, 10, 19/20 and 27, Sola kommune, Rogaland (Oppdragsrapport Vol. 2012/22). Universitetet i Stavanger Arkeologisk Museum, Stavanger

Fyllingen H (2015) Arkeologiske undersøkelser på gårdene Bjorhaug gnr 25 og Bø/Bøhagen gnr 26 i Hå kommune: et bidrag til innblikk i gårdstrukturen på Jæren i perioden 1500 f.Kr. til 800 e.Kr (Oppdragsrapport Vol. 2015/04). Universitetet i Stavanger Arkeologisk Museum, Stavanger

Fyllingen H, Armstrong NJ (2012) Arkeologisk utgravning på Tjora, gnr 10 bnr 5, 17 og 19, Sola kommune, Rogaland (Oppdragsrapport Vol 2012/23B). Universitetet i Stavanger Arkeologisk Museum, Stavanger

Gil TB (2012) Arkeologisk utgravning av lokalitet Alvasteinen på Myklebust gnr. 1134, Sola kommune (Oppdragsrapport Vol. 2012/12). Universitetet i Stavanger Arkeologisk Museum, Stavanger

Gjerland B (1990) Arkeologiske undersøkingar på Haugsneset og Ognøy i Tysvær og Bokn kommunar, Rogaland (Oppdragsrapport Vol. 5). Universitetet i Stavanger Arkeologisk Museum, Stavanger

Glørstad H (2003) Svinesundprosjektet: Bind 2. Varia 55. Universitetets kulturhistoriske museer, Fornminneseksjonen, Oslo

Haaland R (2006) Africa and the Near East: Pot and porridge, bread and oven–two food systems maintained over 10,000 years. In: Wotzka H-P (ed) Grundlegungen: Beiträge zur Europäischen und Afrikanischen Archäologie für Manfred KH Eggert. Francke, Tübingen, 243–254

Halvorsen LS, Hjelle KL (2017) Prehistoric agriculture in western Norway—Evidence for shifting and permanent cultivation based on botanical investigations from archaeological sites. J Archaeol Sci 13:682–696

Hansson AM, Heiss AG (2014) Plants used in ritual offerings and festive contexts. In: Chevalier A, Marinova E, Peña C (eds) Plants and people: Choices and diversity through time. Vol. 1. Oxbow, Oxford, 311–384

Hjelle KL, Solem T (2008) Botaniske undersøkelser–Ormen Lange Nyhamna (Rapport Ormen Lange—Nyhamna). Tapir Academic Press, Trondheim, 483–557

Hjelle KL, Hufthammer AK, Bergsvik KA (2006) Hesitant hunters: A review of the introduction of agriculture in western Norway. Environ Archaeol 11(2):147–170

Hjelle KL, Prøsch-Danielsen L, Soltvedt EC (2016) Potential and recommendation: Agrarian botanical data from western Norway. In: Iversen F, Petersson H, Denham SD, Sundman F (eds) The agrarian life of the North 2000 BC-AD 1000: Studies in rural settlement and farming in Norway 293–342. Portal, Kristiansand

Holst E (1997) Soma—Gudenes drikk? Frà haug ok heiðni 1:4–7

Høgestøl M (1995) Arkeologiske undersøkelser i forbindelse med Rennfastprosjektet, Rennesøy k., Rogaland, Sørvest-Norge. AmS-Varia 23. Universitetet i Stavanger Arkeologisk museum, Stavanger

Høgestøl M, Prøsch-Danielsen L (2006) Impulses of agro-pastoralism in the 4th and 3rd millennia BC on the south-western coastal rim of Norway. Environ Archaeol 11(1): 19–34

Høyem Andreasen M (2009) Agerbruget i enkeltgravskultur—Senneolitikum og ældre bronsealder i Jylland belyst ud fra plantemakrofossiler. In: Laursen J, Nielsen I (eds) Årbog for Jysk Arkæologisk Selskab Kuml 2009. Århus: Aarhus Universitetsforlag, pp 9–56

Jensen C, Fredh DE, Ahlquist J (2015) Stratigrafisk undersøkelse av forhistoriske kulturlag på Eik, Sjernarøyene, gnr. 67, bnr. 2, Finnøy k. (Oppdragsrapport Vol. 2015/11). Universitetet i Stavanger Arkeologisk Museum, Stavanger

Johannsen JW (2017) Mansion on the hill—A monumental Late Neolithic house at Vinge, Zealand, Denmark. J Neolithic Archaeol 19:1–27

Kirleis W, Fischer E (2014) Neolithic cultivation of tetraploid free threshing wheat in Denmark and northern Germany: Implications for crop diversity and societal dynamics of the Funnel Beaker Culture. J Quaternary Plant Ecol 23(1):81–96. doi:10.1007/s00334-014-0440-8

Kirleis W, Klooß S, Kroll H, Müller J (2012) Crop growing and gathering in the northern German Neolithic: A review supplemented by new results. Veget Hist Archaeobot 21(3):221–242

Landbruksdirektoratet.no (2017) Https://www.landbruksdirektoratet.no/filserver/statistikkgrafikk/pt-912_2017_11.html

Lid J, Lid DT (2005) Norsk flora. 7th ed. Elven R, Alm T (eds). Samlaget, Oslo

Lilleøren KS, Etzelmüller B, Schuler TV, Gisnås K, Humlum O (2012) The relative age of mountain permafrost—estimation of Holocene permafrost limits in Norway. Global Planet Change 92–93:209–223

Løken T, Pilø L, Hemdorff O (1996) Maskinell flateavdekking og utgravning av forhistoriske jordbruksboplasser—en metodisk innføring (Mechanical topsoil stripping and excavation of prehistoric agricultural sites on cultivated land—a methodological guide): AmS Varia 26, 1–104.

McClatchie M, Bogaard A, Colledge S, Whitehouse N, Schulting R, Barratt P, McLaughlin T (2014) Neolithic farming in northwestern Europe: Archaeobotanical evidence from Ireland. J Archaeol Sci 51: 206–215

Meling T (2015) Arkeologiske utgravingar på Austbø, gnr. 7, bnr. 2. Stavanger kommune, Rogaland i 2000-2001 (Oppdragsrapport Vol. 2006/1). Universitetet i Stavanger Arkeologisk Museum, Stavanger

Meling T (2016) Farm—Manor—Estate: Agricultural landscape and settlement at Hunvåg, Southwest Norway. In: Iversen F, Petersson H, Denham SD, Sundman F (eds) The agrarian life of the North 2000 BC-AD 1000: Studies in rural settlement and farming in Norway. Portal, Kristiansand, pp 293–342

Meling T (2017) Hus fra bronsealder og førromersk jernalder på Sømme i Stavanger. Frà haug ok heiðni 2

Meling T, Jensen C, Soltvedt EC (2020) Arkeologiske og naturvitenskaplige undersøkelser av landbruks- og bosetningsspor fra seinneolitikum til middelalder samt metallhåndtverk fra vikingtid (id 150770,id 150765, id150766) på Sømme (Sømme IV). Sømme gnr. 15, bnr. 5 m.fl. i Sola kommune, Rogaland fylke (Oppdragsrapport Vol. 2020/16) Universitetet i Stavanger Arkeologisk Museum, Stavanger

Moen A (1999) National atlas of Norway: Vegetation. Norges Kartverk, Hønefoss

Musit database (2018) Available at: http://www.unimus.no/arkeologi/forskning/index.php 15.01.2018.

Prescott C, Glørstad H (2012) Becoming European: The transformation of third millennium northern and western Europe. Oxbow, Oxford

Prøsch-Danielsen L (1996) Vegetation history and human impact during the last 11 500 years at Lista, the southernmost part of Norway: Based primarily on Professor Ulf Hafsten's material and diary from 1955–1957. Norsk Geografisk Tidsskrift 50, 85–99

Prøsch-Danielsen L, Soltvedt EC (2011) From saddle to rotary hand querns in south-western Norway and the corresponding crop plant assemblages. Acta Archaeologica 82(S): 129–162

Robinson DE (2003) Neolithic and Bronze Age agriculture in southern Scandinavia—Recent archaeobotanical evidence from Denmark. Environ Archaeol 8(2):145–165

Sandvik PU (1999) Analysar av plantemakrofossilar i jordprøver frå Sørbøtunet, Skei gnr. 43 og Sørbø gnr. 45. 1 og 2–4, Sandnes kommine, Rogaland (Oppdragsrapport Vol. 1999/2). Universitetet i Stavanger Arkeologisk Museum, Stavanger

Sandvik PU (2018) Myklebust bnr.3, Sola k., Rogaland. Analysar av makroskopiske subfossil (Oppdragsrapport Vol. 2018/11) Universitetet i Stavanger Arkeologisk Museum, Stavanger

Sandvik PU, Jensen C (2009) Analyse av makro- og mikrofossil i sedimentprøver frå Hellvik gnr. 60/13, Eigersund k., Rogaland (Oppdragsrapport Vol. 2009/21). Universitetet i Stavanger Arkeologisk Museum, Stavanger

Simonsen A, Prøsch-Danielsen L (2005) Tidlig jordbrukspåvirkning innen kystlyngheibeltet i Sørvest-Norge. Økosystemer i endring (The impact of early farming on forest ecosystems within the costal heathland of southwestern Norway: Changing coastal ecosystems). AmS-Varia. Universitetet i Stavanger Arkeologisk museum, Stavanger

Solberg A (2014) Arkeologiske undersøkelser av boplasser fra yngre steinalder og eldre jernalder på Helganes ved Haugesund lufthavn, Stokkdal 78/1, Karmøy kommune, Rogaland (Oppdragsrapport Vol. 2014/1, del 1). Universitetet i Stavanger Arkeologisk Museum, Stavanger

Solberg B (1993) Western Norway in the Late Neolithic and Early Bronze Age: Can loose finds contribute to our understanding of demography and social stratification? Minneskrift Egil Bakka Arkeologiske skrifter no. 7. Historisk Museum Universitetet i Bergen, Bergen

Soltvedt EC (2000). Carbonised cereal from three Late Neolithic and two Early Bronze Age sites in western Norway. Environ Archaeol 5(1):49–62. doi:10.1179/env.2000.5.1.49

Soltvedt EC (2005) Plantemakrofossiler i konstruksjonsspor: Kilde til økt kunnskap om forhistorisk tid. In: Høgestøl M, Selsing L, Løken T, Nærøy AJ, Prøsch-Danielsen L (eds) Konstruksjonsspor og byggeskikk: Maskinell flateavdekking—metodikk, tolkning og forvaltning. AmS-Varia 43:57–66

Soltvedt EC (2011) Naturvitenskapelige undersøkelser fra forhistoriske hus på Skadberg: Skadberg gnr. 32, bnr. 1 og 2, Sola kommune (Oppdragsrapport 2011/12). Universitetet i Stavanger Arkeologisk Museum, Stavanger

Soltvedt EC, Henningsmoen KE (2016) Agricultural and household activities in Vestfold, southeast Norway, as illustrated by pollen data and the charred remains of crops and wild plants. Environ Archaeol 21(1):11–30. doi:10.1179/1749631414Y.0000000052

Soltvedt EC, Jensen C (2011) Makrosubfossil- og pollenanalyser fra forhistoriske åkerlag, hustomter og graver på Tjora: Tjora gnr. 10, bnr. 5, 17 og 19, Sola kommune (Oppdragsrapport Vol. 2011/24). Universitetet i Stavanger Arkeologisk Museum, Stavanger

Soltvedt EC, Jensen C (2012) Botaniske analyser av makro- og mikrofossiler fra Jåsund gnr 1, Sola kommune (Oppdragsrapport Vol. 2012/24). Universitetet i Stavanger Arkeologisk Museum, Stavanger

Soltvedt EC, Løken T, Prøsch-Danielsen L, Børsheim RL, Oma K (2007) Bøndene på Kvålehodlene Boplass-, jordbruks- og landskapsutvikling gjennom 6000 år på Jæren, SV Norge. AmS-Varia 47:1–215

Stevens CJ, Fuller DQ (2012) Did Neolithic farming fail? The case for a Bronze Age agricultural revolution in the British Isles. Antiquity 86:707–722

Streeter D, Hart-Davies C, Hardcastle A, Cole F, Harper L (2010) Collins flower guide. Harper Collins, London

Sørensen L, Karg S (2014) The expansion of agrarian societies towards the north—new evidence for agriculture during the Mesolithic/Neolithic transition in southern Scandinavia. J Archaeol Sci 51:98–114

Viklund K (2011) Early farming at Umeå in Västerbotten: Charred cereal grains dated to the Bronze Age. Fornvännen 106(3):238–242

Westling S, Overland A (2012) Naturvitenskaplige undersøkelser på Kvia-Motland: Kvia gnr. 19/31, og Motland gnr 20/3, Hå k. (Oppdragsrapport Rogaland 2012/3). Universitetet i Stavanger Arkeologisk Museum, Stavanger

Whitehouse NJ, Kirleis W (2014) The world reshaped: Practices and impacts of early agrarian societies. J Archaeol Sci 51:1–11

Zohary D, Hopf M, Weiss E (2012) Domestication of plants in the old world: The origin and spread of domesticated plants in southwest Asia, Europe and the Mediterranean Basin. Oxford University Press, Oxford

Free-threshing wheat in Danish prehistory

Marianne Høyem Andreasen

Author address:
Department of Archaeological Science and Conservation,
Moesgaard Museum, Denmark

E-mail: mha@moesgaardmuseum.dk

Abstract

Free-threshing wheat appears to have been cultivated throughout Danish prehistory, ever since the introduction of agriculture. It is most often only found in low numbers, in a few samples, at archaeological sites and is only rarely found in high numbers in a sample. The cultivation of free-threshing wheat seems to reach its peak during the Late Bronze Age and the Early Iron Age. The small but constant presence during other periods indicates that it was cultivated on a small scale, perhaps as a 'luxury' crop, as indicated by Viking age and medieval sources. Furthermore, new evidence has shown that free-threshing wheat in Denmark cannot automatically be identified as bread wheat, because free-threshing wheat at the Early Neolithic site of Frydenlund has been identified as durum wheat.

Introduction

Grains from free-threshing wheat occur in archaeobotanical samples in Denmark since the introduction of agriculture (Robinson 1994; Robinson 2003; Robinson et al. 2009). It is most often described in papers dealing with the development of crop species in Denmark as a crop of minor importance. Emmer, barley and rye are the dominant crops during the different periods of Danish prehistory (Robinson 1994; Robinson 2003; Robinson et al. 2009). However, the constant presence of free-threshing wheat throughout the prehistoric periods shows that it must have played a role in prehistoric agriculture and the prehistoric economy. The aim of this chapter is to explore the role of free-threshing wheat in prehistoric Denmark. Cereal nomenclature follows Cappers and Neef (2012).

Danish finds of free-threshing wheat: Criteria and limitations

Prehistoric free-threshing wheat taxa in Denmark

Free-threshing wheat comprises several subspecies, including bread wheat (*Triticum aestivum* ssp. *aestivum*), sometimes called common wheat in Danish ("almindelig hvede"); durum wheat (also known as hard wheat) (*Triticum turgidum* ssp. *durum*); and club wheat (also known as dwarf wheat) (*Triticum aestivum* ssp. *compactum*) (Cappers and Neef 2012, p. 294ff). These subspecies have been found at prehistoric sites in Denmark. Club wheat is mainly identified in older literature and seems to refer to small grains of free-threshing wheat, which, according to Cappers and Neef (2012, p. 316–317), could also derive from either bread wheat or durum wheat with many fertile florets. Furthermore, the grains from bread and durum wheat can vary a lot in size (Cappers and Neef 2012, p. 316). Therefore, the finds of club wheat will be considered as bread/durum wheat in this chapter.

On most prehistoric archaeological sites in Denmark, free-threshing wheat occurs only as grains, whereas rachis fragments generally are found in very limited numbers and in very few locations. This greatly limits the possibilities of identifying the subspecies (Cappers and Neef 2012, pp. 294ff).

Until recently, it has been presumed that only bread wheat could grow in Denmark due to the comparatively cold weather, whereas durum wheat has been considered a Mediterranean crop (Robinson 1994, p. 22; Cappers and Neef 2012, pp. 296–297). New findings from the early Neolithic from the island of Funen have, however, shattered this assumption, as durum wheat was identified in several samples (Kirleis and Fischer 2014; Andreasen 2017).

The dataset

This study is mainly based on data from archaeobotanical screenings and analyses conducted by the Department of Archaeological Science, Moesgaard Museum (unpublished data are available at https://naturvidenskab.moesgaardmuseum.dk) and the Department of Environmental Archaeology and Material Science, National Museum of Denmark, including unpublished data kindly provided by the latter department, together with published data found in the litera-

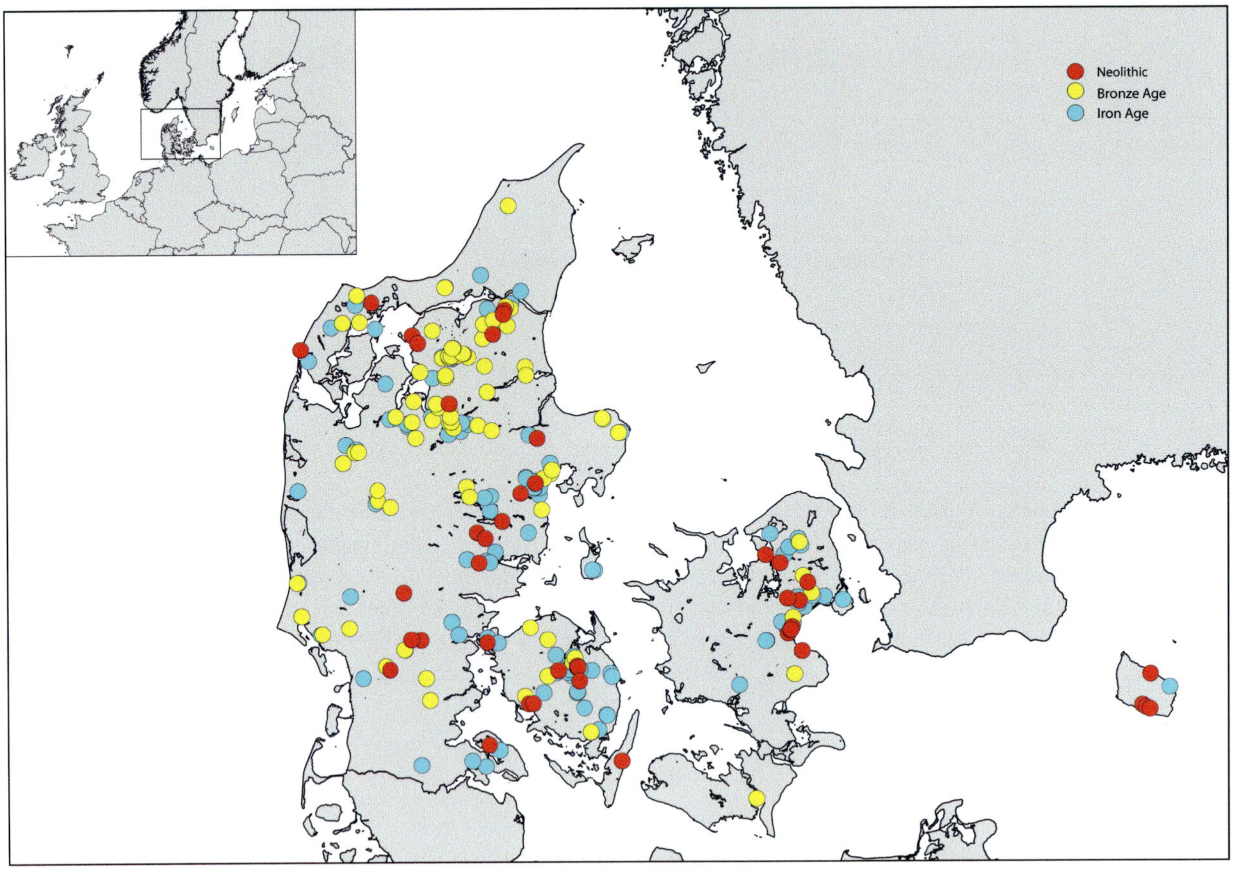

Figure 1. Map showing the sites in Denmark where free-threshing wheat has been identified and which are included in this study. Sites attributed to a transition between periods have been placed in the oldest of the two periods. Details on the sites and references can be found in Electronic Supplementary, see acknowledgements.

ture. It is likely that some free-threshing wheat finds from the literature have escaped my notice and some finds have not been accessible, or the information has not been detailed enough. Therefore, this study is not a complete overview of all the finds of free-threshing wheat in Denmark but, instead, an attempt to evaluate and clarify the role and spread of free-threshing wheat in Danish prehistory.

All samples where the identification of the free-threshing grains is uncertain have been excluded. Data compiling ended on the 31th of July 2017.

Archaeological contexts and preservation conditions

The dataset comprises several types of archaeological contexts, ranging from burnt-down structures (houses, cellars), pits and cultural layers to single postholes and houses where only the postholes are preserved. The contexts are most commonly interpreted as mundane, but a few have been interpreted as ritual. This, of course, makes it difficult to compare the finds, as they reflect many different depositional situations and different stages in the processing, use and abandonment of free-threshing wheat. Here, this lack of comparability is not an issue, as direct comparisons between sites and finds are not the purpose of this chapter.

All the finds of free-threshing wheat in this chapter are carbonised, as plant material on prehistoric settlement sites in Denmark is most commonly only preserved if carbonised. Due to this and other formation processes, the macrofossils in the samples will never provide the full picture of the prehistoric plant economy. Not even a burnt-down house will give the full picture, because what was inside the house at the time of the fire will depend on the time of the year, accidental/intentional burning of the house and other factors. The interpretation of the macrofossil record is further complicated by the fact that some crops may have a greater risk of being exposed to heat/fire. For instance, grain used for beer will be exposed to heat during the roasting process (Stika 2011, p. 44). It is not possible to fully compensate for all taphonomic factors when interpreting the macrofossil record, but it is important to bear them in mind.

Method for data compilation

In this study, I chose to count the number of samples containing free-threshing wheat per site, instead of the number of grains from free-threshing wheat per site. This approach enables me to include screened samples that have not been analysed in detail and for which the number of grains is not known. The aim is to reveal how often free-threshing wheat is found in the samples, not necessarily in what amounts it is found. However, some examples of sites with large numbers of free-threshing wheat are mentioned in the text.

It has not been possible within the scope of this chapter to compare the number of sites or samples containing free-threshing wheat with the total number of sites and samples. That will be left for future work.

The distribution of free-threshing wheat is very heterogeneous due to several factors (Fig. 1). First and foremost, the distribution clusters in areas with recent land-development and intense rescue archaeology. For example, there are many finds where new roads have been built or towns have expanded, whereas there are fewer in areas where building activities and archaeological excavations have been limited. Furthermore, during the years there have been great differences in sampling strategies by excavating museums, which will also cause bias in the data. The absence of sites in the western part of Zealand is caused by a lack of available data, whereas the low number of sites in Lolland-Falster is partly due to a lack of information on the dating of these sites. Therefore, the conclusions based on the distribution of the free-threshing wheat can only be general and not very area specific.

Free-threshing wheat in prehistoric Denmark: An overview

The general picture shows that free-threshing wheat has been cultivated all over Denmark in prehistory and is not confined to certain soils (Fig. 1, Electronic Supplementary: Table Sites, dating and no of samples, see acknowledgements). It is found in the southwestern part of Jutland, where the soil is sandy, in the eastern part of Denmark, where the soil is more clayey, and on Bornholm, where there are rocks in the subsoil covered with clay and sand. Furthermore, free-threshing wheat is known from all the prehistoric periods during which agriculture was practised. Its importance during different periods will be investigated below.

The Neolithic

Free-threshing wheat is found on a relatively large number of Early Neolithic sites, but usually only in limited numbers (Fig. 2; for the definition of archaeological periods in Denmark Table 1; Robinson 1994, p. 22). An exception is the site of Frydenlund, on the island of Funen, where free-threshing wheat is the dominant cereal in most of the samples and is present in large numbers (Kirleis and Fischer 2014; Andreasen 2017). What is also interesting about this site is the extremely large number of rachis fragments from free-threshing wheat, which, for the first time, shows that durum wheat was grown in Denmark in prehistory (Kirleis and Fischer 2014; Andreasen 2017). This site will be described in more detail below.

In both the Middle Neolithic A and the Middle Neolithic B, there are small amounts of free-threshing wheat on a relatively large number of sites (Fig. 2). It appears that the number of sites with free-threshing wheat decreases in the Middle Neolithic B, but this is presumably because there are only few sites from this period that have been sampled and that contained macrofossils (Andreasen 2009).

In the Late Neolithic and by the transition to the Early Bronze Age, free-threshing wheat was still present on a relatively large number of sites, but the number of samples per site containing free-threshing wheat is low (Fig. 2). During this period, the crop seems to have become more widespread in Denmark, but this could partly be due to a general rise in the number of archaeological sites.

The Bronze Age

According to Robinson (1994, p. 22) the Late Neolithic and the Bronze Age mark the apex of the cultivation of free-threshing wheat. And it is obvious from Figure 3 that there is, in fact, a rise in the number of sites where free-threshing wheat has been observed. On some sites, it is also possible to see a rise in the number of samples containing free-threshing wheat and/or the number of grains per sample, for example Bjerre (Henriksen 2003; Henriksen et al. 2018), Egehøj (Rowley-Conwy 1984) and Kongehøj (Andreasen 2011, table 5).

There is an increase in the number of samples containing free-threshing wheat in the Early Bronze Age (Fig. 3). An interesting observation is that, while some sites have many samples, each of which only contain a few grains of free-threshing wheat, e.g. Kongehøj (Andreasen 2011, table 5), other sites have fewer samples, each of which contains many grains. This observation makes it clear that free-threshing wheat was in fact cultivated, for instance at Egehøj (Rowley-Conwy 1984), which will be described below.

The importance of free-threshing wheat increases further in the Late Bronze Age (Fig. 3), when even more sites contain small numbers of free-threshing wheat and a few sites contain larger amounts. An example is Højene (Andreasen 2014), which will be described later. Finds seem to cluster in the area between Viborg and Aalborg, in the northern part of Jutland, which could reflect the situation in the Late Bronze Age. However, this distribution pattern could also be due to a combination of intensive land development and a good chronological control of the Bronze Age versus the Pre-Roman Iron Age in the area, combined with numerous ^{14}C dates on Late Bronze Age material. It is often difficult to distinguish between ceramics from the Late Bronze Age and the Early Pre-Roman Iron Age in Denmark.

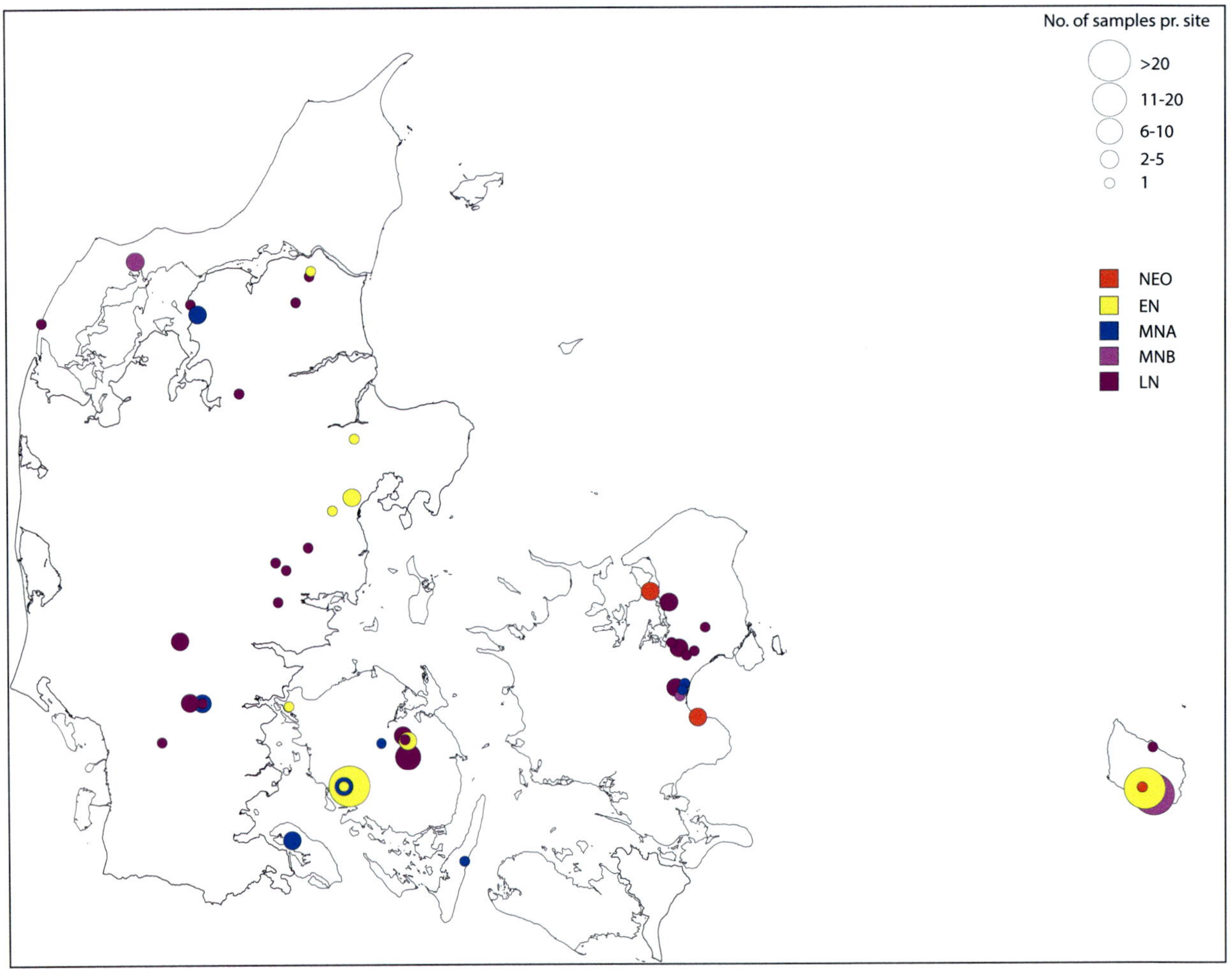

Figure 2. Map showing sites from the Neolithic with free-threshing wheat. Sites attributed to a transition between periods have been placed in the oldest of the two periods. For the period abbreviations and dating, see table 1. The samples are divided into five categories depending on the number of samples that contain free-threshing wheat per site. For more details on the sites, see Electronic Supplementary: Table Sites, dating and no of samples, see acknowledgements.

Table 1. Periods in Danish Prehistory (after Jensen 2001, 2002, 2003, 2004).

Neolithic (NEO)		Early Neolithic	EN	4000–3300 BC
		Middle Neolithic A	MN A	3300–2800 BC
		Middle Neolithic B (Single Grave Culture)	MN B	2800–2400 BC
		Late Neolithic	LN	2400–1700 BC
Bronze Age (BA)		Early Bronze Age	EBA	1700–1000 BC
		Late Bronze Age	LBA	1000–500 BC
Iron Age (IA)	Early Iron Age (EIA)	Pre-Roman Iron Age	PRIA	500 BC–0
		Roman Iron Age	RIA	0–400 AD
	Late Iron Age (LIA)	Germanic Iron Age	GIA	400–800 AD
Viking Age (ViA)		Viking Age	ViA	800–1050 AD

The Iron Age and the Viking age

From the Pre-Roman Iron Age, most sites contain only a few samples with free-threshing wheat, and only a few sites have many samples, but there seems to be no decrease in the number of sites (Fig. 4). At a few sites, the high number of grains indicates that the crop was cultivated, for example at Overbygård (Lund 1979; Henriksen and Robinson 1994, 1996) and Hodde (Jørgensen in Hvass 1985).

There is no decrease in the number of sites containing free-threshing wheat in the Roman Iron Age either. It can be found in all areas of Denmark, but in a particularly large cluster of sites on Funen (Fig. 4). Most of the sites only contain a few samples with free-threshing wheat, with few grains in each sample. However, also during this period there are a few sites that contain samples with larger numbers of grains, for example Alrum (Henriksen and Robinson 1994) and Dragebjerggård Vest (Jensen 2011).

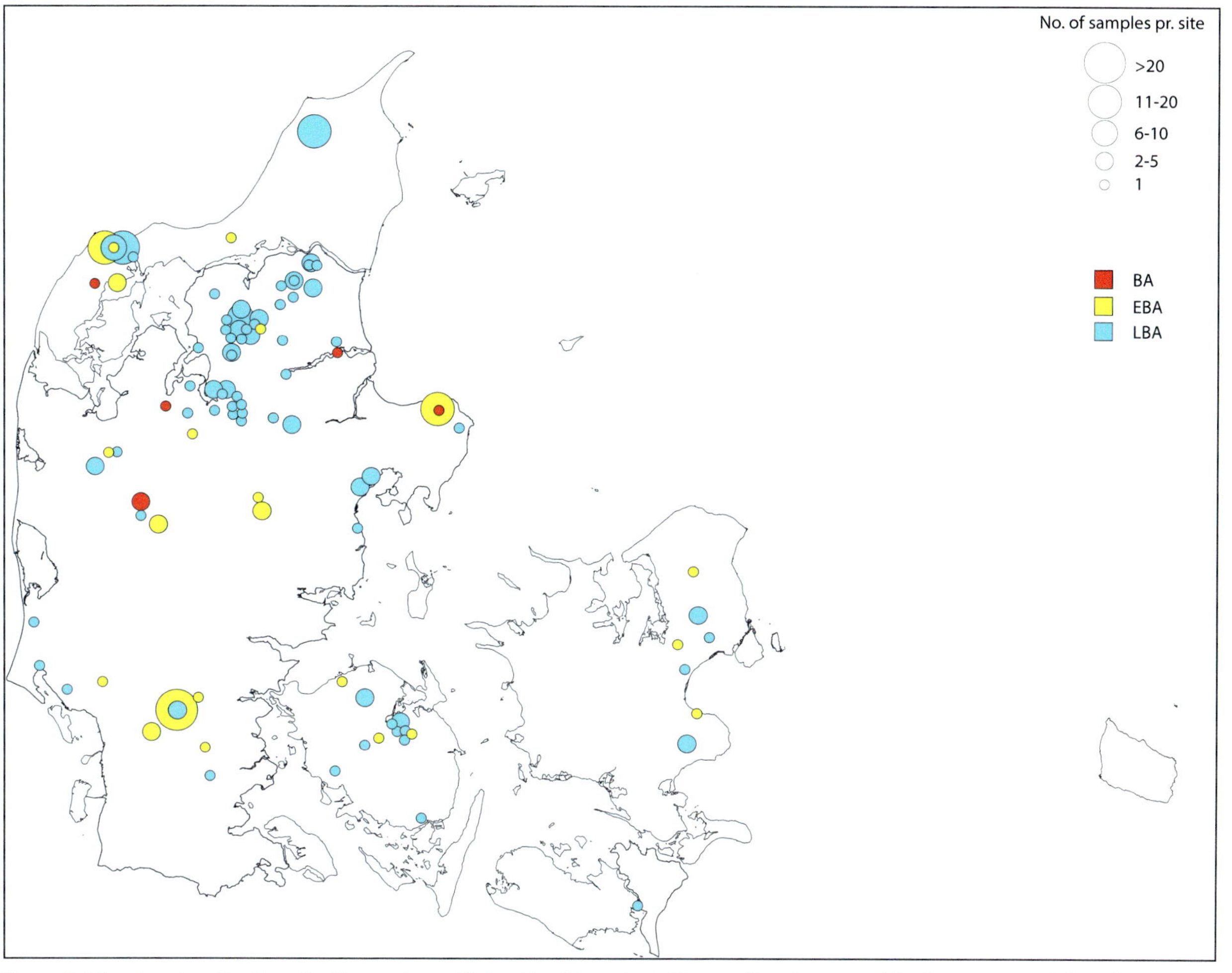

Figure 3. Map showing sites from the Bronze Age with free-threshing wheat. Sites attributed to a transition between periods have been placed in the oldest of the two periods. For the period abbreviations and dating, see table 1. The samples are divided into five categories depending on the number of samples that contain free-threshing wheat per site. For more details on the sites, see Electronic Supplementary: Table Sites, dating and no of samples, see acknowledgements.

In the Germanic Iron Age, there seem to be fewer sites containing free-threshing wheat compared with the Early Iron Age and the Bronze Age (Fig. 5). All the sites contain relatively few samples with free-threshing wheat and only a few grains have been identified in the samples. On the other hand, free-threshing wheat is still found in all parts of Denmark. Especially towards the end of the Germanic Iron Age, there are very few sites with free-threshing wheat (Electronic Supplementary: Table Sites, dating and no of samples, see acknowledgements). This could reflect the prehistoric reality, but more likely reflects a relatively low number of sites from this period in general.

In the Viking age, there are still sites containing free-threshing wheat (Fig. 5), although the number of finds is relatively low. In many parts of Denmark, there is a lack of free-threshing wheat, but whether this reflects prehistoric reality, or a general lack of sites is not clear. What is clear, is that free-threshing wheat is still present in the Viking age.

Examples of prehistoric sites containing free-threshing wheat

In order to reflect on the role of free-threshing wheat through Danish prehistory, some sites from different periods and with different interpretations are presented.

Frydenlund—Early Neolithic durum wheat

Frydenlund is situated on Funen (Fig. 6) and dated to c. 3630–3640 BC (the very precise dating is based on more than one hundred radiocarbon dates). Many samples for macrofossil analysis were taken during the excavation of two Barkær features (a type of long barrow) and two houses underneath them (Andersen 2015). Some of these samples were analysed by Kirleis and Fischer (2014), while others were analysed later, by me (Andreasen 2017).

What sets Frydenlund apart from most other sites where free-threshing wheat has been identified, is that most analysed samples were dominated by ra-

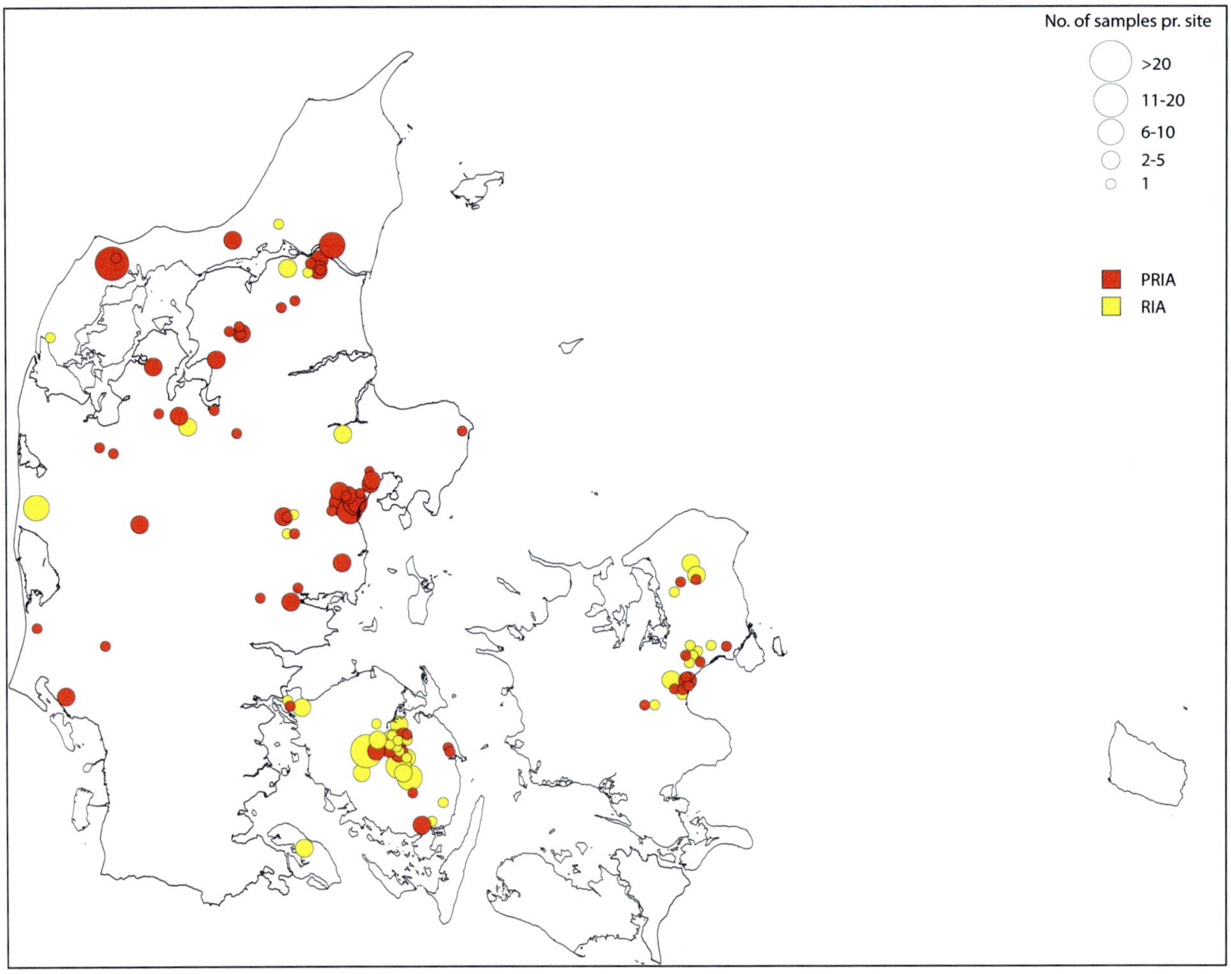

Figure 4. Map showing sites from the Early Iron Age with free-threshing wheat. Sites attributed to a transition between periods have been placed in the oldest of the two periods. For the period abbreviations and dating, see table 1. The samples are divided into five categories depending on the number of samples that contain free-threshing wheat per site. For more details on the sites, see Electronic Supplementary: Table Sites, dating and no of samples, see acknowledgements.

chis fragments, and that almost all rachis fragments were identified as durum wheat (Fig. 7) (Kirleis and Fischer 2014; Andreasen 2017). Possibly, some of the features dominated by rachis fragments reflect threshing sites or features where waste from threshing has been deposited. Other features have different combinations of macrofossils and could reflect storage or perhaps offerings of food, while others again presumably contain waste from cereal cleaning (Andreasen 2017). Because of the connection with the Barkær features in an area where a lot of rituals have presumably taken place, it is not clear whether the macrofossils reflect everyday consumption or ritual use of food/plants (Kirleis and Fischer 2014; Andersen 2015; Andreasen 2017).

In addition to providing information about practices at Frydenlund, the macrofossils broaden our understanding of the cultivation of crops during the Early Neolithic, due to the detection of a hitherto unknown crop in Danish contexts. It seems that durum wheat was cultivated and did quite well in the early phase of agriculture in Denmark, which was probably a relatively warm period (Dreibrodt et al. 2012). This leaves the question whether durum wheat was also cultivated in other periods of Danish prehistory.

Egehøj—An Early Bronze Age house

House III from Egehøj, on Djursland, is dated to the Early Bronze Age period I (Fig. 6). It is a two-aisled longhouse with a sunken floor in the eastern part. In this sunken part, large numbers of carbonised grains were recognised during the excavation. Samples were consequently taken in a profile along the sunken floor and from a pit beneath it (Fig. 8) (Boas 1983; Rowley-Conwy 1984, p. 104). Naked barley, free-threshing wheat (interpreted as bread wheat) and possible emmer were identified in the samples in large quantities, whereas very few weed seeds and no rachis or glume fragments were observed (Table 2). This led to the conclusion that the grains represent cleaned, separately stored crops (Rowley-Conwy 1984, p. 104). The distribution of the different crops throughout the house indicates that all three crops were grown separately and that bread wheat was not just a 'weed'

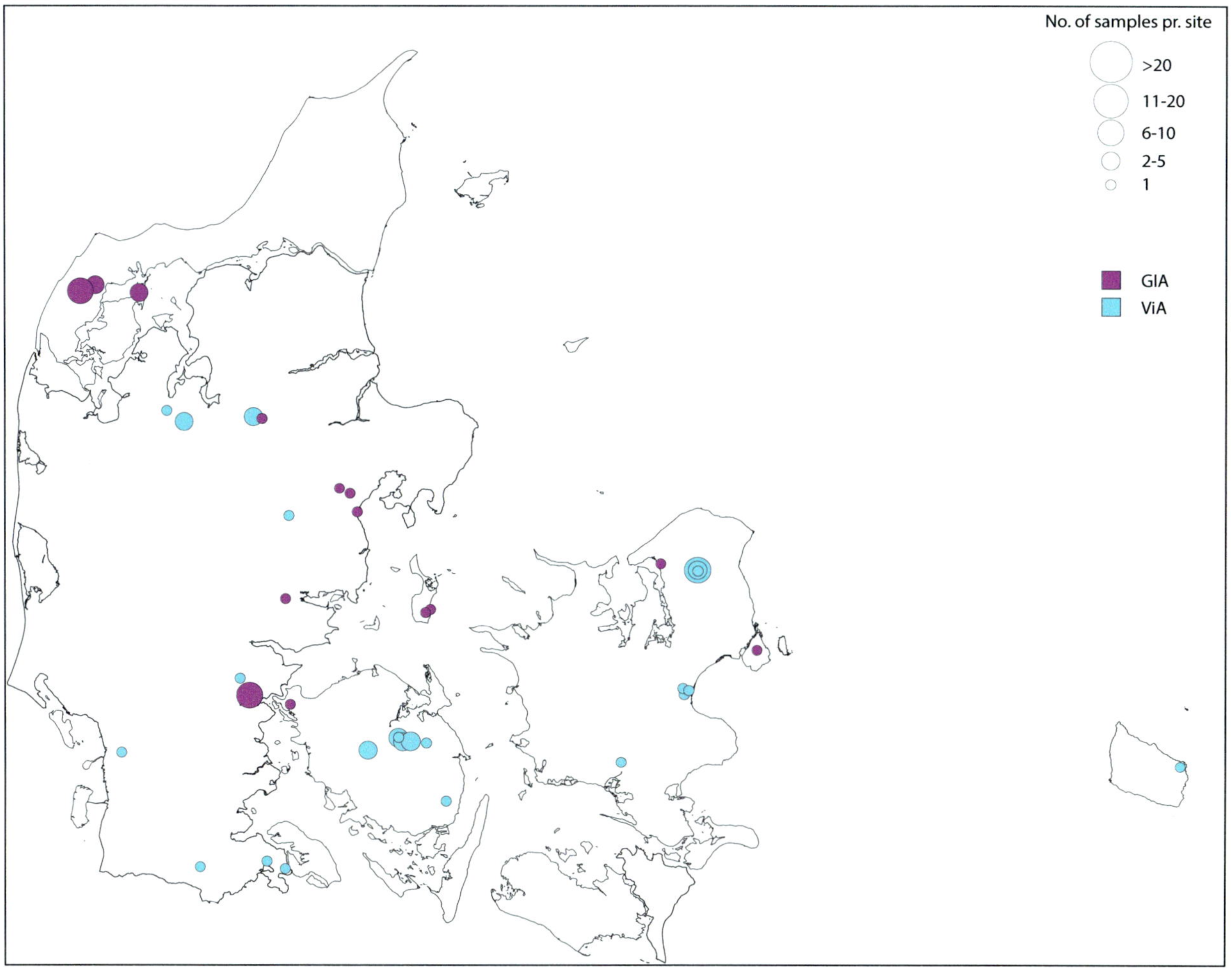

Figure 5. Map showing sites from the Late Iron Age and Viking Age with free-threshing wheat. Sites attributed to a transition between periods have been placed in the oldest of the two periods. For the period abbreviations and dating, see table 1. The samples are divided into five categories depending on the number of samples that contain free-threshing wheat per site. For more details on the sites, see Electronic Supplementary: Table Sites, dating and no of samples, see acknowledgements.

Table 2. Analysed macrofossils from Egehøj, house III (Rowley Conwy 1984, p. 109, table 1).

Excavation co-ordinates	70.80-71.25	72.15-72.50	73.10-73.35	74.00-74.20	74.70-75.00	75.50-75.70	75.70-75.90	75.90-76.10	76.30-76.50	76.70-76.90	77.20-77.50	78.30-78.60	79.20-79.50	pit *cca*	total
Naked 6 row barley (*Hordeum vulgare*)	4	43	232	55	50	26	11	26	43	62	13		3	36	604
Bread wheat (*Triticum aestivum s.l.*)		171	267	63	32	17	6	15	13	41	3	3	2	21	654
Emmer type (*Triticum cf. diccocum*)	1	9	37	15	21	17	4	21	20	34	5	2	2	18	206
Uidentified wheat (*Triticum sp.*)	2	9	22	16	16	7	4	6	11	27	3		2	9	134
Unidentified cereals (*Cerealia*)	5	103	239	59	72	44	14	34	50	88	11	12		41	772
Cereal grains, total	12	335	797	208	191	111	39	102	137	252	35	17	9	125	2370
Cereal grains per litre of deposit (approx)	5	134	241	42	38	15	13	12	10	30	7	2	2	50	
Weeds, total		1		1	1	4				3		1	1	1	13

Figure 6. Map showing the locations of the archaeological sites of Frydenlund, Egehøj and Højene.

Figure 7. Photo of rachis fragments of durum wheat from the site of Frydenlund (Photo: Foto/Medieafdeling, Moesgaard Museum, Rógvi N. Johansen).

among other crops (Rowley-Conwy 1984, p. 105). At the time when Rowley-Conwy wrote about the grains from Egehøj, there were only a few small finds of bread wheat in Denmark. He therefore suggested that bread wheat could have been a rare crop, something that not every household would grow. Alternatively, he suggested, if bread wheat was sown in the autumn and eaten shortly after harvest, it would only have been stored for a very short period in the house, and therefore our chances of finding it would be very small (Rowley-Conwy 1984, p. 108–109). However, evidence suggests that winter crops were not intro-

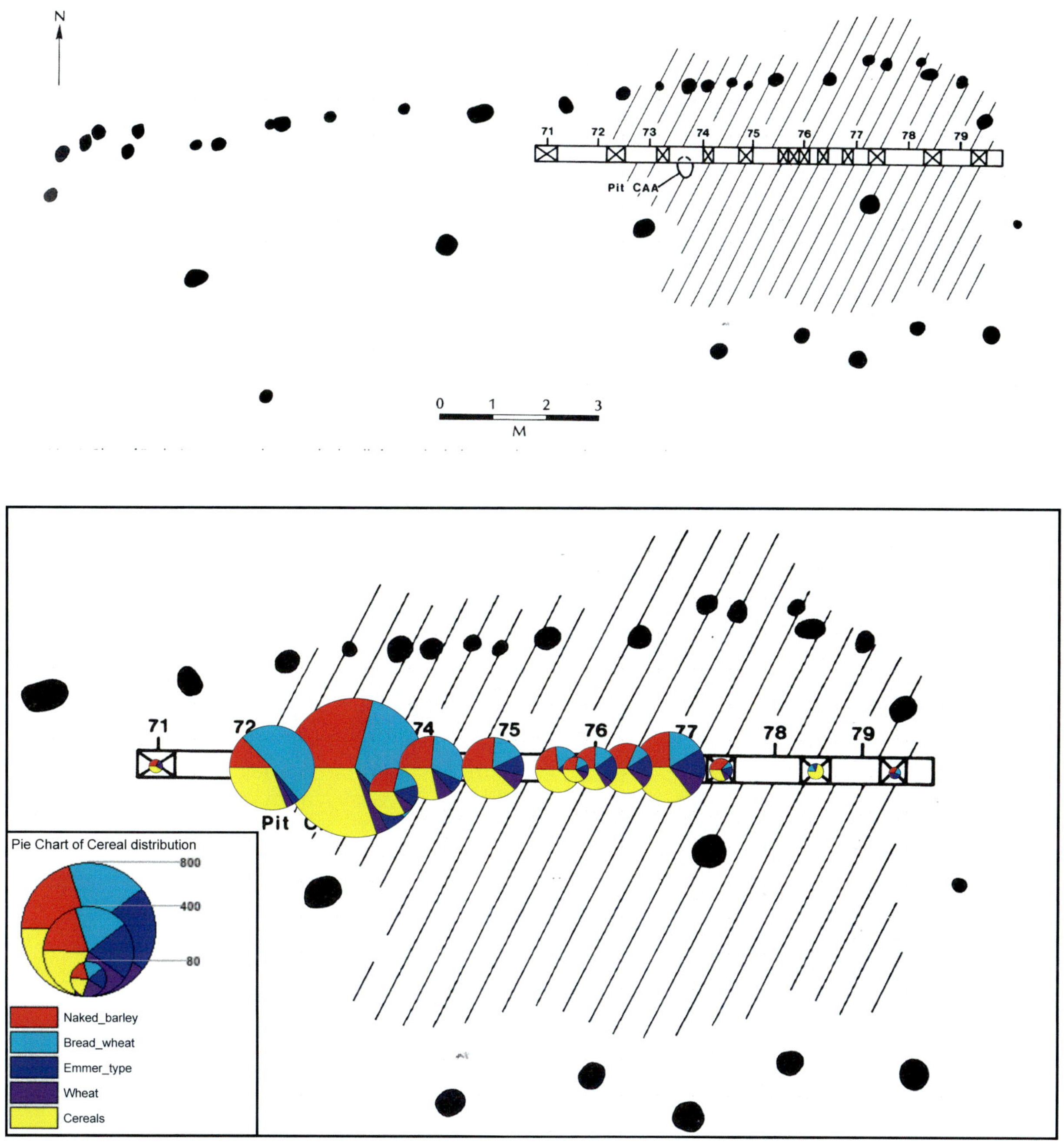

Figure 8. Plan of Egehøj house III, showing the baulk from where the samples were taken (Rowley-Conwy 1984, p. 105, fig. 1) (upper figure) and the distribution of the different crops (lower figure).

duced in Denmark until the Late Roman Iron Age (Mikkelsen and Nørbach 2003, pp. 217–225), and therefore Rowley-Conwy's last assumption is not in agreement with the current knowledge.

Højene—A Late Bronze Age pit

Pit A2595 from the Late Bronze Age site of Højene, in the northern part of Jutland (Fig. 6), could be interpreted either as a drying kiln or as an offering (Marring 2014; Andreasen 2014). Altogether 8.8 litres of carbonised material were found in the pit, and it was dominated by cereal grains. More than 99 per cent of the cereal grains were identified as wheat, and 69–89 per cent of the wheat grains were identified as free-threshing wheat. A few rachis fragments from free-threshing wheat were found, and these were identified as bread wheat. In addition to bread wheat, the site also yielded emmer/spelt and naked and hulled barley, which are also presumed to have been cultivated at the site (Table 3). It is not possible to conclude based on the archaeological or archaeobotanical evidence whether the bread wheat should be regarded as a crop that was used as part of the daily economy or, instead, as a special crop that was grown

Table 3. Analysed macrofossils from Højene. Only some of the floated archaeobotanical samples have been analysed. Numbers in brackets mark the actual number of grains/seeds identified and counted; numbers outside the brackets mark the estimated number of grains/seeds in the entire archaeobotanical sample. For example, if only a quarter of the sample has been analysed, the counted number (inside the brackets) was multiplied by 4 to arrive at the estimated number (outside the brackets). Charcoal is subjectively estimated on a scale of x: very little charcoal to xxxxx: large amounts of charcoal. Only those cereal fragments larger than 2 mm have been counted. Note: the two rootbulbs in x209 are connected.

x-nr	195	197	199	201	203	205	209
Sample volume before flotation (l)	2	3	10	10	8	5	2
Archaeobotanical sample after flotation (l)	0,5	0,25	0,3	0,6	0,5	1	1,2
Avena sp.		(3) 12	(2) 8	(1) 4	(2) 8		
cf. Avena sp., "neddles"		(1) 2					
Hordeum vulgare var. nudum			(16) 64	(1) 8	(1) 8		
Hordeum vulgare var. vulgare				(1) 8			(1) 16
Hordeum vulgare				(2) 16	(2) 16		
cf. Hordeum vulgare		(1) 4	(9) 36			(1) 8	
cf. Secale cereale ssp. cereale			(1) 4	(1) 8			
Triticum aestivum ssp. aestivum	(2395) 18580	(1872) 7488	(1763) 7052	(1897) 14576	(2128) 16592	(1844) 29744	(1702) 29442
Triticum turgidum ssp. dicoccum/ aestivum ssp. spelta	(96) 556	(139) 556	(247) 988	(144) 1056	(146) 1008	(243) 3456	(200) 3200
Triticum cf. monococcum ssp. monococcum							(2) 32
Triticum sp.	(219) 1484	(201) 804	(389) 1556	(331) 2532	(256) 2012	(423) 6448	(595) 9520
Triticum aestivum ssp. aestivum, rachis	(6) 18						
Triticum cf. aestivum ssp. spelta, glumebase (number/equals number of grains)						(1/8) 8/16	
Triticum sp., glumebase (number/equals number of grains)	(2/4) 4/8	(1/2) 2/4			(4/4) 16/16	(1/2) 8/16	
Cerealia indet	(7) 48	(5) 20	(29) 116	(7) 44	(9) 68	(5) 80	(6) 96
Cerealia indet fragment	(324) 2592	(205) 820	(574) 2296	(306) 2448	(350) 2800	(417) 6672	(591) 9456
Cerealia indet, rachis	(34) 88	(18) 36		(7) 28	(42) 168	(1) 8	(1) 8
Cerealia indet, rachis/glumebase			(4) 16				
cf. Cerealia indet, straw fragment						(3) 48	
Arrhenatherum elatius var. bulbosum							(2) 32
Camelina sativa	(1) 2	(3) 6	(1) 4	(2) 8		(2) 16	(3) 24
Linum usitatissimum						(1) 8	
Asteraceae	(1) 2	(1) 2	(1) 4				
Bromus sp.	(5) 18			(2) 12	(10) 56	(4) 32	(1) 16
Chenopodium album	(60) 120	(101) 202	(21) 84	(6) 24	(12) 48	(22) 176	(2) 16
Chenopodium sp.				(20) 80	(72) 288	(33) 264	(33) 264
Fallopia convolvulus	(2) 8		(2) 8		(4) 16	(6) 48	(1) 16
Persicaria maculosa/lapathifolium	(2) 4		(4) 16	(8) 32	(2) 8	(6) 48	(3) 32
Plantago lanceolata		(1) 2					
Poaceae	(100) 346	(198) 396	(46) 184	(31) 232	(64) 256	(47) 376	(20) 160
Polygonum aviculare	(1) 4						
Polygonaceae	(1) 2						
Ranunuculus sp.			(1) 4				
Rumex acetosella	(4) 12	(3) 6	(4) 16	(1) 4	(6) 24		
Rumex sp.	(2) 4						
Spergula arvensis	(2) 4	(1) 2	(6) 24		(5) 20		
Veronica sp.				(2) 16			
Indet	(8) 18	(13) 26	(29) 116	(5) 20	(10) 40	(7) 56	(4) 32
Slagg from organisk materiel				(6) 48	(7) 56	(7) 112	(4) 64
Charcoal	x	x	x	x	x	x	x
Place	Lag a + b	Lag a + b	Lag a	Lag a	Lag a + b	Lag a + b	Lag b

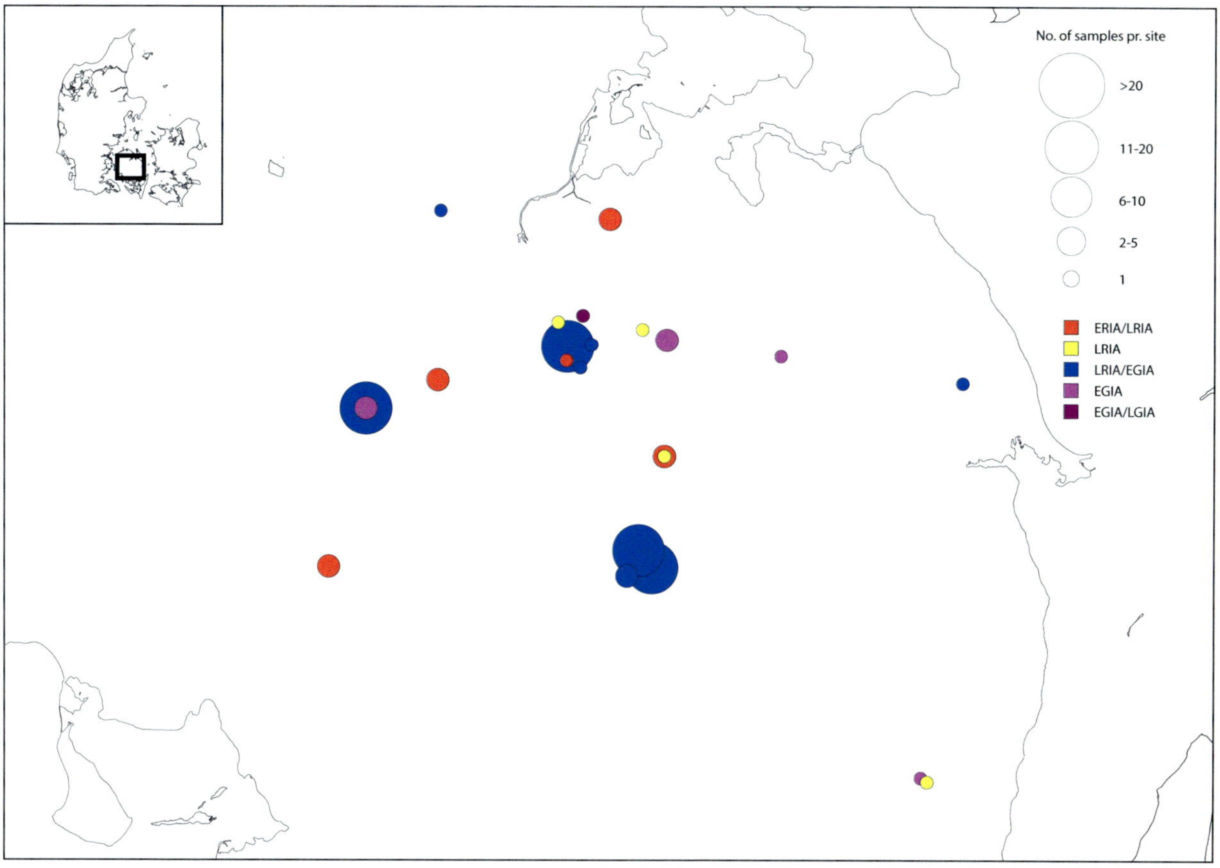

Figure 9. Map showing the sites on Funen from the Early Roman Iron Age/Late Roman Iron Age–Early Germanic Iron Age/Late Germanic Iron Age. For more details on the sites, see Electronic Supplementary: Table Sites, dating and no of samples, see acknowledgements.

for special occasions or could be used as an offering (Andreasen 2014).

Funen in the Late Roman and Early Germanic Iron Age

As seen in Figure 9, there are many sites containing free-threshing wheat clustered in one area on Funen. At most of these sites, free-threshing wheat was recognised in only a few samples, most often in fewer than five (Table 4). Only six sites out of 21 have more than five samples containing free-threshing wheat, even though on most of the sites a lot of samples were taken and screened. Most of the samples contained fewer than 5 grains of free-threshing wheat, while 28 of the samples contained between 6 and 50 grains per sample. Free-threshing wheat was not the dominant cereal in any of these samples. The remaining five samples contained between 56 and 340 grains (Fig. 10, Table 4). Only in the two samples containing 173 and 259 grains of free-threshing wheat was it the dominant crop.

Free-threshing wheat appears on many of the sites on Funen in the Late Roman and Early Germanic Iron Age and therefore seems to be quite widespread in the area at that time, but only a few grains were found on most sites. This is a situation which already seems to arise during the Pre-Roman and Early Roman Iron Age (Jensen and Andreasen 2011, p. 139). On the other hand, the high number of grains from free-threshing wheat on a few sites, and in some cases their dominance in a few samples, makes it clear that it was grown as a separate crop. When we put the evidence together, it seems clear that free-threshing wheat was a fairly widespread crop on Funen in the period. However, it seems likely that it was not generally grown on a large scale, but, rather, as a secondary crop, perhaps a sort of 'luxury'. Free-threshing wheat is in some respects more difficult to grow than other cereals (Körber-Grohne 1995, p. 28), and for this reason, perhaps the inhabitants did not want to depend on it.

Discussion

As demonstrated above, free-threshing wheat can be found in most parts of Denmark from the Neolithic to the Viking age. In the Early and Middle Neolithic, it is found on only a limited number of sites (Fig. 11), which most likely reflects the archaeological reality, as comparably few sites from these periods have been excavated and sampled. The distribution of free-threshing wheat expands a bit in the Late Neolithic and Early Bronze Age, but it is not until the

Table 4. Samples from Funen from the ERIA/LRIA-EGIA/LGIA shown as number of grains or other remains of free-threshing wheat per sample (the sample numbers are not indicated, as they are different from site to site).

Sites	Dating	Grains or other remains from free-threshing wheat per sample														
Fladbjerg Syd	RIA	16	1	2	4	4	56	12	340	100						
Ågård Nord	RIA	2 + 3cf	1cf	4cf	32	4cf	1cf									
Tuekjærsholm	RIA	10	1													
Motorvej S	ERIA/LRIA	1														
Skovgård, etape 2, del 1 2010	ERIA/LRIA	1	2	3	1	<5	1	<5								
Munkebjergvejens forlængelse	LRIA	1														
Skovgård, etape 2, del 1 2010	LRIA	7	1	1cf												
Nørrebjerg	LRIA	<5														
Nyt Universitetshospital / OUH	LRIA/EGIA	1														
Motorvej S	LRIA/EGIA	<5														
Hjulby	LRIA/EGIA	<5														
Villestofte Skovvej	LRIA/EGIA	<10														
Hestehaven 51	LRIA/EGIA	1cf	<10	3	6	5	14	2	2							
Kielbergvej	LRIA/EGIA	13	13	2	12 + 4cf	6 + 2cf	5	26	1	2						
Kielbjergvej Nordvest	LRIA/EGIA	1cf	18	1	17	1	1	4	1	<20						
Harrekærgård Øst I	LRIA/EGIA	<5cf														
Dragebjerggård Vest	LRIA/EGIA	2	173	8	3	8	21	14	4	3 rachis fragments	259	1	3	9	19	4
Tuekjærsholm	EGIA	<10														
Natskyggen	EGIA	<20	<5	<5	<5											
Dragebjerggård Vest	EGIA	<20	<5	<50	<50											
Campus, etape 1 & 2 & 4	EGIA/LGIA	1														

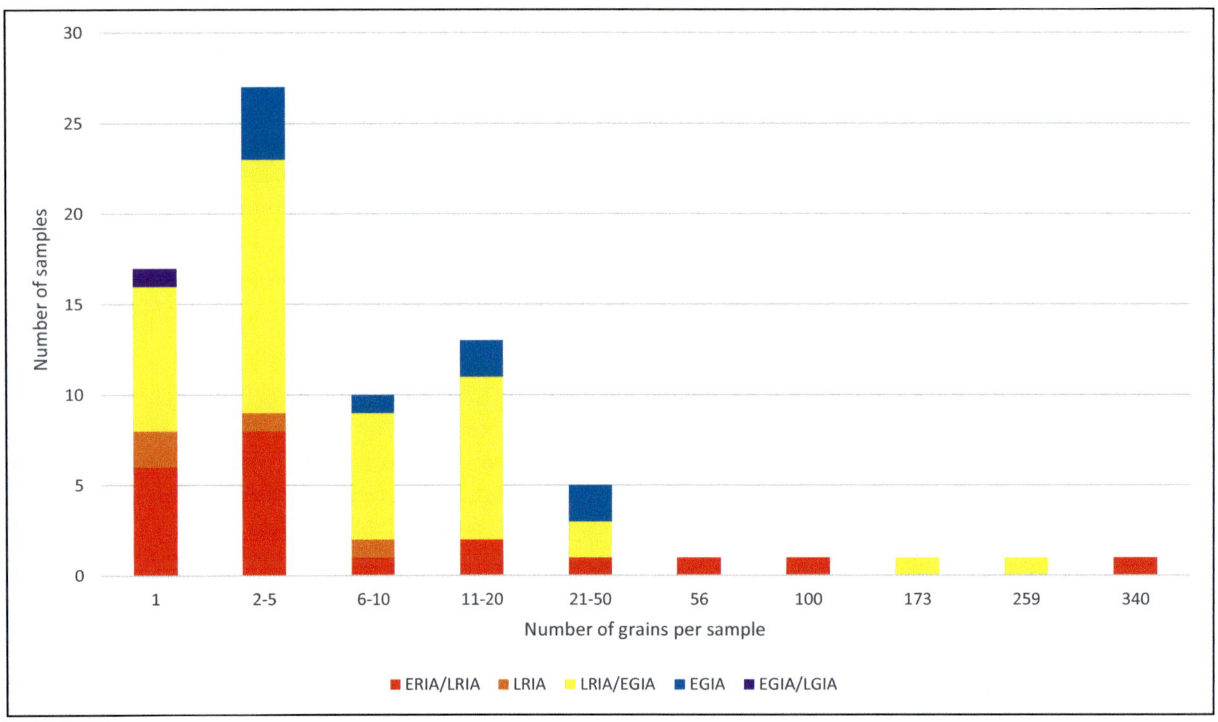

Figure 10. Samples from Funen from the Early Roman Iron Age/Late Roman Iron Age–Early Germanic Iron Age/Late Germanic Iron Age shown as number of grains per sample.

Late Bronze Age and Early Iron Age that it reaches its widest distribution, with relatively many finds per 100 year (Fig. 11). Finds of free-threshing wheat seem to become less common towards the end of the prehistoric period (Fig. 11), either due to a dwindling of its role in agriculture or due to a decrease in the number of excavated and sampled sites.

On most sites, free-threshing wheat is found only in a few samples and in low numbers, but from the Early Neolithic until the Early Germanic Iron Age, there is

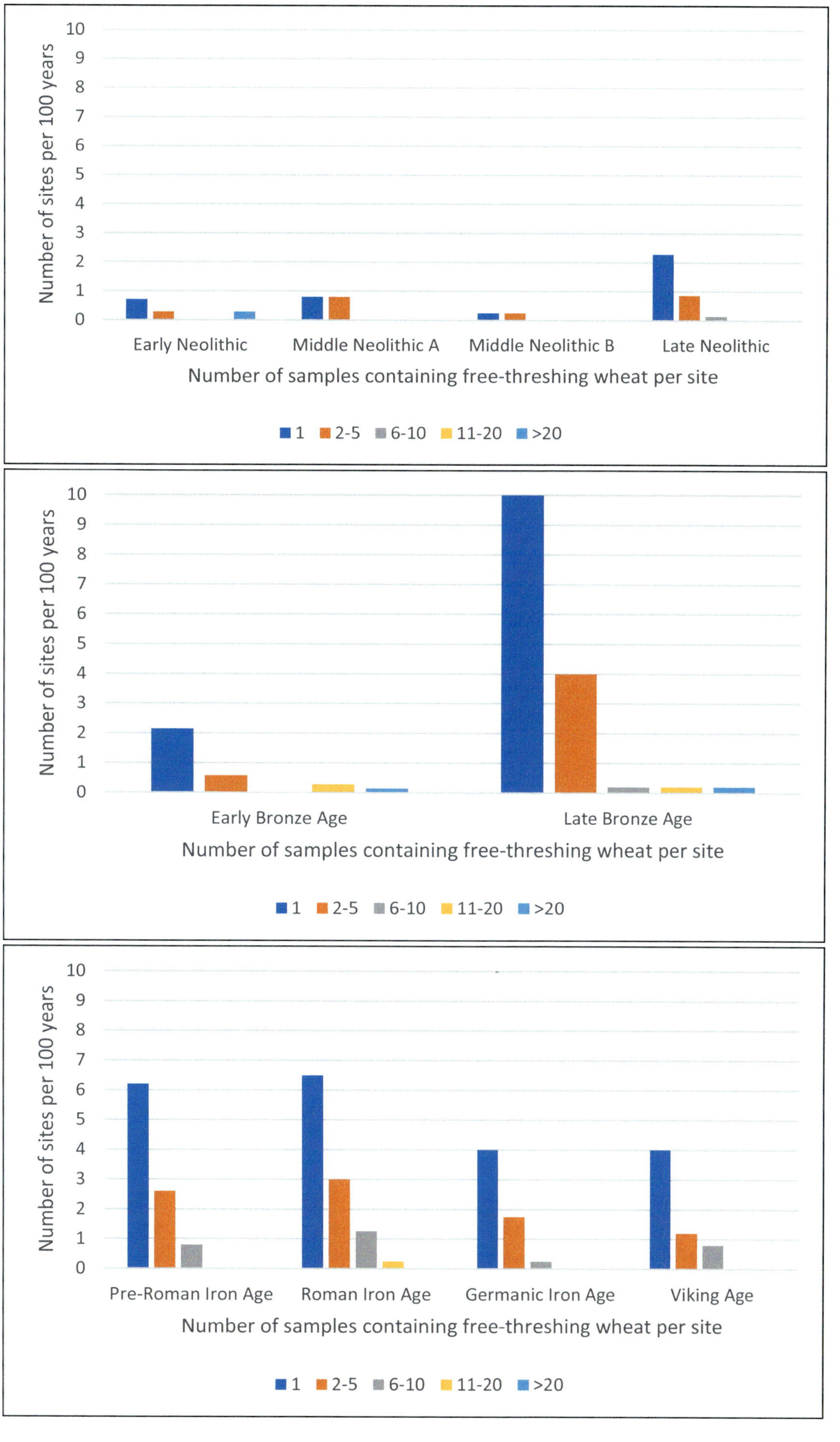

Figure 11. Number of samples per site containing free-threshing wheat per 100 years.

always one or a few sites where free-threshing wheat is either dominant or found in large numbers in one or more samples. This pattern is reflected on a smaller scale on Funen during the Late Roman and Early Germanic Iron Age.

The distribution of free-threshing wheat would have appeared much more restricted if the review had been based only on sites with large numbers of carbonised grains. By also including sites that have only been screened but not analysed, it has been possible to present a more comprehensive picture of its distribution in both time and space. It is therefore of vital importance that all stages of analysis of macrofossils are available for use when making an overview of the different cereals and their importance.

What was the role of free-threshing wheat in prehistoric Denmark? Often when a cereal is represented by only one or a few grains in samples that are dominated by other cereals, it will be interpreted as a sort of weed or a leftover from earlier cultivation. However, in the case of free-threshing wheat, we have a cereal that is usually found only in small numbers in the samples but on a relatively large number of sites. It is dominant only on a few sites and in a few samples, but these sites are from many different periods. Most of the sites containing many grains from free-threshing wheat have been interpreted as ordinary settlement sites, for example Egehøj (Rowley-Convy 1984), Overbygård (Lund 1979) and Hodde (Hvass 1985). However, there are also a few sites where a ritual context cannot be excluded and could, perhaps, even be likely, for example Frydenlund (Kirleis and Fischer 2014; Andersen 2015; Andreasen 2017) and possibly Højene (Marring 2014, Andreasen 2014).

So why is it that free-threshing wheat is most often only found in small numbers? Have certain activities or a lack thereof in the past resulted in grains from free-threshing wheat being underrepresented in the samples? Or does the record reflect the original relationships among the different crops? In order to get closer to answering this question, it is necessary to look at some different aspects of the agricultural processes.

It is often assumed that hulled crops have a greater risk of becoming carbonised, as they require parching to make dehusking easier (Peña-Chorarro and Zapata 2014, p. 226). However, according to the classical literature, ethnographic studies and experiments, parching appears not to have been necessary for a successful dehusking of hulled wheats (Peña-Chocarro and Zapata 2014). Other possible explanations for grains' contact with fire, which would affect both hulled and free-threshing wheats, are drying for better preservation; roasting for better taste or in connection to beer brewing; and burning of awns (Peña-Chorarro and Zapata 2014, pp. 228–229).

Of course, parching of hulled wheat and barley in Danish prehistory cannot be excluded, but is it necessarily the main reason for the carbonisation of grains? If so, it is difficult to understand why naked barley is so dominant or well represented in many periods (Robinson 1994, 2003). Naked barley can, of course, have been used for making beer, which would involve roasting. However, there are only few signs of sprouting, and it is therefore difficult to find positive evidence of beer making, except on very few occasions (Klassen 2005, pp. 39–42; Henriksen 2006, p. 7). Therefore, beer making cannot explain the large amount of naked barley, and parching cannot be exclusively used to explain the differences between the different crops.

As mentioned earlier, macrofossils may also derive from the burning of garbage, either in connection with food preparation or by accident. A burnt-down house gives a unique opportunity to get a snapshot of what was present in the house. It is therefore often assumed that a burnt-down house will give an insight into the ratio between the different crops. But what is inside a burnt-down house depends on several factors. The burnt-down constructions referred to in this article, which were presumably burnt by accident, show a varied picture. Some, such as Egehøj and Overbygård, have a high percentage of free-threshing wheat, but in both cases, naked barley is either similarly or more strongly represented in the samples (Rowley-Conwy 1984; Lund 1979). On other sites, such as Hestenhaven and Alrum, free-threshing wheat apparently played only a minor role (Henriksen and Robinson 1994; Jensen et al. this volume). On the sites that are clearly dominated by free-threshing wheat, i.e. Frydenlund and Højene, special circumstances apply. At Frydenlund, the samples dominated by durum wheat came from an apparent threshing site, that could either be connected to the site's ritual features or be a mundane threshing site for durum wheat (Andreasen 2017). At Højene, the samples all reflect one harvested crop (Andreasen 2014). These two sites, therefore, cannot tell us about the ratio between the different crops, only that free-threshing wheat was cultivated at some sites, which can also be concluded at Egehøj and Overbygård.

If free-threshing wheat was cultivated on a large scale, we can assume that it would be found in fairly large amounts in several samples from ordinary sites (not just burnt-down constructions or contexts representing one single harvest). However, this is not the case. Instead, free-threshing wheat is found in small amounts on many sites. This seems to indicate that it had a widespread distribution but was a crop of minor importance.

As suggested for Funen in the Late Roman and Early Germanic Iron Age, free-threshing wheat was perhaps a 'luxury' crop. The large finds of durum wheat and bread wheat support the interpretation that free-threshing wheat was, in fact, cultivated as a monoculture, and the large finds from a possible ritual context (Frydenlund) also suggest that in some cases free-threshing wheat played a special role. The role of free-threshing wheat as a luxury food or food for special occasions is known from Viking age and medieval sources. In the Nordic Edda sagas, a story is told about Rig's (Heimdal in disguise) travels among the humans. In short, Rig visits humans from three different social levels, and only at the highest social level is white bread made from wheat served (Larsen 1926). From the medieval period, it is known that bread from wheat was mainly served on special occasions, such as holidays and feasts, and then only in small amounts, and that only limited amounts of free-threshing wheat were grown (Kjersgaard 1978, pp. 33, 129–130). Although it is not possible to draw direct parallels from the Viking age and medieval period with earlier periods, the distribution pattern seems to support the idea that food made of either bread wheat or durum wheat could have been a luxury food, served at special occasions.

Conclusion

Free-threshing wheat (bread wheat and/or durum wheat) was widely distributed in Denmark during all periods from the Neolithic to the Viking age, but most samples contain only a limited number of grains. Large amounts are found at only a few sites. This seems to suggest that free-threshing wheat was cultivated as a separate crop, Although most likely only in limited amounts. This way of cultivation could perhaps indicate that free-threshing wheat was not a crop that the farmers were dependent on, but a special crop that was used on special occasions or as a luxury.
Regarding the find of durum wheat at Frydelund (Kirleis and Fischer 2014; Andreasen 2017), the many rachis fragments show that it is necessary to reconsider our identification of free-threshing wheat. In contrast to what has been believed for many years, the find shows that durum wheat was cultivated in Denmark, at least during some parts of prehistory, and that it could thrive despite the northern climate. Therefore, it must be concluded that, unless a find contains rachis fragments, it *cannot* be resolved whether the grains are bread wheat or durum wheat and, further, that grains from free-threshing wheat therefore should be identified as bread wheat/durum wheat (*Triticum aestivum* ssp. *aestivum*/*Triticum turgidum* ssp. *durum*), or for convenience *Triticum aestivum/durum* as suggested by Cappers and Neef (2012, p. 316).

Acknowledgments

This overview of free-threshing wheat in Danish prehistory could not have been completed without the help of many people. I would therefore like to thank the museums that allowed me to use their material and all the archaeologists who put a lot of work into giving me the correct dates of the free-threshing wheat grains. I would also like to send a huge thank you to Peter Steen Henriksen at the Department of Environmental Archaeology and Materials Science, National Museum of Denmark, and David E. Robinson for allowing me to use their huge dataset, published as well as unpublished, and for all their kind help. My thanks go also to Carsten Riisager, formerly of the Moesgaard Museum, who helped with MapInfo and Casper Skaaning Sørensen, Moesgaard Museum for his huge help with transforming the figures into something that could be published. Finally, I would like to give a huge thank you to the editors and peer reviewers and to my colleagues Peter Mose Jensen and Peter Hambro Mikkelsen for their constructive suggestions in connection with this article. I also thank the latter two for their many good discussions on many aspects of archaeobotany, among which the idea for this article first appeared.
Electronic supplementary material for the paper is accessible at the data exchange platform of the Johanna Mestorf Academy at Kiel University (https://www.jma.uni-kiel.de/en/research-projects/data-exchange-platform).

References

Andersen NH (2015) Frydenlund: Early Neolithic settlement and 'barkaer' structures in the Sarup area. In: Brink K, Hydén S, Jennbert K, Larsson L, Olausson D (eds) Neolithic diversities (Perspectives from a conference in Lund, Sweden). Acta Archaeol Lundensia Ser in 8°, 65:117–127

Andreasen MH (2009) Agerbruget i enkeltgravsskultur, senneolitikum og ældre bronzealder i Jylland: Belyst ud fra plantemakrofossiler. Kuml 2009:9–55

Andreasen MH (2011) HBV 1302, Kongehøj I og HBV 1275, Kongehøj II (FHM 4296/670 og 690): Makrofossilanalyser fra en røse/gravhøj fra senneolitikum/ældre bronzealder og en række hustomter fra ældre bronzealder periode II (Rapport for Afdeling for Konservering og Naturvidenskab nr. 3). Moesgaard Museum, Højbjerg

Andreasen MH (2014) VHM 00282, Højene (FHM 4692/1147): Arkæobotanisk analyse af en kornfyldt grube fra yngre bronzealder (Rapport for Afdeling for Konservering og Naturvidenskab nr. 6). Moesgaard Museum, Højbjerg

Andreasen MH (2017) FSM 6973, Frydenlund (FHM 4296/1119): Makrofossilanalyse af anlæg fra tidligneolitikum (Rapport for Afdeling for Konservering og Naturvidenskab nr. 24). Moesgaard Museum, Højbjerg

Boas NA (1983) Egehøj: A settlement from the Early Bronze Age in east Jutland. J Danish Archaeol 2:90–101

Cappers RTJ, Neef R (2012) Handbook of plant palaeoecology. Barkhuis, Groningen

Dreibrodt S, Zahrer J, Bork H-R, Brauer A (2012) Witterungs- und Umweltsgeschichte während der norddeutschen Trichterbecherkultur: Rekonstruiert aus Basis mikrofazieller Untersuchungen an jahresgeschichteten Seesedimenten. In: Hinz M, Müller J (eds) Siedlung, Grabenwerk, Grosssteingrab:

Studien zu Gesellschaft, Wirtschaft und Umwelt der Trichterbechergruppen im nördlichen Mitteleuropa (Frühe Monumentalität und soziale Differenzierung 2). Dr. Rudolf Habelt GmbH, Bonn, pp 145–158

Henriksen PS (2003) Arkæobotaniske analyser af materiale fra Bjerre plads 7, en hustomt fra yngre bronzealder i Thy (Nationalmuseets Naturvidenskabelige Undersøgelser rapport nr. 6) National Museum of Denmark, Copenhagen

Henriksen PS (2006) Hundborg, Thy: De arkæobotaniske undersøgelser (Nationalmuseets Naturvidenskabelige Undersøgelser rapport nr. 17) National Museum of Denmark, Copenhagen

Henriksen PS, Robinson DE (1994) Ældre jernalderens agerbrug: Arkæobotaniske analyser af kornfundene fra Overbygård, Østerbølle, Fjand og Alrum (Nationalmuseets Naturvidenskabelige Undersøgelser rapport nr. 12) National Museum of Denmark, Copenhagen

Henriksen PS, Robinson DE (1996) Early Iron Age agriculture: Archaeobotanical evidence from an underground granary at Overbygård in northern Jutland, Denmark. Veg Hist Archaeobot 5:1–11

Henriksen PS, Robinson DE, Kelertas K (2018) Bronze Age agriculture, land use and vegetation at Bjerre Enge based on the results of archaeobotanical analyses. In: Bech J-H, Eriksen BV, Kristiansen K (eds) Bronze Age settlement and land use in Thy, northwest Denmark, vol. 2. Aarhus University Press, pp 387–458

Hvass S (1985) Hodde: Et vestjysk landsbysamfund fra ældre jernalder (Arkæologiske studier 7). Akademisk forlag Universitetsforlaget i København, Copenhagen

Jensen J (2001) Danmarks oldtid. Stenalder 13.000-2.000 f.Kr. København

Jensen J (2002) Danmarks oldtid. Bronzealder 2.000-500 f.Kr. København

Jensen J (2003) Danmarks oldtid. Ældre Jernalder 500 f.Kr.-400 e.Kr. København

Jensen J (2004) Danmarks oldtid. Yngre Jernalder og Vikingetid 400-1050 e.Kr. København

Jensen PM (2011) OBM 4538, Dragebjerggård Vest (FHM 4296/769). Arkæobotanisk analyse af to hustomter fra slutningen af yngre romertid og ældre germansk jernalder. Rapport for Afdeling for Konservering og Naturvidenskab nr. 7, 2011

Jensen PM, Andreasen MH (2011) Det levede man af (Afsnit om agerbruget). In: Nielsen, MH (ed) Lundø MB, Therkelsen KG Fyn i Fortiden: Det levede liv 500 f.Kr–150 e.Kr. Forlaget Odense Bys Museer, pp 127–152

Kirleis W, Fischer E (2014) Neolithic cultivation of tetraploid free threshing wheat in Denmark and northern Germany: Implications for crop diversity and societal dynamics of the Funnel Beaker culture. Veg Hist Archaeobot 23(Suppl. 1):81–96

Kjersgaard E (1978) Mad og øl i Danmarks Middelalder. Nationalmuseet, København

Klassen L (2005) Refshøjgård: Et bemærkelsesværdigt gravfund fra enkeltgravskulturen. Kuml 2005:17–59

Körber-Grohne U (1995) Nutzpflanzen in Deutschland von der Vorgeschichte bis heute: Das kompetente Nachschlagewerk. K. Theiss, Stuttgart

Larsen T (1926) Nordens Gudekvad: Overs. fra Oldnordisk af Thöger Larsen. Lemvig

Lund J (1979) Tre førromerske jordkældre fra Overbygård. Kuml 1979:109–139

Marring R (2014) VHM 00282, Højene. Sct. Olai Sogn, Vennebjerg Herred, Hjørring Kommune Fund og Fortidsminder 100612-89: Arkæologisk udgravning af bebyggelse fra yngre bronzealder, jernalder og sen vikingetid/tidlig middelalder. Beretning Vendsyssel Historiske Museum

Mikkelsen PH, Nørbach LC (2003) Drengsted: Bebyggelse, jernproduktion og agerbrug i yngre romersk og ældre germansk jernalder. Jysk Arkæologisk Selskab, Højbjerg

Peña-Chocarro L, Zapata L (2014) Parching and dehusking hulled wheat. In: van Gijn A, Whittaker JC, Anderson PC (eds) Exploring and explaining diversity in agricultural technology. Oxbow Books, Oxford, pp 226–232

Robinson, DE (1994) Crop plants in Danish prehistory / Dyrkede planter fra Danmarks forhistorie. Arkæologiske udgravninger i Danmark 1993. Det Arkæologiske Nævn, København, pp 20–39

Robinson DE (2003) Neolithic and Bronze Age agriculture in southern Scandinavia: Recent archaeobotanical evidence from Denmark. Environ Archaeol 8:145–165

Robinson DE, Mikkelsen PH, Malmros C (2009) Agerbrug, driftsformer og planteressourcer i jernalder og vikingetid (500 f.Kr.–1100 e.Kr.). In: Odgaard B, Rømer JR (eds) Danske landbrugslandskaber gennem 2000 år: Fra digevoldinger til støtteordninger. Aarhus Universitetsforlag, Aarhus, pp 117–142

Rowley-Conwy P (1984) The Egehøj cereals: Bread wheat (*Triticum aestivum* s.l.) in the Danish Early Bronze Age. J Danish Archaeol 3:104–110

Stika H-P (2011) Early Iron Age and late mediaeval malt finds from Germany: Attempts at reconstruction of early Celtic brewing and the taste of Celtic beer. Archaeol Anthrop Sci 3:41–48

A Bronze Age house at Hestehaven: An early example of storage and cultivation of hulled barley (*Hordeum vulgare* var. *vulgare*) in Denmark

Peter Mose Jensen[1], Livija Ivanovaitė[2], Anja Vegeberg Jensen[3]

Author addresses and e-mail:
[1] Dept. of Archaeological Science and Conservation,
Moesgaard Museum, Denmark, pmj@moesgaardmuseum.dk
[2] Institute for Prehistoric and Protohistoric Archeology,
Johannes Gutenberg University, Germany, livanova@uni-mainz.de
[3] Museum Skanderborg, Denmark, avj@museumskanderborg.dk

Abstract

This chapter presents the archaeobotanical data from a unique Late Neolithic and Early Bronze Age site at Hestehaven, eastern Jutland, Denmark. Of the four building constructions identified, the late Neolithic house K4 and the Early Bronze Age house K1 were archaeobotanically investigated. The burnt-down, two-aisled house K1 contained a sunken floor which provided exceptional archaeobotanical data. Here a relatively undisturbed 8–10 cm thick cultural layer was identified at the bottom of the floor consisting mainly of carbonised remains of grain stored in the house at the time of the fire. This find is the largest archaeobotanical assemblage from the Early Bronze Age in Denmark. Systematic sampling and subsequent archaeobotanical investigation provided a detailed view of the grain store, which was found to contain several different types of crops, including the earliest large concentration of hulled barley (*Hordeum vulgare* var. *vulgare*) found in Denmark. The investigation also gave new information about grain storage, as different crops were found in separate concentrations, likely reflecting individual, now-decayed containers.

Based on the analyses from houses K4 and K1, the chapter discusses agricultural practices and the general agricultural situation during the Late Neolithic and Early Bronze Age in Denmark, with a special focus on the storage situation in K1 and the early exploitation of hulled barley.

Introduction

In recent times, a number of papers have investigated the general developments and changes in agricultural practices during the Late Neolithic (c. 2400–1800 cal BCE) and Early Bronze Age (c. 1800–1100 cal BCE) in Denmark (Robinson 1994, 2000, 2003; Andreasen 2009). Although these studies have provided general overviews of agrarian practices, they have been limited by the uneven distribution of analysed grain finds in Denmark. Many regions in Jutland with light soils have been relatively well investigated, while eastern parts of Jutland and the isles of Funen and Zealand, which are generally characterised by heavier soils, are represented by few finds. Thus, there is a need for new archaeobotanical data, especially from areas with heavier soils, in order to obtain a comprehensive overall view on the late Neolithic and Early Bronze Age agricultural situation in Denmark as a whole.

A recent archaeobotanical investigation of the site of Hestehaven (Fig.1) is significant for expanding our view of the agricultural practices in eastern Jutland. Here, the remains of four buildings dated to Late Neolithic–Early Bronze Age were excavated. The rich archaeobotanical assemblage found in a burnt two-aisled Early Bronze Age house has provided important new data about the agricultural practices. In the sunken floor area of the house, a relatively undisturbed bottom layer was identified consisting mainly of carbonised cereals, which were stored in the house at the time of the fire. After the layer of carbonised grain was identified in the excavation, a number of samples from not only the floor layer but also from other archaeological features from this and another, Late Neolithic house on the site were sampled and later analysed archaeobotanically.

The analyses provided important new information about agricultural practices, such as crop handling and storage patterns, in the Late Neolithic to Early Bronze Age.

Figure 1. Map showing location of the two sites, Hestehaven (blue star) and Hjulby (red star), Denmark, discussed in this chapter.

Materials and methods

The archaeological site of Hestehaven was excavated in 2011 by the local Museum Skanderborg (Fig. 1) (Jensen A 2012, Jensen & Bagge 2019).
During excavation, structural evidence for four houses was investigated (Fig. 2). Three of the houses (named K1, K2 and K4) were of the two-aisled house type, while K3 was a less well preserved and relatively unclear three aisled building of unknown date (Jensen A 2012). Radiocarbon dates from the three two-aisled constructions show that the houses belonged to the transitional period from the Late Neolithic to the Early Bronze Age (Table 1 and Fig. 3). All calibrated ages are presented as the two standard deviations (2σ) range.
The most prominent house in terms of both archaeological and archaeobotanical finds was house K1, dating to the Early Bronze Age. This house, which showed signs of having burnt down, was oriented east–west, measured approx. 10 × 5 m and contained a sunken floor area (5 × 5.5 m) in the eastern part of the building. During excavation, a number of archaeological find groups, including pottery and quern stone fragments were found, together with tools. A pressure-flaked dagger provided an archaeological dating to the Late Neolithic or Early Bronze Age. Pottery with Bell Beaker decorations was also found

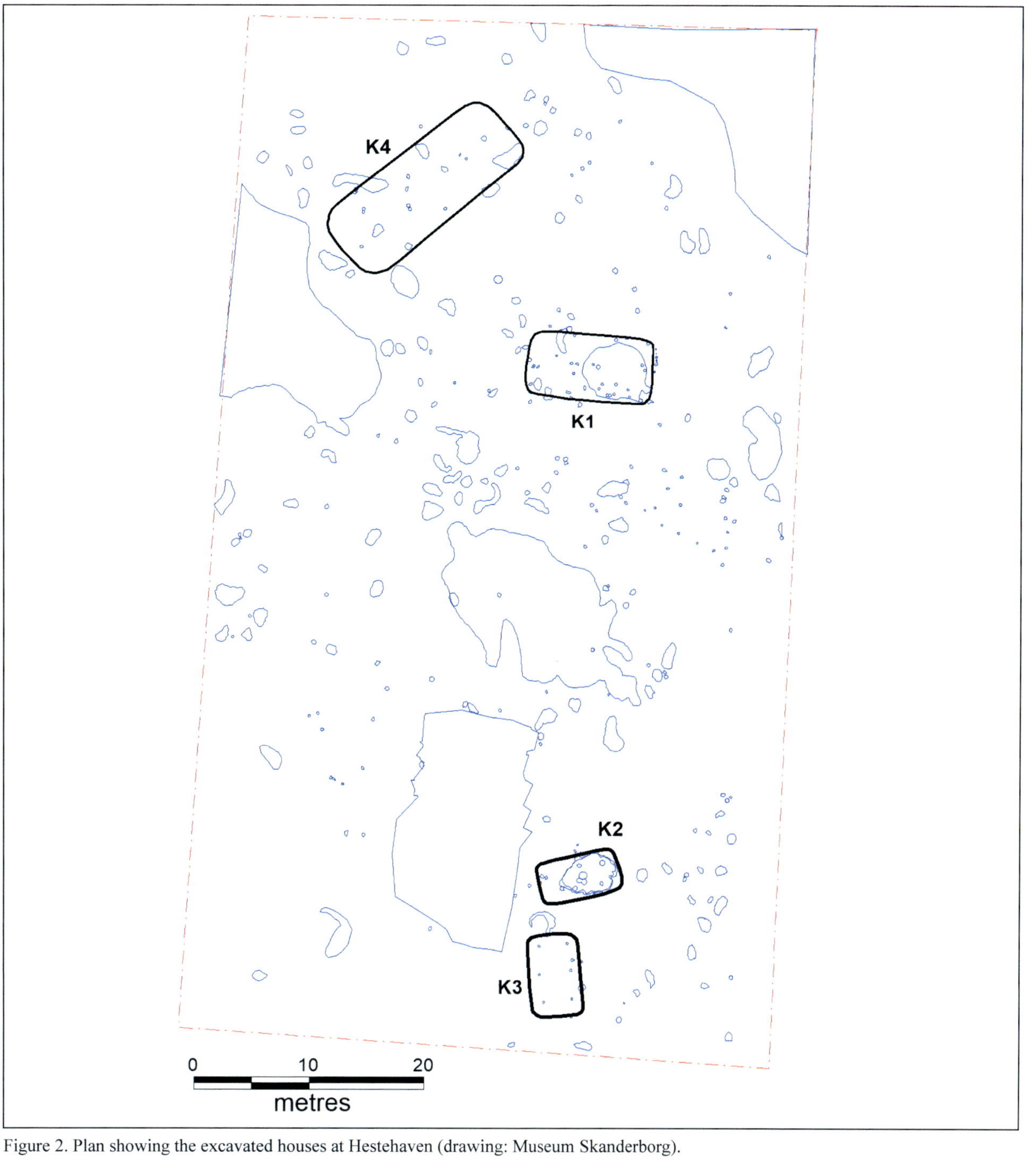

Figure 2. Plan showing the excavated houses at Hestehaven (drawing: Museum Skanderborg).

(Fig. 4), representing a very late appearance of the Bell Beaker phenomenon, considering the dating of the house to the Early Bronze Age. However, the most remarkable discovery was encountered in the sunken floor area. At the bottom of the sunken floor, an 8–10 cm thick charred layer (A212) (Fig. 5) was found, which contained huge amounts of carbonised cereal grains. The grain layer was lying directly on top of the relatively undisturbed, sandy floor layer/subsoil and was interpreted as the possible remains of a grain store that had been preserved when the house burnt.

During the excavation, a large number of soil samples for archaeobotanical analysis were taken from different features, such as postholes, but especially from layer A212. All soil was sampled from this layer in 0.5 × 0.5 m squares, with each square being further subdivided into a bottom layer sample and 1 or 2 overlying samples, apart from the corner squares, where the grain layer was relatively thin. The reason for taking overlaying samples was to ascertain whether the layer consisted of a massive layer of the same kind of grains or whether there were differences in species composition between layers, which may indicate different depositional events.

In order to conduct the archaeobotanical analysis, all soil samples were treated in a flotation machine, whereby the carbonised material was washed through

Table 1. ^{14}C dates from Hestehaven.

Sample number	Sample context	Sample type	^{14}C age (BP)	Calibrated age (1 sigma)	Calibrated age (2 sigma)
AAR-18535	House K4, A282, sample 1	Charred acorn (*Quercus*)	3567±25	1946-1887 BC	2015-1782 BC
AAR-18536	House K4, A282, sample 2	Charred barley (*Hordeum vulgare var. nudum*)	3634±26	2030-1956 BC	2126-1920 BC
AAR-18537	House K2, A386, sample 1	Charred grain (*Cerealia*)	3313±25	1623-1533 BC	1668-1520 BC
AAR-18538	House K2, A386, sample 2	Charred hazelnut shell (*Corylus avellana*)	3390±45	1741-1631 BC	1872-1533 BC
AAR-18539	House K2, A386, sample 3	Charred hazelnut shell (*Corylus avellana*)	3401±25	1742-1670 BC	1752-1627 BC
AAR-18540	House K1, A212, JP13 (2:3) sample 1	Charred barley (*Hordeum vulgare var. vulgare*)	3286±25	1608-1524 BC	1622-1501 BC
AAR-18541	House K1, A212, JP13 (2:3) sample 2	Charred barley (*Hordeum vulgare var. vulgare*)	3279±25	1608-1517 BC	1619-1498 BC
AAR-18542	House K1, A212, JP57 (2:2) sample 1	Charred barley (*Hordeum vulgare var. nudum*)	3226±25	1517-1455 BC	1602-1432 BC
AAR-18543	House K1, A212, JP57 (2:2) sample 2	Charred barley (*Hordeum vulgare var. nudum*)	3301±40	1622-1524 BC	1687-1497 BC

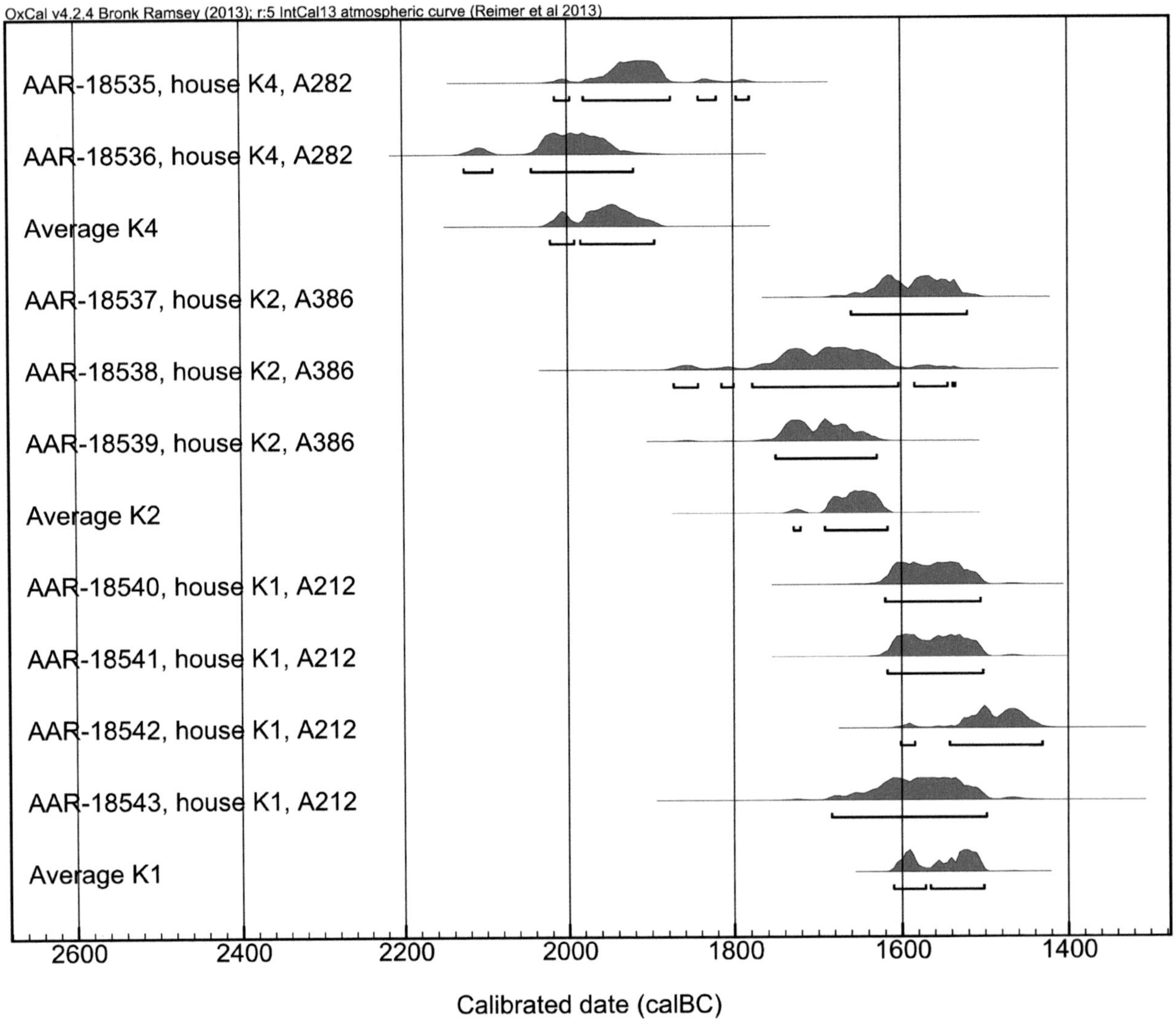

Figure 3. Calibrated radiocarbon ages of houses K1 and K4 at Hestehaven, using OxCal v 4.2 (Bronk Ramsey 2009) and the IntCal 13 calibration curve (Reimer et al. 2013). The average probability curves for houses K1 and K4 were produced with OxCal's R-Combine function.

Figure 4. Pressure-flaked dagger and two ceramic sherds with Bell Beaker ornamentation from K1 at Hestehaven, Denmark (photo: Museum Skanderborg and Moesgaard Museum.

Figure 5. The sunken floor area in house K1 at Hestehaven, Denmark, divided into sections, with soil layers seen in the profile. The black grain layer, A212, is visible at the bottom of the profile, directly on top of the light subsoil (photo: Museum Skanderborg).

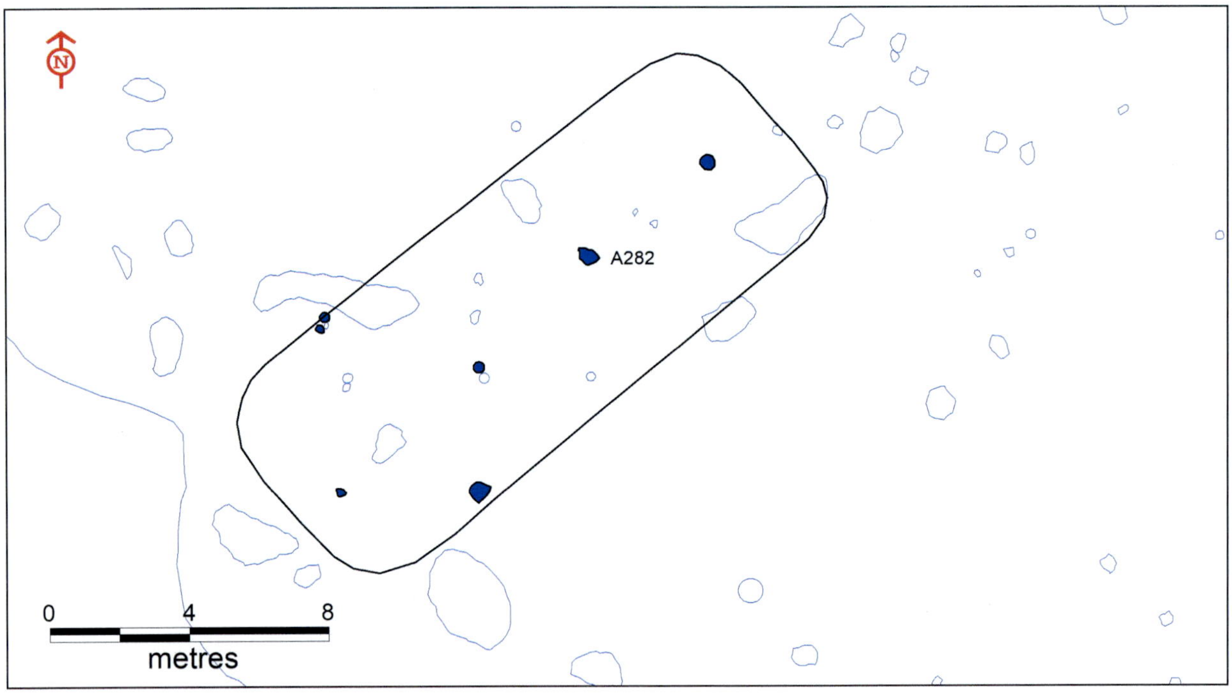

Figure 6. Plan of House K4 at Hestehaven, showing the analysed posthole A282 (drawing: Museum Skanderborg).

a 0.25 mm mesh and dried. Afterwards, all samples were screened under a microscope using up to 40× enlargement to evaluate the carbonised material.

Based on the amount of carbonised material and species distributions in the samples, as estimated due to the screening, a number of samples from houses K1 and K4 were selected for a more detailed analysis. No samples from houses K2 and K3 were archaeobotanically investigated due to lack of carbonised cereals and weeds from the postholes related to these structures.

From the Late Neolithic house K4, only a single sample was selected for analysis. This sample came from the fill of a roof-bearing posthole (A282) from the north-eastern part of the building (Fig. 6).

All floated samples from grain layer A212 in house K1 together are estimated to contain more than 37 litres of carbonised grain. This made analysis of 100% of the samples too time and resource consuming, and for this reason 13 samples from the lowest level of the grain, the layer directly above the sunken floor level, were selected for analysis.

In addition to the samples from A212, five additional samples, belonging to postholes from house K1, were also selected for analysis (Fig. 7). These include three samples from outside the sunken area and two from postholes under layer A212. These postholes were only visible under the grain layer, and apparently did not cut through A212. This suggests that the posts most probably date from the time around or before the fire that caused the formation of the overlying burnt layer.

Regarding sample sizes, 300 or more plant remains have been analysed from all samples from Hestehaven when possible, the only exceptions being a few smaller posthole samples from K1 (see Figs. 8 and 9).

Results

The results from the archaeobotanical analyses from Hestehaven will be presented in the following section. Throughout the text, the English plant names will be used; a list of all identified plant taxa from the site, together with their scientific names, can be found in Table 2, below.

The Late Neolithic house K4

The grains from K4 were mainly barley (*Hordeum vulgare*) (about 1100 grains), most probably naked barley (*Hordeum vulgare* var. *nudum*). A single glume base fragment of emmer (*Triticum turgidum* ssp. *dicoccon*) was also present, indicating exploitation of this crop. A few seeds from fat-hen (*Chenopodium album*) were present, which in all likelihood represent remains of weeds associated with the crops. In addition to the remains of cereal grains and associated field weeds, the sample also yielded a relatively high proportion of remains of likely collected food plants. These include an estimated 162 fragments of nuts and 80 fragments of shell of acorns (*Quercus* sp.), about 15 pips and 80 fruit fragments of crab apple (*Malus sylvestris*), and 10 fragments of nutshells from hazel (*Corylus avellana*) (Jensen 2013).

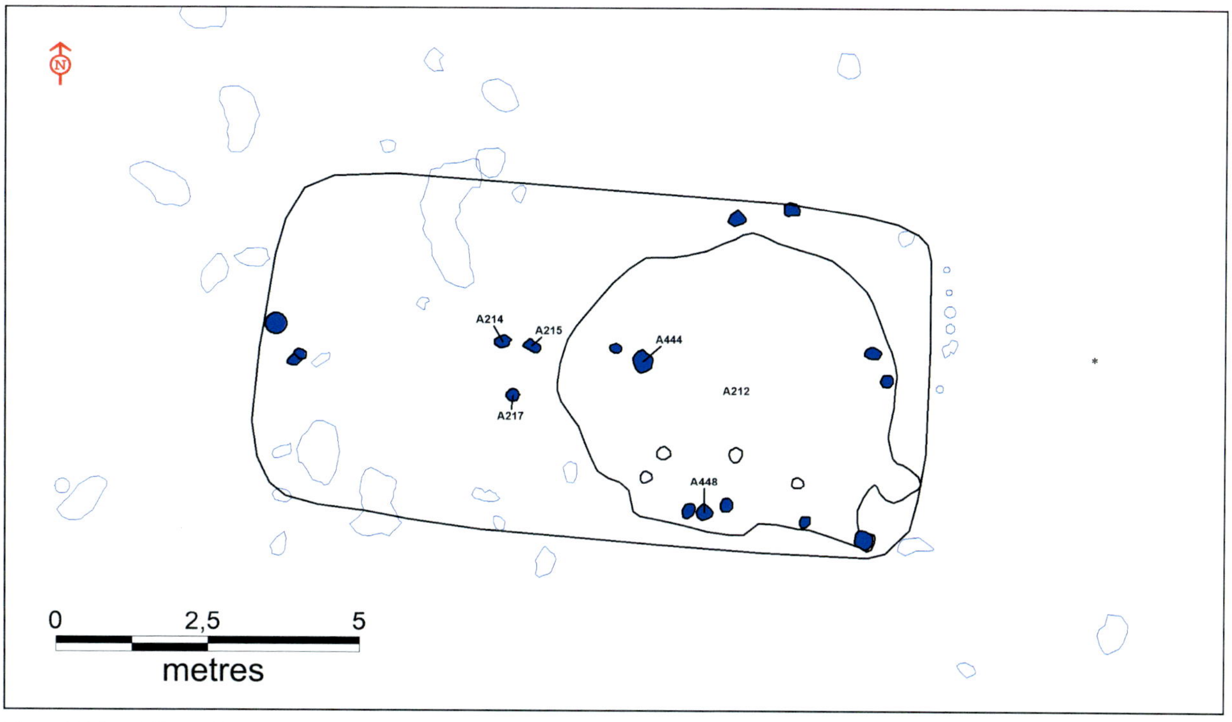

Figure 7. Plan of House K1 at Hestehaven, showing postholes (blue fill) and the outlines of the house and layer A212 (black. Analysed postholes are labeled) (drawing: Museum Skanderborg).

The Early Bronze Age house K1

The macrofossils from the postholes in the central part of house K1

As seen in Figure 8, the samples from the postholes in the central part of house K1 contained comparatively few carbonised plant remains compared to the overwhelming amounts of carbonised grain in the sunken area to the east. The central posthole samples mainly contained cereals, represented by a few grains of barley, probably mainly naked barley, 17 grains of emmer or spelt (*Triticum turgidum* ssp. *dicoccon/ Triticum aestivum* ssp. *spelta*) and 6 half glume bases from emmer. In addition to cereal remains, a few probable weed seeds from redshank or pale persicaria (*Persicaria maculosa/lapathifolium*) were found in this part of the house. Finally, the samples contained a few remains from likely collected food plants, in the shape of a single nutshell fragment from hazel and a pip of crab apple (Jensen 2013).

The macrofossils from the sunken floor area in the eastern part of house K1

The analysed samples from the sunken area of K1 mainly come from the grain-rich layer A212, but a few samples of fill from the postholes found underneath the layer were also analysed, as mentioned above (Figs. 8 and 9). The distribution of plant species in the postholes to a large degree reflects the distribution in the overlying grain layer samples. This implies that the postholes contained material that had fallen down from the grain store above, either during or after the fire.

As the screenings from A212 already clearly showed, apart from small amounts of charcoal, weed seeds and occasional hazel nutshell fragments, the samples consisted almost exclusively of carbonised cereal grains. The screenings also revealed that, while a clear majority of samples were composed of several cereal species mixed together to a varying degree, some of the samples were dominated by one or a few crop types. We chose which samples to analyse on the basis of the distribution patterns obtained from the screenings.

The analysed samples primarily contained a mixture of hulled barley (*Hordeum vulgare* var. *vulgare*), naked barley, and wheats (Fig. 10). The barley samples were heavily dominated by grains alone and only sporadically contained awn fragments. In comparison to this, the relationship between grains and awns was much more varied within the wheat group, represented by both bread wheat or durum wheat (*Triticum aestivum* ssp. *aestivum/Triticum turgidum* ssp. *durum*), emmer and spelt. Among the wheats, bread wheat/durum wheat was present in only a few samples and was exclusively represented by grains. Most wheat samples in A212 apparently contained a relatively even mixture of emmer and spelt, with both grains and glume bases present. Carbonised glume bases from emmer and spelt are fork shaped and often break into two halves. The relationships between glume base halves and grains varied a lot between samples. It was as low as 1:6 or 1:7 in some samples and more than 1:1 in others, with no clear distribution

Table 2. The identified plant taxa in the analysed contexts from house K4 and K1 at Hestehaven.

		House K4	House K1					
		A282	A212	A214	A215	A217	A444	A448
Scientific names	**English names**							
Cultivated species	**Cultivated species**							
Avena sp.	Oat		X					
Hordeum vulgare	Barley	X	X	X	X	X	X	X
Hordeum vulgare var. nudum	Naked Barley	X	X		X	X	X	X
Hordeum vulgare var. vulgare	Hulled Barley		X				X	X
Triticum aestivum ssp. *aestivum* / *Triticum turgidum* ssp. *durum*	Bread Wheat/Durumwheat		X					
Triticum turgidum ssp. *dicoccon*	Emmer	X	X			X	X	
Triticum turgidum ssp. *dicoccon* / *Triticum aestivum* ssp. *spelta*	Emmer/Spelt		X			X	X	
Triticum monococcum ssp. *monococcum*	Einkorn		X					
Triticum aestivum ssp. *spelta*	Spelt		X				X	
Triticum sp.	Wheat		X	X		X	X	X
Wild/collected species	**Wild/collected species**							
Bromus sp.	Brome		X					
Amaranthaceae	Amaranth family		X					
Caryophyllaceae	Carnation family		X					
Chenopodium album	Fat-hen	X	X			X	X	X
Chenopodium sp.	Goosefoot		X					
Corylus avellana (nutshells)	Hazel (nutshells)	X	X			X		
Epilobium palustre	Marsh Willowherb		X					
Fabaceae	Legume family		X					
Fallopia convolvulus	Black Bindweed		X				X	
Galium sp.	Bedstraw		X					
Malus sylvestris	Crab Apple	X		X				
Persicaria maculosa/lapathifolium	Red Shank/Pale Persicaria	X	X	X	X	X	X	X
Plantago lanceolata	Ribwort Plantain		X					
Poaceae	Grass family		X					
Polygonaceae	Knotweed family		X					
Quercus sp. (nuts and nutshells)	Oak (nuts and nutshells)	X						
Ranunculus sp.	Buttercup		X					
Rumex acetosella	Sheep's Sorrel	X	X					
Rumex sp.	Dock		X					
Silene sp.	Campion		X					
Stellaria sp.	Chickweed		X					

patterns throughout the layer. But since a clear distinction between emmer and spelt can only be made on the basis of the glumes, not the grains, the high number of glume bases in many of the samples was fortunate indeed.

Despite the mixture of several crop species in the samples from the grain layer, a clear pattern could still be seen in the distribution of samples that were dominated by specific cereal species. Especially close to the edges of the sunken area, i.e. at the walls, the samples were more or less clearly dominated by either hulled barley, naked barley or different wheats (Fig. 10), indicating separate storage of the crops. However, due to the large degree of mixing in the samples, the volumes of the individual stores could not be estimated.

In addition to barley and wheats, a few grains of oat (*Avena* sp.) and a single grain and glume base of einkorn were also found in the layer. Oat and einkorn are normally very rare in Danish Early Bronze Age contexts (Robinson 1994, 2000, p. 283, 2003, p. 148; Robinson & Kempfner 1988), and due to their scarce appearance at Hestehaven we think they likely represent weeds or other pollutants among the other crops. For this reason, the two species will not be further discussed in this chapter.

Cereal remains clearly dominated over weed seeds in all analysed samples (Fig. 11). This indicates that the grain on the sunken floor had been at least partially cleaned, which is typical for grain put in storage (Hillman 1981, pp. 132–137, 1984, pp. 4–6; Jones 1984, p. 44; Viklund 1998, pp. 90–107). By far the most common weed seeds identified were from pale persicaria or redshank and black bindweed (*Fallopia*

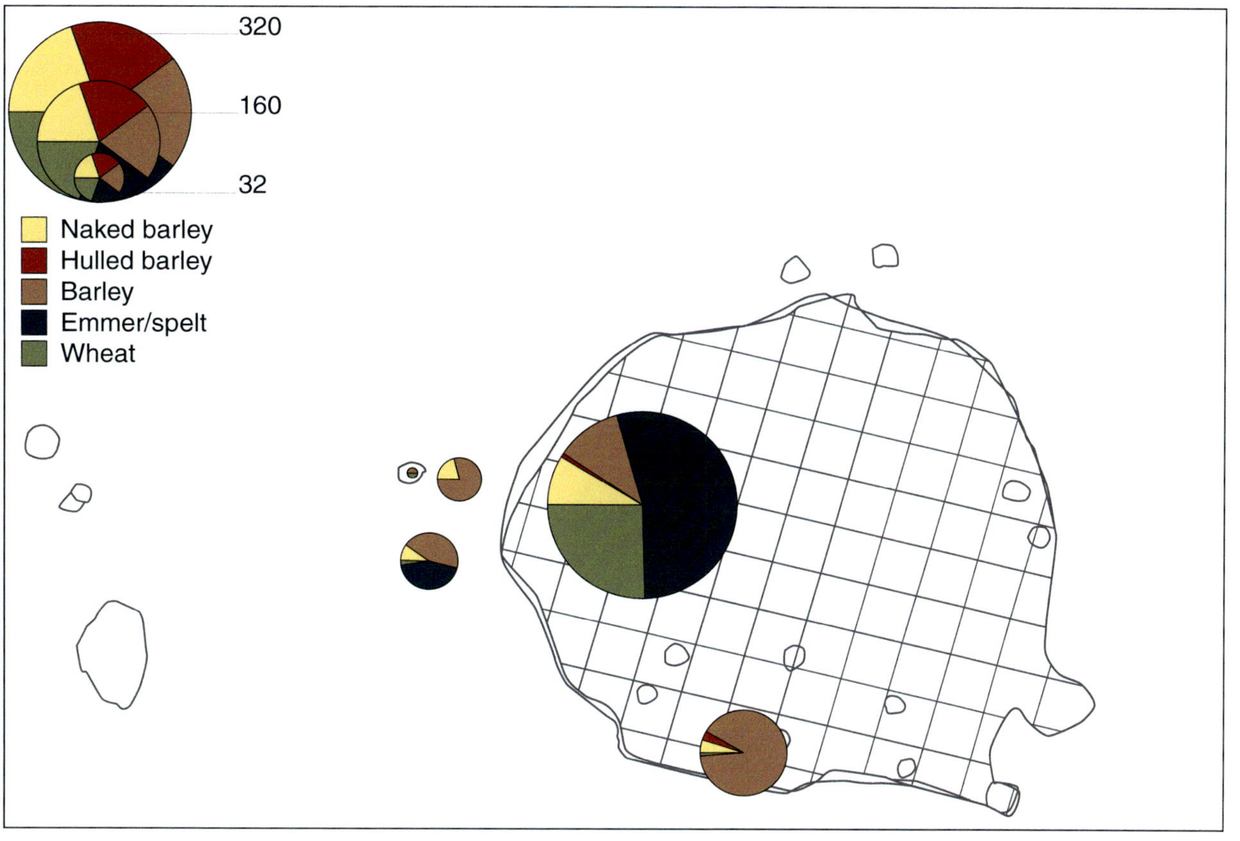

Figure 8. Plan showing distribution of whole grains from cultivated species in the posthole samples from the central and eastern parts of house K1 at Hestehaven.

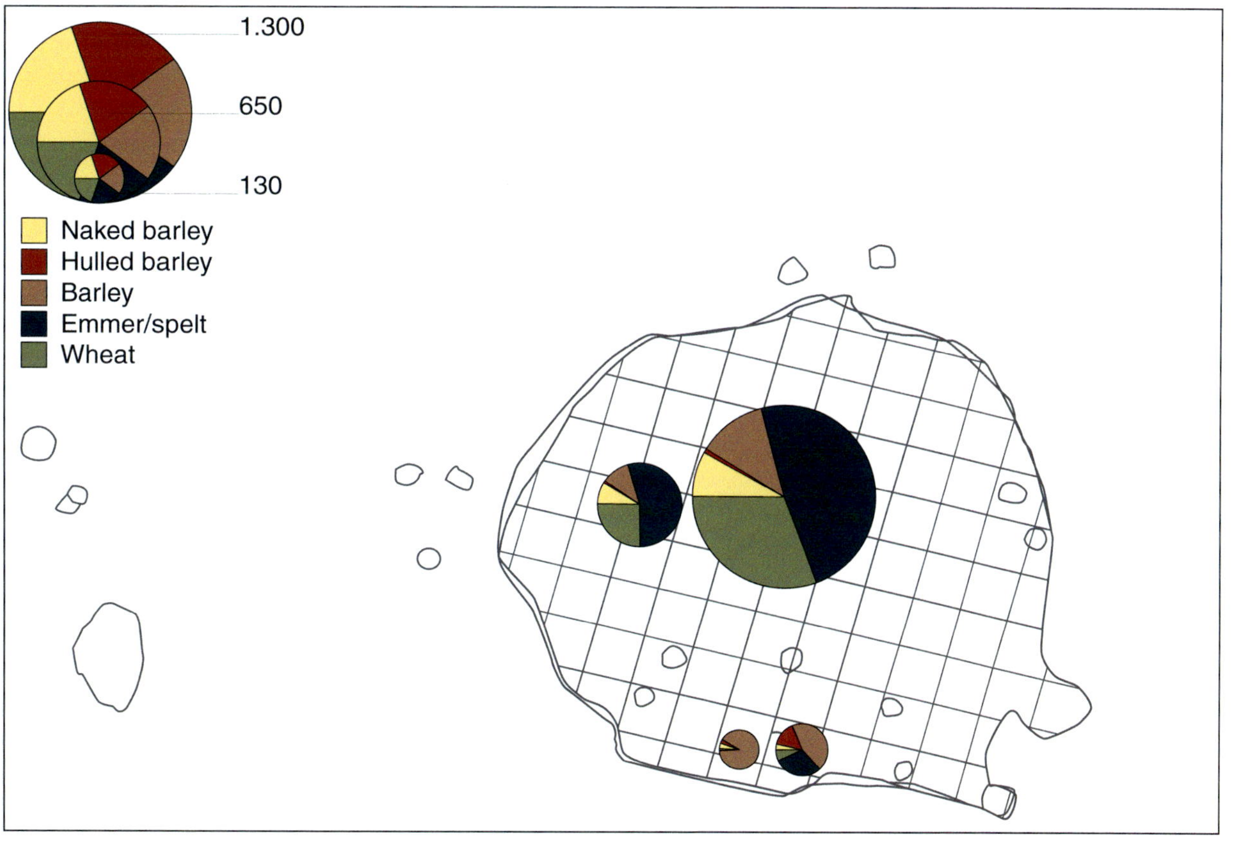

Figure 9. Plan showing distribution of whole grains from cultivated species in postholes from underneath layer A212 at Hestehaven and the corresponding distributions in the overlying layer.

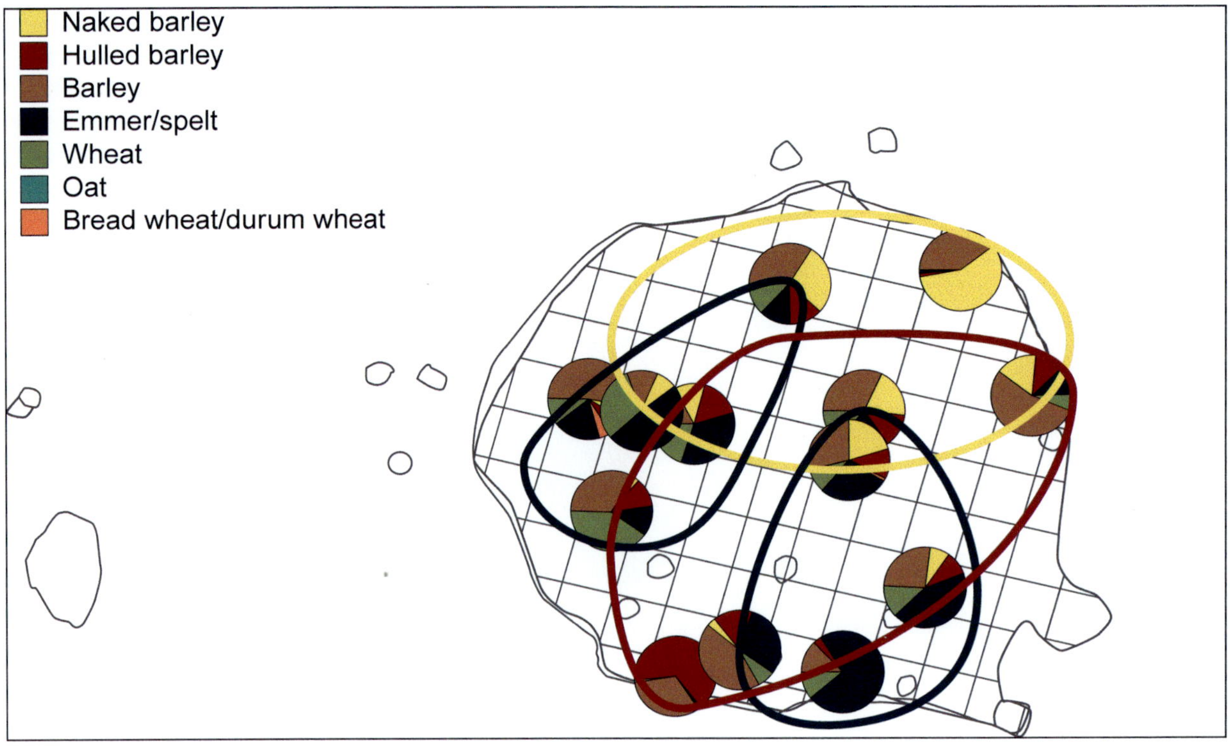

Figure 10. Plan showing percentage distribution of whole grains from cultivated species in layer A212 at Hestehaven and suggested storage areas for naked barley (yellow line), hulled barley (red line), and wheats (black line), as suggested by the analyses and the screenings.

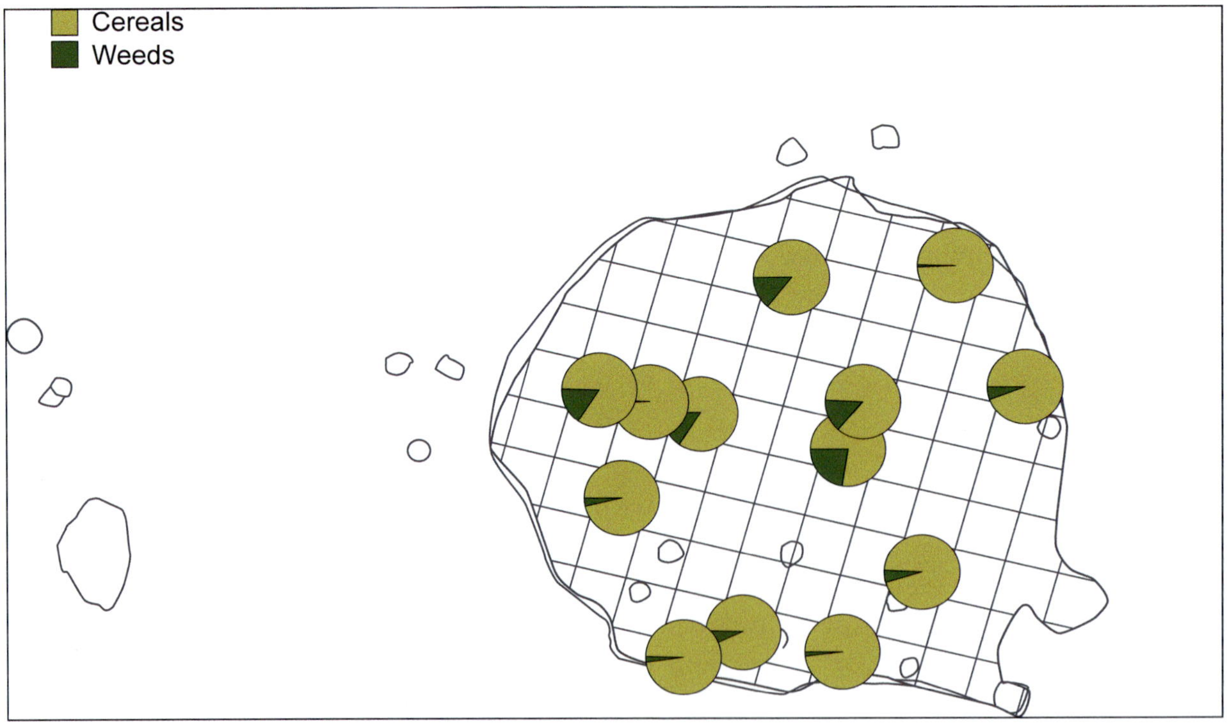

Figure 11. Plan showing percentage distribution of whole grains from cultivated species compared with weed seeds in layer A212 at Hestehaven.

convolvulus). These species typically occur in fields and other frequently disrupted soil types (Hansen 1993, pp. 191–192, 194; Mossberg et al. 1994, pp. 70, 72) and probably represent remains of cereal-related weeds still present among the grains after the initial cleaning. Apart from this interpretation, the weeds are difficult to evaluate further due to the low numbers of analysed seeds, and it is impossible to make detailed conclusions about agricultural practices on the basis of the weed seed distributions found in the samples.

Discussion

The Late Neolithic–Early Bronze Age site at Hestehaven has provided valuable new archaeobotanical data, not only in terms of overall agricultural trends, but also in terms of local onsite development. When comparing the archaeobotanical data from the Late Neolithic house K4 and the Early Bronze Age house K1, the relationship between cultivated and collected plants especially warrants discussion.

The strong presence of collected food plants, in the form of acorns, apples, and hazelnuts, in house K4 is remarkable, although it is not unknown in Danish contexts, as acorns, apples, and hazel nutshell fragments are relatively common in settlements from both the later parts of the Neolithic and the Early Bronze Age. One example is the assemblage from a burnt house from Vorbasse, in the southern part of Jutland (Jørgensen 1977, pp. 233–234), dated to around 2800–2400 cal BCE, which, just like K4, contained acorns, apples and hazelnuts. While apples and hazelnuts are typical human food plants, acorns were in historical times widely used as fodder for pigs that were normally brought into oak forests to feed on the nuts in the autumn (Brøndegaard 1978, pp. 286–290; Jørgensen 1977, p. 238). This, however, was clearly not the practice in the Neolithic and Bronze Age, where acorns were collected and often stored with other human food plants, which indicates human rather than animal consumption (Jensen 2008; Jensen & Pedersen 2008). The large number of collected acorns, apples and hazelnuts in house K4 thus indicates a strong importance of human plant food gathering in this Late Neolithic settlement.

K4, with its strong presence of collected plant foods, contrasts very much with the cereal-dominated house K1, which is more than 300 years younger. This contrast implies differences in subsistence strategies between the two houses, and perhaps also a local change over time towards an increasing reliance on cultivated over collected plant foods at the site. While it is certainly possible that subsistence changes are reflected in the finds from the two houses, we must be careful when making conclusions, because they result from two very different taphonomic situations. The Neolithic house, K4, is probably unburnt, and this means that the charred material in the house may derive from everyday activities that involved fire and heat throughout the habitation period. In contrast, Bronze Age house, K1, is definitely burnt, and the assemblage probably stems from one catastrophic event. Furthermore, the carbonised material in K1 is dominated by the many samples from layer A212. If only the plant species composition from the postholes from the central parts of the two houses are compared, a certain similarity with K4 can be observed, as remains of crab apples and hazel nutshells also were found in K1. This implies that the greatest differences between the houses are caused simply by the preserved huge cereal store in the eastern part of house K1.

Crop handling and storage

The samples from the postholes in the central part of the Bronze Age house, K1, differed from the samples from the sunken area to the east by containing lower amounts of carbonised plant food remains. Obviously, a direct comparison between posthole samples and samples from a grain-rich layer is problematic. But even when comparing the posthole samples from different areas of the house, we noted that the amounts of carbonised material in the posthole samples outside the sunken part of K1 are considerably lower than the amounts found in the postholes from within the sunken area (Fig. 8). This mixing of many different remains indicates that the central part of the house was probably not a major grain storage area, and this hypothesis is further confirmed by the diverse species composition in the samples from the central house area. A presence in this part of the house of cereal remains, together with a few weed seeds from redshank or pale persicaria, a single nutshell fragment from hazel and a pip of crab apple, indicate mixed remains from a number of food preparations, crop-cleanings or other everyday activities, rather than a single store. As no samples exist from the western part of K1, no conclusions can be made about this area, but still the analyses from the eastern and central parts of the house strongly suggest that the storage area for cereals was situated in the sunken eastern part.

Although the amount of charred grains and the composition of the archaeobotanical assemblage in the grain layer from the eastern part of house K1 is unique, we note that it is not the first Danish grain find of this type from the period. Several large cereal finds from sunken floors in houses dated to the later part of the Neolithic or the Early Bronze Age have been investigated, and they are typically interpreted as grain stores that have become carbonised due to house fires (Andreasen 2009; Jensen P 2012; 2013a; 2013b; Jensen in press; Møbjerg et al. 2007). Even though these finds show that house fires and the storage of large quantities of grain in houses may have been relatively common, the systematic sampling in house K1 at Hestehaven still provides a unique chance to investigate the detailed composition of such a grain store.

Regarding the crop distribution in layer A212, the analysis indicates that naked barley was concentrated in the northern half of the sunken area, while hulled barley was mainly present in the south-western part of the sunken area. Although many samples contained a mixture of species, a few samples from the two barley concentrations are still remarkably pure (Fig. 10). This clearly shows how naked and hulled barley were stored as separate crops in different parts of the sunken area. The wheat-dominated samples, consisting mainly of emmer and spelt, were particularly concentrated in the south-eastern part of the sunken area, indicating wheat storage here. The western part of the sunken floor also contained a relatively large amount of wheat, suggesting secondary wheat storage in this area (Fig. 10).

The relatively even mixture of emmer and spelt in many samples indicates that these two species were stored together. The very varied occurrences of emmer and spelt glume bases in relation to grain in the different samples is more difficult to explain. One reason for the variation could be that some samples contain few identifiable wheat remains, which makes these wheat samples less representative. Other likely reasons are differences in preservation conditions between grains and glumes (Boardman & Jones 1990) and depositional differences between different plant remains in different samples. All in all, the variations among samples mean that the amount of glume bases of emmer and spelt in relation to grains is impossible to compare correctly. In several large wheat samples, the ratio between glume base halves and grains is more or less 1:1. As each glume base holds two grains, the relationship in these samples makes it likely that the emmer and spelt in A212 had at least to some extent been stored as spikelets rather than as cleaned grains. When stored as spikelets, the grain would have been more protected against attacks by insects and microorganisms, and as the removal of the grain from the spikelet is generally quite rough, the process can also damage the viability of the grain and destroy its use as seed corn (Møbjerg et al. 2007, p. 26). There are thus several reasons to store emmer and spelt as spikelets, and a number of finds indicate that this storage method happened relatively often in Late Neolithic and Early Bronze Age Denmark (Møbjerg et al. 2007, p. 26). Spikelet storage is also well known ethnographically from the traditional cultivation of emmer and spelt, where the grains are typically removed from the spikes on a day-to-day basis prior to food preparation (Hillman 1981, pp. 132–137, 1984, pp. 4–6).

Archaeologically there were no clear traces of storage containers in layer A212. It is also difficult to infer whether the grain had been stored at floor level or higher up. It should be noted, however, that the sunken area contained several postholes that could have supported a loft used for storage (Jensen A 2012, p. 10). The sunken floor itself may also indicate a loft, as the lowering of the floor would have resulted in increased standing height in this part of the building, which again would have been an advantage if a loft was present. It is therefore very likely that the grain was originally stored in a loft, from where it fell down when the house burnt. A relatively long fall, for instance from a loft, would also explain why the carbonised grain was apparently evenly spread out on the floor level in the sunken area. If the grain had been stored close to the floor, we would have expected it to concentrate in separate piles. This situation is known from several burnt houses from the Early Iron Age (Henriksen & Robinson 1994). A long fall may also explain the high degree of mixing of crops in the samples, although post-depositional mixing from activities on the house plot after the fire could also have played a role. That the mixing resulted from storage of different crops together seems less likely though. The reason for this is that the mixed samples concentrated in the central parts of the sunken area, where they would have been relatively exposed to mixing, while all of the more clean samples came from close to the walls of the house, where they were probably less exposed to pollution from other samples.

Agricultural changes and cultivation of hulled barley in Late Neolithic and Early Bronze Age Denmark

The archaeobotanical assemblage from house K1 at Hestehaven is a significant find not only because of the presence of the preserved store, but also because of the taxonomic composition of the grain. While both emmer, spelt, bread wheat and naked barley are common cereals in Late Neolithic and Early Bronze Age Denmark (Robinson 2000; 2003), the high amounts of hulled barley are highly unusual and require further discussion.

Although both hulled and naked barley were introduced in Denmark from the beginning of the Neolithic (Robinson 1994, 2000, 2003), the naked variety is clearly the most common barley type, while hulled barley appears only sporadically and remains at low levels for a very long time (Hjelmquist 1975, pp. 216–217; Robinson 1994). At some sites in Jutland, a slight increase is registered in the archaeobotanical assemblages in the Early Bronze Age (Andreasen 2009, p. 23), but still, naked barley remains the dominant barley type in many regions of Denmark until the beginning of the 1st millennium AD. Around this time, the importance of hulled barley increases relatively quickly, after which it remains the dominant barley type. At the same time, the importance of naked barley decreases markedly, and it is at low levels in the majority of finds from this time onwards (Henriksen 2003; Jensen & Andreasen 2011, p. 142; Robinson et al. 2009, pp. 128–129. This decrease in naked barley in relation to hulled barley from the Pre-Roman/Roman Iron Age and onwards is observed in many other parts of Europe as well, but in other areas, hulled barley is more prominent than the naked variety. This is, for instance, seen in Germany and England, where hulled barley has a strong presence on many sites starting in the Neolithic period (Lister & Jones 2013).

More recent analysis has also indicated that there were probably regional variations in barley cul-

Figure 12. Hulled barley from house K1 at Hestehaven (photo: Moesgaard Museum).

tivation in different areas of Denmark (Jensen & Andreasen 2011, p. 142). Naked barley thus apparently remained the dominant barley type in some areas of Jutland until the Roman Iron Age (Henriksen 2002; Henriksen et al. 2009; Jensen & Andreasen 2011, p. 142). In contrast, hulled barley probably became dominant already in the Late Bronze Age in eastern Zealand, where the change corresponds more or less with the rise of hulled barley cultivation in neighbouring southern Sweden (Grabowski 2011; Henriksen 2003, p. 30).

Several reasons have been advanced to explain the increase in hulled barley during the Late Bronze Age to Early Iron Age in southern Scandinavia (Grabowski 2011, p. 481; Lister & Jones 2013; Mikkelsen 2003, pp. 149–151). One suggested reason is that hulled barley reacts more positively to fertilising and that the change could be connected with an assumed increase in the use of fertiliser at the time (Grabowski 2011, p. 481; Gustafsson 1998; Jensen & Andreasen 2011, p. 142; Lagerås & Regnell 1999). Other suggested reasons are changes in food habits (Skoglund 1999) or an increased use of barley for animal fodder rather than for human food (Grabowski 2011). A further suggested reason is that hulled barley was more resistant to diseases in a supposedly deteriorating climate around the beginning of the Iron Age (Helbæk 1957, p. 266). A final suggested reason is that, because the grains of hulled barley are more firmly attached to their spikes than are those of naked barley grains, harvesting and post-harvest handling of hulled barley could have been done more roughly, and with less grain loss than was the case for naked barley (Mikkelsen 2003, pp. 150–151).

Although hulled barley generally becomes dominant in southern Sweden during the Late Bronze Age, an exception from the overall situation can still be seen. On the site of Fosie IV, in south-western Sweden, close to Zealand in Denmark, the dominance of hulled barley over naked barley is recorded already in the Late Neolithic (Gustafsson 1995, pp. 2–5). The archaeobotanical find from Hestehaven therefore corresponds to an emerging pattern, indicating regional differences in agricultural practices and a possible greater importance of hulled barley towards the east, a pattern that can be supported by another recent find from Hjulby, on the eastern Danish island of Funen (Fig. 1). Here, an Early Bronze Age pit containing an estimate of about 16.000 carbonised cereal grains was sampled for palaeobotanical analysis. While the majority of the grains were probably spelt, the remaining identified grains (about 200) were almost exclusively hulled barley, and not a single grain of naked barley was identified among the barley grains (Jensen et al. 2011, pp. 72–73).

Thus, two recent Early Bronze Age finds from the eastern parts of Denmark have come to light now that show an increasing focus on hulled barley cultivation in the period before hulled barley becomes the most dominant barley type in Scandinavia.

Generally, the species distribution in Hestehaven house K1 corresponds well with the agricultural patterns seen during the Early Bronze Age, at least in Jutland, where Andreasen (2009) has recorded two trends in the archaeobotanical finds. The first is an overall increase in the range of cultivated species encountered, both on a general level and on many individual sites, which is an ongoing trend already seen in the late part of the Neolithic. The second trend is a tendency towards a very high occurrence of either emmer, spelt or bread wheat compared with other species on some Early Bronze Age sites compared with earlier times, where the different cultivated species tend to appear in more equal proportions (Andreasen 2009, Rowley-Conwy 1984). The high amounts of hulled barley in the samples from Hestehaven and Hjulby therefore indicate that now there was an increased focus on the cultivation of specific crops not only within the wheat types, but also within the barley types on some sites of this period.

When providing possible explanations for the abovementioned Early Bronze Age agricultural changes, Andreasen (2009) suggests that a reliance on many different cultivars, as opposed to a few, would probably be advantageous as it would help to ensure a good overall harvest by spreading the risks. Jensen (2000, pp. 92–93) suggests that an advantage could be the ability to better exploit soil nutrients by changing between species on a given field. When providing possible explanations for the increase in cultivars during the Single Grave period, Klassen (2005, pp. 5–6) suggests that the increased range of cultivated species and the increased focus on specific crops on some sites may simply reflect a growing tendency to cultivate on new and more diverse soil types, as a consequence of a gradual rise in population. In our opinion, this reason could also be used to explain the agricultural changes in the Early Bronze Age. The soils in eastern Denmark and Jutland are generally heavier than the lighter and sandier soil types found in many areas of western Jutland. As hulled barley has a higher demand for nutrients than naked barley, it is quite possible that the reason for both the increased focus on hulled barley at specific sites in the Early Bronze Age and the earlier general dominance of the hulled variety in the Late Bronze Age in eastern Denmark can, to a large degree, be attributed to differences in soil types.

Conclusions

The two investigated houses, K4 and K1, from Hestehaven have provided significant new data from a previously less researched region of eastern Jutland, from the Late Neolithic and Early Bronze Age. Especially the preserved grain store in the burnt-down Early Bronze Age house K1 has given a unique insight into the composition of stored grain from the time of the fire, an insight that to a large extent was gained through a systematic sampling of the storage layer in 0.5 × 0.5 m squares and in different layers. Subsequent screenings of all samples showed a high degree of crop species mixing in many samples, but the screenings were also able pinpoint a few cleaner samples close to the walls of the house. These indicated that emmer and spelt had probably been stored together, while naked and hulled barley had likely been stored as separate, clean crops. They also indicated that the many mixed samples were caused by events related to the house fire. Regarding future analyses of houses from this period, the observations gained from the screenings and analyses from the storage layer A212 are thus very significant, as they show how systematic sampling and sampling close to the house walls can be vital for a thorough understanding of grain store compositions.

The large amount of hulled barley in layer A212 makes Hestehaven unique among contemporaneous Early Bronze Age sites. Coupled with data from the site of Hjulby on Funen, it suggests that this grain species may have played a larger agricultural role in regions with heavier soil in Denmark than previously thought. The scarcity of finds from the period makes any statements about regional patterns uncertain, but the analyses from Hestehaven nevertheless provides important new information about hulled barley cultivation in the Early Bronze Age that can be further investigated in the future.

Acknowledgments

A number of samples from layer A212 at Hestehaven K1 were analysed as parts of various supervised student projects at the Department of Archaeological Science and Conservation at the Moesgaard Museum. These were executed by Dominika Kofel and Rehne Vokstrup, to whom grateful thanks are extended for helping with the analysis of the samples used for this article.

References

Andreasen MH (2009) Agerbruget i enkeltgravsskultur, senneolitikum og ældre bronzealder i Jylland: Belyst ud fra plantemakrofossiler. Kuml 2009:9–55

Boardman S, Jones G (1990) Experiments on the effects of charring on cereal plant components. J Archaeol Sci 17(1):1–12

Brøndegaard VJ (1978) Folk og Flora, Dansk Etnobotanik vol. 1. Rosenkilde og Bagger, Copenhagen

Grabowski R (2011) Changes in cereal cultivation during the Iron Age in southern Sweden: A compilation and interpretation of the archaeological material. Veg Hist Archaeobot 20:479–494

Gustafsson S (1995) Fosie IV: Jordbrukets förändring och utveckling från senneolitikum till yngre järnålder (rapport nr. 6). Stadsantikvariska avdelningen, Malmö Museer, Malmö

Gustafsson S (1998) The farming economy in south and central Sweden during the Bronze Age: A study based on carbonized botanical evidence. Current Swedish Archaeol 6:63–71

Hansen K (1993) Dansk feltflora, Gyldendal, Copenhagen

Helbæk H (1957) Bornholm plant economy in the first half of the 1st millennium AD. In: Klint-Jensen O (ed) Bornholm i folkevandringstiden og forudsætningerne i tidlig jernalder (Nationalmuseets Skrifter, Større beretninger II). National Museum of Denmark, Copenhagen, pp. 259–277

Henriksen PS (2002) Smedegård, en byhøj fra Thy. Arkæobotaniske undersøgelser. Nationalmuseets Naturvidenskabelige Undersøgelser rapport nr. 20. National Museum of Denmark, Copenhagen

Henriksen PS (2003) Arkæobotaniske undersøgelser af materiale fra Kroppedal Museum for Astronomi, Nyere tid, Arkæolog (Nationalmuseets Naturvidenskabelige Undersøgelser rapport nr. 12) National Museum of Denmark, Copenhagen

Henriksen PS, Robinson DE (1994) Ældre jernalders agerbrug: Arkæobotaniske analyser fra Overbygård, Østerbølle, Fjand og Alrum (Nationalmuseets Naturvidenskabelige Undersøgelser rapport nr. 12) National Museum of Denmark, Copenhagen

Henriksen PS, Harild JA, Jensen PM (2009) De arkæobotaniske analyser af materiale fra Nr. Hedegård. In: Runge M (ed) Nørre Hedegård: En nordjysk byhøj fra ældre jernalder. Jysk Arkæologisk Selskabs Skrifter, Højbjerg, pp. 247–252

Hillman G (1981) Reconstructing crop husbandry practices from charred remains of crops. In: Mercer R (ed) Farming practice in British prehistory. Edinburgh University Press, Edinburgh, pp. 123–162

Hillman G (1984) Interpretation of archaeological plant remains: The application of ethnographic models from Turkey, In: van Zeist W, Casparie WA (eds) Plants and ancient man. A.A. Balkema, Rotterdam, pp. 1–41

Hjelmquist H (1975) Getreidearten und andere Nutzpflanzen aus der Frühneolitischer Zeit von Langeland. In: Skaarup J (ed) Stengade: Ein Langeländischer Wohnplatz mit Hausresten aus der Frühneolitischer Zeit. Langelands Museum, Rudkøbing, pp. 211–219

Jensen AV (2012) SBM 1271 Hestehaven, Skanderborg (Excavation report). Skanderborg Museum, Skanderborg.

Jensen AV, Bagge M (2019) Two-aisled houses in the Skanderborg area from the Late Neolithic and Early Bronze Age. In: Sparrevohn LR, Kastholm OT, Nielsen PO (eds) Houses for the living: Two-aisled houses from the Neolithic and Early Bronze Age in Denmark. The Royal Society of Northern Antiquaries, Copenhagen, pp. 279-286

Jensen PM (2000) Agerbruget i yngre bronzealder belyst ved arkæobotaniske analyser af kornfund. MA dissertation, Aarhus University

Jensen PM (2008) Agern og korn fra Gilmosevej: En senneolitisk grube med agern og korn fra Gilmosevej (HEM 4086 rapport nr. 2). Department for Archaeological Science and Conservation, Moesgaard Museum, Højbjerg

Jensen PM (2012) HEM 4357, Sjællandsvej V (FHM 4296/369): Arkæobotanisk analyse af kornprøver fra bundlag i hus fra tidlig bronzealder (rapport nr. 2). Department for Archaeological Science and Conservation, Moesgaard Museum, Højbjerg

Jensen PM (2013) Makrofossilanalyse fra SBM 1271, Hestehaven (FHM 4296/1059): Arkæobotanisk analyse af to neolitiske huse fra Skanderborg (rapport nr. 5). Department for Archaeological Science and Conservation, Moesgaard Museum, Højbjerg

Jensen PM (in press) Cereal grain deposits from Late Neolithic and Early Bronze Age two-aisled houses in east Denmark. In: Sparrevohn LR, Kastholm OT, Nielsen PO (eds) Houses for the living: Two-aisled houses from the Neolithic and Early Bronze Age in Denmark. The Royal Society of Northern Antiquaries, Copenhagen

Jensen PM, Pedersen VJ (2008) Agern og korn fra Gilmosevej: Nyt om udnyttelse af planter i tidlig bronzealder. Midtjyske Fortællinger, Herning Museum pp.103–114

Jensen PM, Nielsen MH, Runge M (2011) Kornkamre fra Bronzealderen. In: Henriksen MB, Runge M (eds) Arkæologiske netværk og tværvidenskabelige samarbejdsprojekter: Eksempler fra den Fynske arbejdsmark. Fynske Minder 2011. Odense Bys Museer, Odense, pp.72–75

Jensen PM, Andreasen MH (2011) Det levde man af. In: Nielsen MH (ed): Fyn i Fortiden: Det levede liv 500 f.Kr–150 e.Kr. Forlaget Odense Bys Museer, Odense, pp. 127–152

Jones GEM (1984) Interpretation of archaeological plant material: Ethnographic models from Greece. In: van Zeist W, Casparie WA (eds) Plants and ancient man. A.A. Balkema, Rotterdam, pp. 43–61

Jørgensen G (1977) Acorns as food source in the later Stone Age. Acta Archaeologica 48:233–238

Klassen L (2005) Zur Bedeutung von Getreide in der Einzelgrabkultur Jütlands. jungsteinsite.de

Lagerås P, Regnell M (1999) Agrar förändring under sydsvensk bronsålder: En diskussion om skenbara samband och olösta gåtor. In: Olausson M (ed) Spiralens öga: Tjuga artiklar kring aktuell bronsålderforskning. Avdelningen för Arkeologiska Undersökningar, Skrifter 25. Riksantikvarieämbetet, Stockholm, pp. 263–276

Lister DL, Jones MK (2013) Is naked barley an eastern or a western crop? The combined evidence of archaeobotany and genetics. Veget Hist Archaeobot 22:439–446

Mikkelsen PH (2003) Agerbruget. In: Mikkelsen PH, Nørbach LC (eds) Drengsted: Bebyggelse, jernproduktion og agerbrug i yngre romersk og ældre germansk jernalder (Jysk Arkæologisk Selskabs Skrifter, 43). Jysk Arkæologisk Selskab, Højbjerg, pp. 115–225

Mossberg B, Stenberg L, Ericsson S (1994) Den store nordiske flora. G.E.C. Gads Forlag, Copenhagen

Møbjerg T, Jensen PM, Mikkelsen PH (2007) Enkehøj: En boplads med klokkebægerkeramik og korn. Kuml:9–45

Reimer PJ, Bard E, Bayliss A, Beck JW, Blackwell PG, Bronk Ramsey C, Buck CE, Cheng H, Edwards RL, Friedrich M, Grootes PM, Guilderson TP, Haflidason H, Hajdas I, Hatté C, Heaton TJ, Hoffmann DL, Hogg AG, Hughen KA, Kaiser KF, Kromer B, Manning SW, Niu M, Reimer RW, Richards DA, Scott EM, Southon JR, Staff RA, Turney CSM, van der Plicht J (2013) IntCal13 and marine13 radiocarbon age calibration curves 0–50,000 years cal BP. Radiocarbon 55(4):1869–1887

Robinson DE (1994) A catalogue of crop plant remains from Danish prehistoric sites (Nationalmuseets Naturvidenskabelige Undersøgelser, rapport 13). National Museum of Denmark, Copenhagen

Robinson DE (2000) Det slesvigske agerbrug i yngre stenalder og bronzealder. In: Ethelberg P, Jørgensen E, Robinson DE (eds) Det sønderjyske landbrugs historie: Sten- og Bronzealder. Haderslev Museum, Haderslev, pp. 281–298

Robinson, DE (2003) Neolithic and Bronze Age agriculture in southern Scandinavia: Recent archaeobotanical evidence from Denmark. Environ Archaeol 8:145–165

Robinson DE, Kempfner D (1988) Carbonized grain from Mortens Sande 2. J Danish Archaeol 6:125–129

Robinson DE, Mikkelsen PH, Malmros C (2009) Agerbrug, driftsformer og planteressourcer I jernalder og vikingetid (500 f.Kr.–100 e.Kr.): In: Odgaard B, Rydén Rømer J (eds) Danske landbrugslandskaber gennem 2000 år: Fra digevoldinger til støtteordninger. Aarhus Universitetsforlag, Aarhus, pp. 117–142

Rowley-Conwy P (1984) The Egehøj cereals. J Danish Archaeol 3:104–110

Skoglund P (1999) Diet, cooking and cosmology: Interpreting the evidence from Bronze Age plant macrofossils. Current Swedish Archaeol 7:149–160

Viklund K (1998) Cereals, weeds and crop processing in Iron Age Sweden. Archaeology and Environment 14. University of Umeå, Umeå

Tracing pioneer agriculture in northern Norway

Christin Eldegard Jensen

Author address:
Arkeologisk Museum, University of Stavanger, NO-4036 Stavanger, Norway

E-mail: christin.jensen@uis.no

Abstract

New empirical evidence, including archaeobotanical data obtained from archaeological excavations, supports earlier pollen-based hypotheses about a Bronze Age and Early Iron Age farming culture north of the Arctic Circle in Norway.

Since the late 1970s, pollen studies have been undertaken on mire sediments close to the unexcavated sites of prehistoric settlements in northern Norway. These have revealed a general pattern of a cultural landscape development similar to that in coastal areas farther south in Norway and indicate the presence of animal husbandry and cereal cultivation from at least the Late Bronze Age onwards. But this conclusion has been difficult to test empirically due to a lack of large-scale excavations of farm sites in northern Norway. During the last decade, however, mechanical topsoil stripping and on-site archaeobotanical analysis, coupled with direct ^{14}C dating of cereal grains and bones, have begun to be employed on archaeological excavations, with useful and interesting results.

Data from some recent interdisciplinary investigations in the coastal region between 68 and 69° N are presented here and discussed. Extensive field layers containing charred cereals and settlements corresponding to the southern Scandinavian house typology date from the Late Bronze Age to the Merovingian period, including intervening phases of abandonment. This shows that, where feasible, people in the north practised farming at an early stage in prehistory. Early, permanent, coast-bound settlements were possibly sustained by a farming economy based on animal husbandry (primarily sheep and/or goat) and small-scale cereal cultivation, in addition to the exploitation of marine resources.

Introduction and current research status

Botanical remains recovered from archaeological excavations are extremely useful when it comes to linking local agricultural activity to a specific site. This is particularly evident in regions with a low incidence of archaeological excavations of potential ancient farm sites. Development-led archaeology, regulated by the cultural heritage act, is the main source of empirical research data from prehistoric human settlements in Norway. The magnitude and quality of these data therefore relate primarily to modern socio-political patterns and may not always be fully adequate for the interpretation of past regional settlement patterns and practices. Northern Norway is a region with a low level of urbanisation and industrial development in direct conflict with cultural heritage management. Relative to southern Scandinavia, topsoil stripping, and accompanying more extensive application of archaeobotanical methods, came late to northern Norwegian excavation practice. Prior to the first large-scale archaeological excavation employing mechanical topsoil stripping at Kveøya, in 2008–2009 (Arntzen & Sommerseth 2010; Sjögren & Arntzen 2013), the empirical foundation for archaeological research into early agricultural activity in northern Norway mainly comprised stray finds of artefacts corresponding to the established southern Scandinavian farming typology. Flint sickles with traces of wear, dating from the Late Neolithic and Early Bronze Age, are the oldest artefacts that can be directly related to cereal production in Scandinavia, but there are only three finds of this kind from northern Norway, all from the southern part of Nordland. There are, however, other finds indicating contact with southern Scandinavia, such as flint daggers, axes and bronzes; see, for example, reviews by Valen (2007) and Rønne (2011). Even so, the overall impression from the archaeological record regarding material culture and the available resources is markedly different from that in southern Scandinavia. Consequently, the term Early Metal Age is used for the period corresponding to the Nordic Bronze Age (1800–500 BC) and Pre-Roman Iron Age (500 BC–AD 1) in northern Fennoscandia (Carpelan 1979; Olsen 1993; Jørgensen 1986). Nevertheless, for ease of comparison, I will follow the general southern Scandinavian terminology in this presentation (Vandkilde et al. 1996).

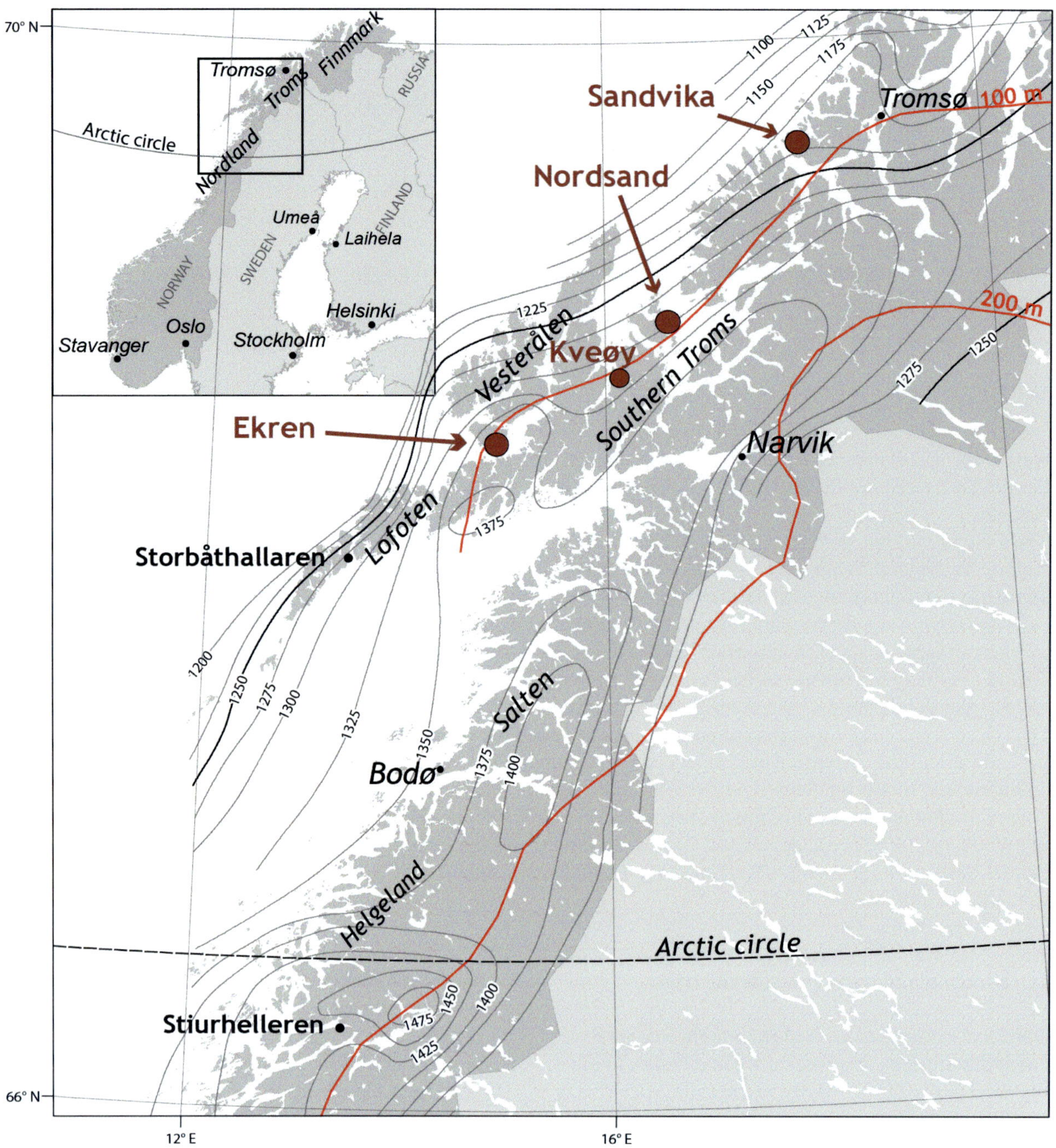

Figure 1. Map of the Nordland and Troms regions of Norway showing the study region and other locations mentioned in the text. The red contours mark the 100 m and 200 m altitudinal limit a.s.l. of the mid-boreal vegetation zone (after Moen 1999), while the degree-day isotherms are marked with grey contours (after Fjærvoll 1961). Insert: Map showing the location of the study region in Scandinavia.

Although virtually no archaeological excavations of ancient farm sites had been undertaken in northern Norway until the past decade, several palaeoecological investigations of wetlands have revealed anthropogenic signals indicating the existence of a farming culture.

The first major interdisciplinary research project to employ pollen analysis as the main method for tracing farming history was the *-staðir Farm Project*, which included investigations at 16 farms (Vorren et al. 1990; Nilssen 2010). This was a local spin-off from a Nordic research project on abandoned farms, initiated by the historian A. Holmsen in the 1970s. This aimed to investigate the demographic crises of the Late Middle Ages from a long-term perspective, and employed a retrospective methodology based on farm name classes to date the time of farm establishment. The '-staðir farms' of northern and central Norway represent a large, heterogeneous group in terms of their size, topography and archaeology. The project demonstrated a 2000-year span of farm establishment, from c. 1100 BC until c. AD 1100 (Vorren et al. 1990). In 9 out of 15 possible cases, an interruption in continuity was observed during the Merovingian period, concentrated around AD 660, followed by one or two hundred years of abandonment. Maximum agricultural activity was evident in the period from AD 160–390 (Late Roman Iron Age) to AD 780–920 (Viking age). Four farms showed continuity during the Merovingian depression. One of these was Bøstad, in the Lofoten islands, located c. 1 km from the Viking chieftain's residence at Borg (Munch et al. 2003). Slight impact from grazing and cultivation is evident from the Early Bronze Age onwards, before increasing during the Early Iron Age and reaching a maximum during the Viking age and Early Middle Ages. Local cereal cultivation was confirmed by finds of hulled barley (*Hordeum vulgare* var. *vulgare*) within the Borg site, which included some immature grains (Soltvedt 1984; Hansson 2003). Several subsequent pollen studies have confirmed the main trends in settlement history documented by the *-staðir Farm Project* and substantiated the hypothesis about animal husbandry and cereal cultivation being a part of the economy during the Bronze Age and Pre-Roman Iron Age. There is no complete review of all relevant pollen studies from northern Norway, but some publications present regional syntheses of and/or reflections on prehistoric farming in the north (Jensen 2012; Jensen & Arntzen 2016; Josefsson et al. 2014; Høeg 2000; Nilssen 2010; Sjögren & Arntzen 2013; Vorren 1986; Vorren 2005; Vorren et al. 1990).

Knowledge of cereal cultivation may have reached the northernmost part of Norway from south-eastern cultures, as well as from southern Norway. Very early finds of cereal pollen are documented in northern Sweden and Finland as well (Mökkönen 2010; Josefsson et al. 2014). An excavated Early Bronze Age farm settlement at the mouth of the Umeå River yielded finds of barley grains and bones of sheep or goat (Heinerud & Larsson 2013; Lindqvist & Granholm 2016; Östman 2014; Ellinor Johansson, pers. comm. 2016). At about the same latitude in Finland, grains of hulled barley from a hearth and a cultivation layer have been ^{14}C dated to the Late Bronze Age (Holmblad 2010) and an overlapping cultivation phase with barley pollen was identified from a nearby mire (Holmblad 2010; Wallin 2009).

Some of the pollen data from northern Norway are from early investigations with poor temporal resolution. Archaeobotanical analyses, and other scientific proxy methods applied to archaeological sites, are therefore welcome to confirm or refute the results of the pollen studies. Since the Kveøya investigation, three further excavations of ancient farm sites involving topsoil stripping and on-site archaeobotanical analysis have been carried out (Fig. 1). Two of these were development-led investigations: Ekren in Hadsel (Jensen & Westling 2013) and Nordsand in Harstad (Cerbing 2016; Jensen & Ahlqvist 2015). Sandvika in Tromsø was a research excavation (Arntzen 2015; Jensen & Arntzen 2016).

Results from interdisciplinary investigations of excavated ancient farm sites

A summary of the four relevant archaeological excavations is presented below, highlighting the archaeobotanical results. All calibrated ages are presented as the two standard deviations (2σ) range. The taxonomic nomenclature follows Fægri & Iversen (1989) and Moore et al (1991).

Kveøya (Hunstadneset)

The investigation at Kveøya (Kvæfjord municipality of southern Troms, excavated in 2008–2009), which remains the largest excavation of its kind undertaken in the region, provided evidence of Bronze Age and Iron Age agrarian settlement (Arntzen & Sommerseth 2010; Sjögren 2010; Sjögren & Arntzen 2013).

A single carbonised grain of barley (*Hordeum vulgare*) was directly ^{14}C dated to the Late Neolithic (Table 1). It originated from an oblong pit containing a charcoal-rich sandy fill, overlain by massive Pre-Roman Iron Age field layers. The authors considered the find context and the ^{14}C date to be uncertain due to the lack of corresponding dates from any other area of the excavation (Sjögren & Arntzen 2013). However, there are no obvious sources of error in relation to the ^{14}C date, and the possible presence of barley on the site during the Late Neolithic cannot be totally dismissed. As this was the only directly dated cereal grain from the site, a further grain of barley was ^{14}C dated as part of a later research project (Jensen & Arntzen 2016). This came from a posthole belonging to a Late Bronze Age house (House 1). The date obtained for the barley grain proved, however, to be much later; it calibrated to AD 420–575 (Merovingian period) (Table 1). Even so, the large number of features dated to the Late Bronze Age found during the excavations at Kveøya, including cooking pits, possible flat-field graves, field layers and postholes, do indicate the presence of a permanent settlement with an agrarian economy. ^{14}C-dated birch charcoal from three discontinuous layers of palaeosol underlying the Pre-Roman Iron Age field layer indicate two phases of activity, one during the Early Bronze Age (c. 1260–1040 BC) and one during the Late Bronze Age (c. 900–850 BC). No plant macroremains other than charcoal were recorded from these palaeosols, but unidentified charred cereal grains were found in the fill of a posthole belonging to the Late Bronze Age House 1—in addition to the Merovingian-period barley grain and charred grains of barley (*Hordeum* sp.) and naked barley (*Hordeum vulgare* ssp. *nudum*) in a clearance cairn. Few arable weeds were recorded from a Bronze Age context, namely, common chickweed (*Stellaria media*) and goosefoot (*Chenopodium*). Pollen analysis of the Late Bronze Age palaeosols revealed a transition from possibly grazed, birch-dominated forest to

Table 1: Direct AMS ^{14}C dates for carbonised cereal grains and bone/tooth of domestic animals. All ^{14}C dates were calibrated using OxCal version 4.3 (Bronk Ramsey 2009) and the calibration curve IntCal13 (Reimer et al. 2013).

Site	Context	Material	Lab.ref.	^{14}C age BP	Cal. age (95,4 %)	Reference
Stiurhelleren	Rock shelter	Sheep/goat, tooth	University of Arizona	3135±60	1530-1230 BC	Hultgreen et al. 1985, Johansen 1990
Stiurhelleren	Rock shelter	Barley, 1 grain	University of Arizona	3060±70	1495-1115 BC	Hultgreen et al. 1985, Johansen 1990
Stiurhelleren	Rock shelter	Barley, 1 grain	University of Arizona	2365±60	755-240 BC	Hultgreen et al. 1985, Johansen 1990
Storbåthallaren	Rock shelter	Cattle, bone	unknown	2740±80	1110-790 BC	Utne 1973
Storbåthallaren	Rock shelter	Cattle, bone	unknown	2050±60	340 BC-AD 75	Utne 1973
Kveøy	Pit	Barley, 1 grain	Wk-26504	3936±30	2560-2310 BC	Arntzen & Sommerseth 2010
Kveøy	Posthole	Barley, 1 grain	Beta-399667	1550±30	AD 420-575	Jensen & Arntzen 2016
Sandvika	Hearth	Sheep/goat, toe bones	Beta-399126	3030±30	1395-1135 BC	Arntzen 2015, Jensen & Arntzen 2016
Nordsand	Field layer	Barley (cf.), 1 grain	Beta-407554	1700±30	AD 255-405	Jensen & Ahlqvist 2016
Nordsand	Field layer	Barley (cf.), 1 grain	Beta-406507	1610±30	AD 390-540	Jensen & Ahlqvist 2016
Nordsand	Field layer	Hulled barley, 1 grain	Beta-407555	1600±30	AD 395-540	Jensen & Ahlqvist 2016
Nordsand	Field layer	Barley (cf.), 1 grain	Beta-406506	1560±30	AD 420-565	Jensen & Ahlqvist 2016
Nordsand	Field layer	Barley, 1 grain	Beta-406511	1460±30	AD 550-650	Jensen & Ahlqvist 2016
Nordsand	Field layer	Sheep/goat, toe bone	Beta-406509	1270±30	AD 660-860	Jensen & Ahlqvist 2016
Ekren	Hearth	Barley, 1 grain	Beta-399122	1750±30	AD 220-385	Jensen this paper
Ekren	Posthole	Barley *m*, 1 grain	UBA-24009	1469±30	AD 545- 645	Jensen this paper

more open pasture woodland. The charcoal content, including birch (*Betula*) and willow/aspen (*Salix/Populus*), indicates woodland clearance by fire, and the authors' hypothesis is that this was for cultivation purposes.

The site provided evidence of a full-scale farm structure in the Pre-Roman Iron Age, including a 23-m longhouse (with asbestos-tempered pottery), a utility building, a clay-built oven, several cooking pits and cremation graves, in addition to the clearly defined subfossil field layer, dated to c. 400–100 BC, associated with a lynchet and a clearance cairn. No preserved ard marks were observed, but the homogenous quality of the soil substrate and its pollen content indicate heavy reworking and manuring. Plant macroremains were only recovered from the archaeological structures and features, mainly from postholes, and showed a rather high taxonomic diversity. Charred cereal grains were identified as hulled and naked barley and wheat (*Triticum*), some of which were identified as emmer (*Triticum dicoccum*). The cereals were found together with arable weeds of local origin, for example corn spurrey (*Spergula arvensis*), chickweed and goosefoot, and it seems likely that they had been cultivated on-site. A special feature is the major representation of crowberry (*Empetrum nigrum*) seeds, suggesting some kind of domestic use of this dwarf shrub. Sjögren and Arntzen (2013) suggest that shrub-fallow cultivation was practised during the Pre-Roman Iron Age, involving, for example, cycles of 10–15 years of regrowth, followed by burning and then 2–3 years of cultivation. They base this view primarily on the lithology and pollen content of the clearance cairn and on the soil layer, which indicates periods with grazed fallow land.

Nordsand

The site of Nordsand (at Sandsøy, Harstad municipality of southern Troms, excavated in 2013) yielded evidence of cereal cultivation and animal husbandry (impact of grazing and manuring, in addition to bones of sheep or goat) during the time from the Late Roman Iron Age to the transition between the Migration and Merovingian periods, recovered from an extensive field layer found between distinct, homogenous sand layers interpreted as sand drift (Cerbing 2016; Jensen & Ahlqvist 2016).

The excavation was undertaken in connection with the construction of a new road, and permission was only given to investigate part of this extensive ancient farm site. This, together with the fact that the existing main road and a water pipe ran through the main settlement area, made it impossible to obtain a full picture of all the structures. In particular, the reconstruction of complete house sites was hampered. Four house complexes can possibly be outlined (Fig. 2). A three-aisled longhouse (House 2) can be tentatively dated as Pre-Roman Iron Age, or possibly even very Late Bronze Age, according to its typology and ^{14}C dates for charcoal from a posthole and an adjacent pit. No empirical evidence of farming has so far been linked with this early settlement. Soil samples for analysis of plant macroremains were taken from the archaeological structures, but due to the limitations of a modest budget, these remain to be analysed. The majority of the Nordsand material derives from the later stages of the Iron Age. This includes Houses 1, 3 and 4, which all have partially overlapping ^{14}C dates extending from the middle of the Roman Iron Age until the transition between the Migration period and the Merovingian period. These partly excavated

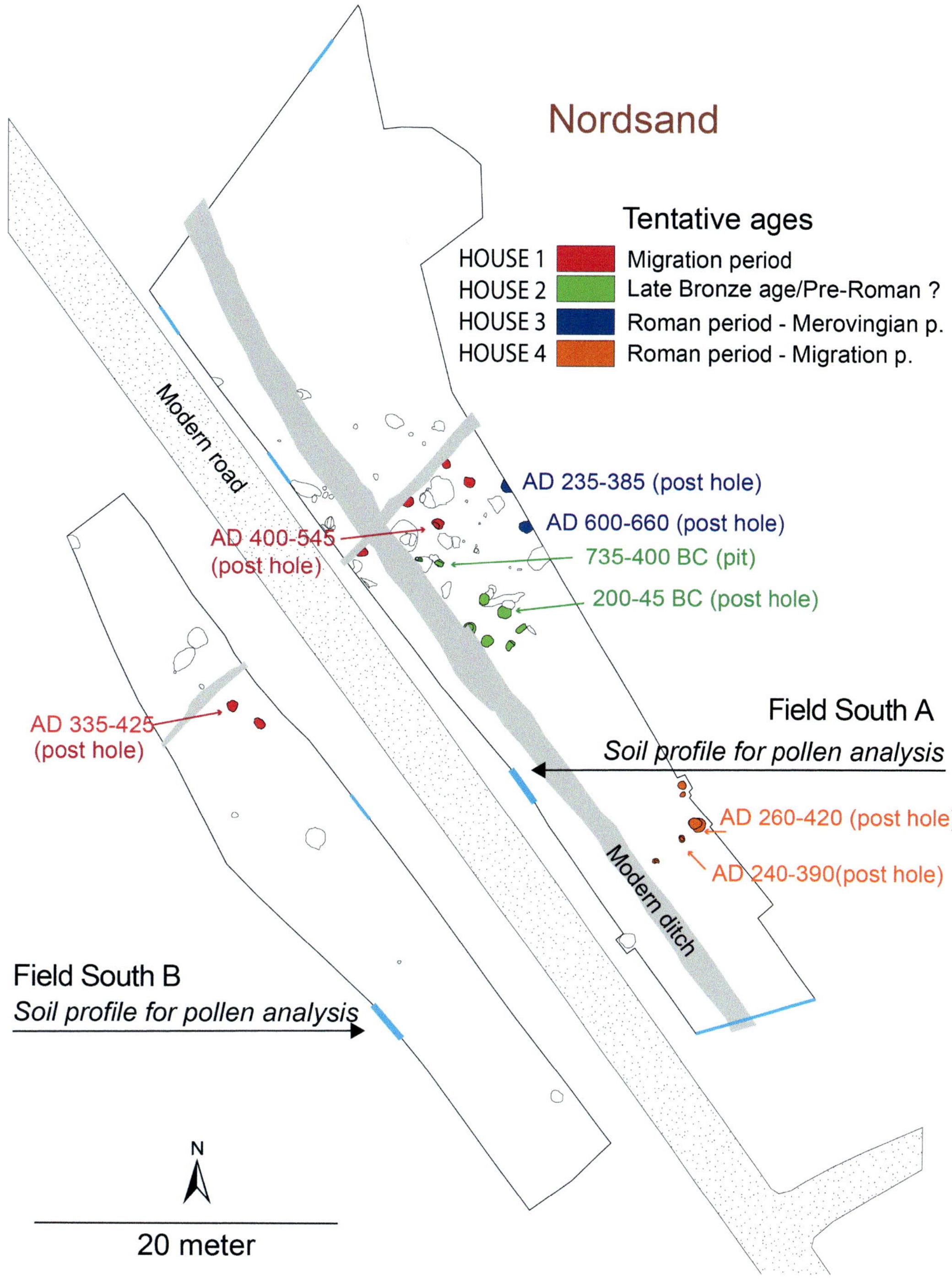

Figure 2. Plan view of the two excavated subareas at Nordsand, Norway, showing remains of house structures and locations of soil profiles analysed for pollen and plant macroremains (drawing: Tromsø University Museum).

houses, and remains of four hearths, form a cluster within a radius of 20 m at the highest point of elevation. The fields were located close to the settlements. No clearance cairns were observed, but there may have been no need for clearance of stones from the sandy arable fields.

A continuous subfossil field layer covered most of the two excavated subareas of c. 4000 m^2 in total, located between 9 m and 13 m above sea level (Fig. 3). The thickness of the field layer was generally 20–30 cm, and it had distinct ard marks at its lower boundary. Six direct ^{14}C dates obtained for carbonised grains of barley from three separate soil profiles show that cultivation extended from the Middle or Late Roman Iron Age (AD 255–405) until early in the Merovingian period (AD 550–600) (Table 1). One of these barley grains was identified as hulled barley (*Hordeum vulgare var. vulgare*) and was dated to AD 395–540. The soil substrate showed a major degree of reworking. The ^{14}C dating of two separate barley (*Hordeum* sp.) grains to the Migration period (AD 390/395–450) indicates cereal cultivation dur-

Figure 3. View of the excavated subareas of Nordsand, Norway, shown in Figure 2 looking north-west. Another excavated subarea lies to the north of these, close to the marine shore; it has a similar soil stratigraphy, including a subfossil cultivation layer and ard marks.

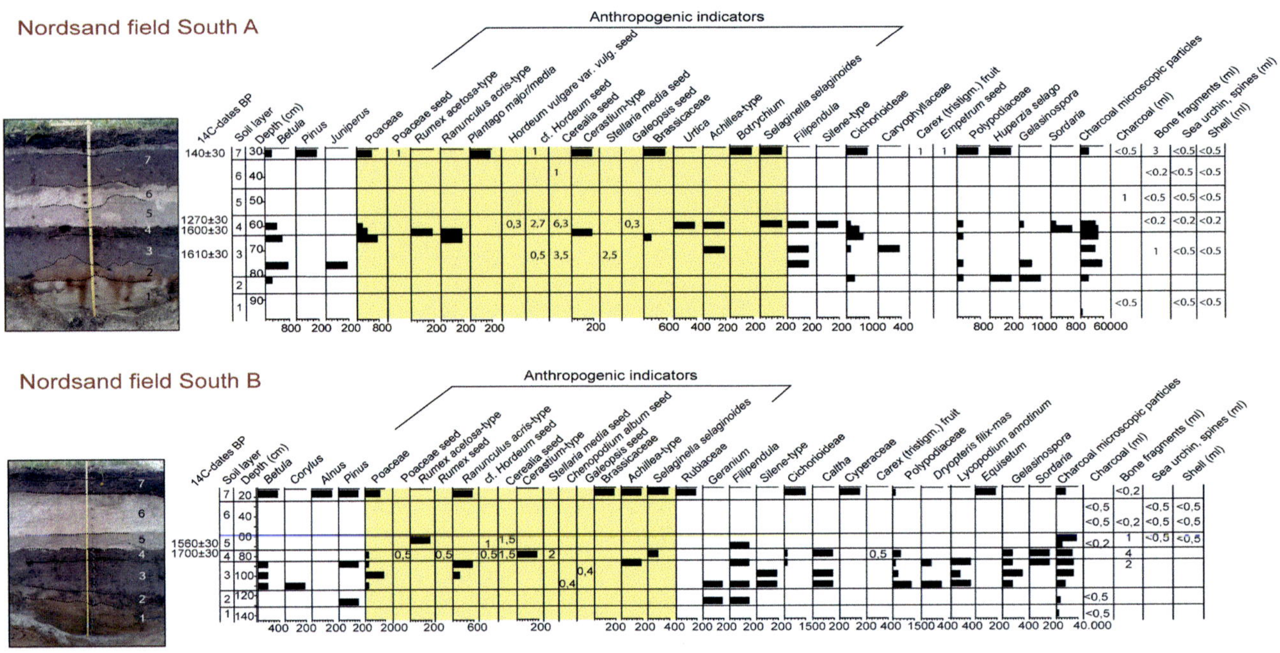

Figure 4. Concentration (number per volume) of plant macroremains, pollen, spores and microscopic charcoal fragments in samples from the soil profiles in two excavated subareas at Nordsand, Norway. Soil descriptions, from bottom to top: 1 grey sand; 2 greyish brown humic sand with ard marks; 3 and 4 dark brown, strongly humic sand with macroscopic charcoal and some gravel, cultivated soil; 5 grey sand with charcoal fragments, aeolian deposit; 6 light grey sand, aeolian deposit; 7 dark grey humic sand with roots; 8 grey sand with roots, aeolian deposit; 9 strongly humic soil with roots, modern cultivation; 10 topsoil with living plants.

ing the entire time span. The weed flora associated with the cereals was very sparse, and only a few charred seeds of common chickweed, goosefoot, hemp nettle (*Galeopsis*) and sorrel *(Rumex)* were recorded (Fig. 4). No Chenopodiaceae pollen was observed, and goosefoot may not have been a prominent part of the local weed flora. The pollen samples mainly contained a variety of taxa reflecting heaths and grassland, dominated by grasses (Poaceae), buttercup (*Ranunculus acris*–type), yarrow or mayweed (*Achillea*-type), meadowsweet (*Filipendula*), and pink family (Caryophyllaceae), and by fern spores. Spores of dung fungus (*Sordaria*) support the assumption that manure was used as a fertiliser, although these spores were found in the upper part of the field layer and may derive from a later context. Spores of another ascomycete, *Gelasinospora*, were also observed; this genus is associated with charred wood. The pollen samples from the field layer contained abundant microscopic charcoal particles, and these may have been transported from the households by the wind. Alternatively, they may well originate from ash used to enrich and improve the soil. Sea urchin spines and shell and bone fragments incorporated into the field soil may also derive from domestic waste which was included in a deliberate practice of soil improvement. The fields were, however, located not far from the seashore, and they may have been exposed to wind erosion of material from the open, sandy beach. A toe bone of a sheep or goat found in the upper part of the field layer was dated to the Late Merovingian period/Early Viking age (Table 1), i.e. contemporaneous with the bottom layer of a cooking pit cut into the sand drift nearby.

The settlement history of Nordsand can therefore be briefly summarised as follows: Sporadic activity during the Bronze Age until the establishment of a more permanent settlement during the Late Bronze Age or Pre-Roman Iron Age, possibly with farming involving animal husbandry. Establishment of a major farm site during the Roman Iron Age, with animal husbandry and barley cultivation, lasting until the middle of the Merovingian period, when it was abandoned. The reason for this abandonment may be the massive sand drift in the area. Only occasional activity is observed from then until recent/sub-recent farming practices. Remains from the Late Iron Age and the Middle Ages may have been destroyed by this more recent activity.

Sandvika

Sandvika (Tromsø municipality, excavated in 2013) was probably a short-term settlement, inhabited during the Late Bronze Age and with an economy based on animal husbandry and the exploitation of marine resources. The main feature of this site comprised the remains of a three-aisled longhouse, constructed between 1000 and 800 BC on grassland close to the former marine shoreline (Arntzen 2015).

In addition to the house construction type, several other archaeological features and artefacts associated with Nordic Bronze Age settlements were documented. A thin-walled soapstone vessel is the northernmost find of its kind so far in Norway. The vessel is bowl-shaped, and charred food crust covering one of the sherds gave a ^{14}C date of 896–802 BC, which is consistent with the assignation of this type of vessel to the Late Bronze Age in Norway (Pilø 1989). Asbestos-tempered potsherds constitute the most abundant artefact category associated with the house. The pottery is generally poorly preserved, but some of the vessels/sherds have tempering and a rim shape similar to the Risvik type (Jørgensen & Olsen 1988), which is linked to coastal Bronze Age settlements in the area between southern Nordland and the northern part of Troms (Arntzen 2015). Marine fishing and hunting are confirmed by the find of a large sinker for hand-line fishing and the presence of numerous bones of fish and seal.

The area is heavily influenced by sand drift, and erosion may be one cause of farm abandonment. There is no clear empirical evidence for crop cultivation, but there are indications that the site lay exposed for several hundreds of years after its abandonment. Consequently, the very low content of plant macroremains and pollen may be explained by taphonomic loss due to biological decay and chemical degradation, in addition to material being washed out by rainfall. The only charred plant macrofossils recovered from features associated with the house are a seed of common chickweed from the fill of a soapstone vessel, a crowberry from a hearth, and seeds of orache (*Atriplex*) from the same hearth and a posthole. The hearth and a refuse pit also contained a considerable quantity of burnt bone fragments identified as sheep or goat, small mammal, fish, bird and seal. The bones of sheep or goat show cutmarks attributable to butchering. Phalanges and metapodials of sheep or goat have been ^{14}C dated to the Early Bronze Age (1390–1335 BC). This date may, however, be a bit too old due to residual carbonate derived from old wood used as fuel.

Two soil profiles, extending down to the marine shore substrate, were analysed for pollen and plant macroremains (Jensen & Arntzen 2016). One was from within the central part of the house and the other from the outskirts of the settlement area. The floor layer of the demolished house and the corresponding stratigraphic level outside the house contained a similar pollen assemblage, reflecting grassland influenced by human activity. The combination of buttercup, sorrel, ribwort plantain (*Plantago lanceolata*), members of the pink family, moonwort (*Botrychium*) and lesser clubmoss (*Selaginella selaginoides*) is connected with northern grazed meadow communities, and the presence of common plantain (*Plantago major*) indicates ruderal anthropogenic habitats (e.g. Vorren 1986; Fremstad 1997; Lid & Lid 2005). A similar pollen assemblage is recorded in a pollen diagram from a mire located 450 m south-west of the Bronze Age house (Tveraabak & Alm 1997), following a peak in charcoal dust that may be associated with the settlement and initial burning of the woodland (Tveraabak & Alm 1997; Jensen & Arntzen 2016). Furthermore, a pollen diagram from the Dypingen mire at Brensholmen, 3–4 km south of Sandvika, reflects grazing impact and possible cereal cultivation (Cerealia-type pollen) during the period 1150–550 BC, overlapping with the settlement in Sandvika, followed by a period of abandonment until c. AD 200 (Vorren 2005). The combined results from these sites imply a local Bronze Age farming culture based on fishing and marine hunting, together with animal husbandry, possibly only sheep or goat, and perhaps some cereal cultivation.

Ekren

Ekren is located in the Hadsel municipality of Vesterålen and was excavated in 2012. This south-

Figure 5. View of the site of Ekren, Norway (marked with a white arrow), seen from the outfield looking south-east.

facing site is located on a moraine, with a view towards the Lofoten islands (Fig. 5). Although no field layers were found here, barley grains and arable weed seeds recovered from postholes in a three-aisled house (15–18 m long) and an unidentified circular construction, as well as from hearths and cooking pits (Fig. 6), indicate that local cereal cultivation may have taken place (Jensen & Westling 2013). A few ^{14}C dates for charcoal extend back to the Pre-Roman Iron Age, but the main period of settlement was during the Late Roman Iron Age and the Early Migration period (AD 140–580). This is confirmed by the direct ^{14}C dating of two barley grains, from a hearth and a posthole (Table 1). Any fields were most probably located just outside the excavation area, where well-drained, south-east-facing slopes provided suitable cultivation conditions. Palaeosols had, however, been lost from this part of the site prior to excavation, due to modern agricultural activity and development. The weed remains found associated with the barley grains were charred seeds of common chickweed, corn spurrey, goosefoot (*Chenopodium album*), common knotgrass (*Polygonum aviculare*), hemp nettle and sheep's sorrel (*Rumex acetosella*). Other seeds were from meadow herb taxa, such as buttercups, cinquefoils (*Potentilla*), pink family, vetches (*Vicia*) and grasses. Crowberries were present in great abundance in one of the hearths. Burnt bone fragments, some of which were of sheep or goat, were found in several of the structures, but none have been directly ^{14}C dated. Thc nitrophilous character of the weeds may suggest manuring with the dung of domestic animals. A pollen study at the neighbouring farm of Lekang, located 2 km north-east of Ekren (Jensen 1991), revealed an expansion in farming with animal husbandry and indications of barley cultivation during the same period as the Ekren settlement, as well as during the Pre-Roman Iron Age. A special feature of the archaeological remains at Ekren is probably the largest collection of cooking pits (>100) found in northern Norway to date. It prompts the assumption that Ekren may have been a site for social, ritual or political gatherings during the Roman Iron Age and possibly also later. The Ekren and Lekang sites lie within 3 km of Hadsel Prestegård. The large farm mound, longboat houses, grave fields and rich archaeological finds at Hadsel suggest that this was a chieftain's residence and a place of religious ritual during the Iron Age. Such events would require a substantial supply of food and beer, and some self-sufficiency in terms of local farming activity seems plausible. The archaeological interpretation of the Ekren site is, however, yet to be completed, and no final conclusions on the type of settlement here have yet been drawn.

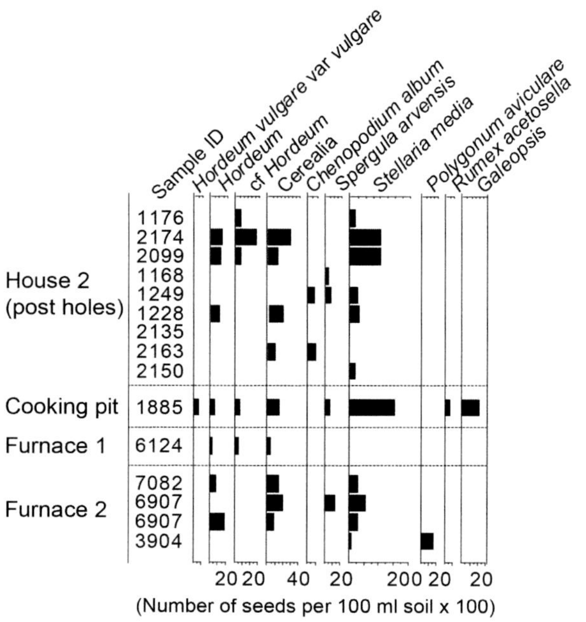

Figure 6. Volume densities of charred cereal grains and weed seeds recovered from archaeological structures and features at the site of Ekren, Norway.

Discussion

The results from the four farm sites show that people of the north of Norway engaged in farming at an early stage of prehistory. Animal husbandry is demonstrated from the Early Bronze Age onwards by the ^{14}C dates for sheep or goat bones, and there is evidence of cereal cultivation from at least the Late Bronze Age onwards. Two of the sites, Sandvika and Nordsand, have been exposed to episodes of extensive sand drift that may have prompted abandonment of the sites. The settlement periods overlap with periods when regional pollen diagrams from mires indicate expansion in farming, and with archaeological data indicating contact with well-established farming cultures.

Numerous investigations of vegetation history have shown that the presence of pastures associated with animal husbandry is characterised by increased openness of the landscape, combined with the presence of herbs favoured by persistent grazing and manuring over decades. The impact of local grazing by reindeer may be reflected in pollen diagrams in a similar way to low grazing impact by cattle and sheep or goat, conditional on there being a high grazing intensity and close proximity to the pollen site. The latter is mostly the case when many animals are deliberately gathered together in a limited space, such as for milking, calf marking and slaughtering, as has been documented in northern Sweden (Aronsson 1994; Kamerling et al. 2017). The extensive grazing pattern of roaming or non-domesticated reindeer is too weak to be traced by pollen analysis. Grazing impact prior to the intensified herding of reindeer during the Late Middle Ages is therefore most probably due to sheep or goats. The ^{14}C-dated bones from Sandvika, Nordsand and rock shelters (Table 1) confirm the presence of domestic animals at coastal sites during the Bronze Age and Iron Age. The increased landscape openness and grassland development evident from the Late Bronze Age/Pre-Roman Iron Age onwards, as well as the observed discontinuity in agricultural indices during the last millennium BC, may be linked to the regional climatic deterioration indicated by several independent climate proxies, which resulted in cooler and more humid conditions (Vorren et al. 2007). This change in climate may have prompted a move towards keeping animals inside during the winter instead of allowing them to roam freely in the open air, which in turn resulted in a need for the harvesting of winter fodder in the form of grass, foliage, dwarf shrubs, etc. The Lofoten/Vesterålen region is presently the northern limit for the heather (*Calluna*) heathland vegetation commonly managed by regular use of fire along the Atlantic coast of Europe (Kaland 1986; Odgaard 1994; Prøsch-Danielsen & Simonsen 2000). A similar process has been documented in Lofoten during the Late Bronze Age and Iron Age (Tveraabak 2004).

The first direct ^{14}C dating of charred cereal grains to the Late Neolithic and Early Bronze Age at Stiurhellaren and Kveøy (Table 1) was undertaken on grains that were slightly out of context, because the stratigraphy of the features from which they were recovered had been disturbed. This makes it difficult to make the connection with known phases of cultivation. Farther south along the west coast of Norway, however, agriculture was already consolidated at this time, with permanent settlements practising animal husbandry and cereal cultivation from about 2400 BC onwards (Soltvedt 2000; Diinhoff 2005; Prøsch-Danielsen & Simonsen 2000; Hjelle et al. 2012).

At Kveøy, two phases of burning have been recognised that may be linked to early cultivation - an early one in the 13th–11th century BC, followed by a main phase in the 10th–9th century BC, prior to the Pre-Roman Iron Age cultivation phase (Sjögren & Arntzen 2013). Burning prior to cultivation was also recorded at the Nordsand site, but no direct evidence of arable cultivation exists prior to the ^{14}C-dated cereal grains, of which the oldest date was AD 255-405 (Table 1). Each field layer documented by visual and tactile examination during fieldwork was generally treated as a single unit. The soil within each field layer was regarded as being mixed to such a degree that greater stratigraphical or temporal resolution was impossible. A single dated cereal grain may therefore not represent the entire time span of the field layer. But the fields at Sandsøy are located close to the main settlement buildings, and the soil of the dominant field layer appears to have been rather strongly reworked by the ard, suggesting that these were permanent fields used over a long period, probably, as documented, from the middle of the Roman Iron Age until the middle of the Merovingian period. A cultivation method involving shifting fields or crop rotation without the use of an ard or plough may be less easy to trace, as there will be a less clear formation of field layers, and the pollen content from crops and weeds will be low, perhaps representing only one season. Nothing is known about shifting cultivation practices during prehistory in northern Norway. One hypothesis relating to the Sandvika settlement is that this was a small sheep farm associated with the larger farm at Brensholmen, where cereals were cultivated, and that people, in this way, made better use of the environmental benefits of the local landscape (Arntzen 2015; Jensen & Arntzen 2016). The climate of the region would have been suitable for cereal cultivation, as it satisfied some general criteria for stable cereal growth in Norway: a mid-boreal bioclimate (Moen 1999) and a degree-day sum of 1250 during the main growing season from 1st June to 30th September (Fjærvoll 1961) (Fig. 1).

Until drainage of peat bogs began in the 18th century, people were dependent on the presence of fertile soil with natural drainage, such as areas of moraine substrate, fluvial deposits or exposed former seabed. Such substrates are vulnerable to nutrient exhaustion over time and to erosion and sand drift. The causality of the sand drift events evident at Sandvika and Nordsand is not clear. Their timing is consistent with periods of climatic deterioration, for example at the transition from the Late Bronze Age to the Pre-Roman Iron Age and in the Migration period. But intensified human impact on the local vegetation, for example trampling and grazing by domestic animals and cereal cultivation, may well have triggered erosion too. These sites fit into a possible pattern of early settlement on sandy soils that is observed along the coast of Nordland and Troms. Some sites stand out due to their multiple settlement phases during the Late Neolithic and Bronze Age and until the Early Iron Age, with finds that may be associated with farming and correlated with the Sandvika site—for example Kolvika, Lofoten (Jørgensen 1989), Bøsanden, Steigen (Moltu 1988) and Hofsøy, Tranøy (Johansen 1976–1980; Lahtiperä 1980; Vorren 1986).

The Viking age and Middle Ages are rarely represented in the on-site stratigraphy, either because later farming activities have removed the deposits from this period or because they have become merged with later cultivation substrates. People have tended to choose the same habitation sites over a long period of time. A favourable local climate and soil properties, as well as proximity to the sea and good logistics, are likely to have been crucial.

The results presented here support and broadly reinforce the existing general view of northern Norwegian farming history. Economic strategies involving the utilisation of multiple resources constitute a strength for the northern coastal societies. The continental shelf and the Gulf Stream are both very close to the coastline in this region of Lofoten/Vesterålen and Troms, as are huge coral reefs used as a breeding area by several fish species. Non-migratory colonies of sea mammals living close to the coastline, such as harbour and grey seals, were hunted up into Modern times. Small, fjord-living whales, such as the killer whale, were occasionally hunted, but it was mainly stranded whales that were killed. In addition to fishing and hunting, farming probably also sustained these early coast-bound permanent settlements, where and when this was possible. In areas where the climate permitted animals to graze all year round, there would also have been time to exploit marine resources, and it would have been possible to subsist even on very small farms. However, the empirical database remains modest, and further excavations, with an interdisciplinary approach, are required to address more subtle research questions regarding farming practices and their socioeconomic significance.

Conclusions

Some general assumptions can be made based on the existing scientific data, and the following main conclusions can be drawn:

Pollen-based theories on farming practices north of the Arctic Circle from the Bronze Age onwards are supported by archaeological evidence resulting from archaeological excavations employing topsoil stripping, intra-site archaeobotanical analysis and direct ^{14}C dating of cereals and bones from archaeological contexts.

Early coast-bound permanent settlements were possibly sustained by a farming economy based on animal husbandry (primarily sheep and/or goat) and small-scale cereal cultivation, in addition to the exploitation of marine resources.

In general, the Late Bronze Age and Pre-Roman Iron Age in coastal northern Norway was characterised by the expansion of pastures and the establishment of new arable fields. But there were also periods of abandonment, which may reflect a high degree of mobility or changes in the subsistence economy. The Pre-Roman Iron Age was a period when cereals appear to have been introduced at most sites in Nordland and Troms and the impact of grazing increased. Barley, mainly of the hulled form (*Hordeum vulgare* var. *vulgare*), was the main crop in northern Norway, but, as documented at Kveøy, naked barley (*Hordeum vulgare* var. *nudum*) and emmer (*Triticum dicoccum*) may also have been cultivated.

The earliest ^{14}C date so far for a cereal grain (*Hordeum*) from a field layer is from the Late Roman Iron Age.

The weeds associated with cereal cultivation were primarily common chickweed (*Stellaria media*), corn spurrey (*Spergula arvensis*), hemp nettle (*Galeopsis*) and goosefoot (*Chenopodium*).

Expansions in agriculture are seen during the Late Bronze Age and Pre-Roman Iron Age, Late Roman Iron Age, Migration period and Late Merovingian period/Viking age. The best documented cereal production is during Early Iron Age, c. AD 200–400, which seems to be the most likely time for the consolidation of agriculture in the southern Troms/northern Nordland region.

Periods of abandonment during the Late Bronze Age, Late Pre-Roman Iron Age/Early Roman Iron Age and Merovingian period may be related to erosion and sand drift. These may have been climatically unstable periods with increased storm activity and/or the effects of grazing/cultivation impact on unstable soils.

Acknowledgements

The results presented from the Nordsand, Sandvika and Ekren sites are from interdisciplinary projects undertaken jointly by the University of Tromsø, The Arctic University of Norway (UiT), and the Arkeologisk Museum (museum of archaeology), University of Stavanger (UiS). Colleagues at both institutions are sincerely thanked for fruitful discussions and help, both in the field and in the laboratory. The UiS research programme *Prehistoric farming at the north-European fringe—Interdisciplinary aspects* financed the ^{14}C dating of cereal grains from the Ekren and Kveøya sites. This chapter was initially presented to the Nordic Archaeobotany Group Meeting 2018, and it has been reviewed by anonymous referees, to whom I am grateful for useful comments.

Literature

Arntzen JE (2015) Sandvika in northern Norway: The northernmost 'Bronze Age' settlement in Scandinavia. Fennoscandia Archaeologica 32:3–35

Arntzen JE, Sommerseth I (eds) (2010) Den første gården i Nord-Norge: Jordbruksbosetting fra bronsealder til jernalder på Kveøy [The first farm in Arctic northern Norway: Agricultural settlement during the Bronze Age and the Iron Age at Kveøy]. TROMURA 39 (in Norwegian with English summary). Tromsø University Museum, UiT The Arctic University of Norway, Tromsø

Aronsson, K-Å (1994) Pollen evidence of Saami settlement and reindeer herding in the boreal forest of northernmost Sweden: An example of modern pollen rain studies as an aid in the interpretation of marginal human interference from fossil pollen data. Rev Palaeobot Palyno 82(1–2):37–45

Bronk Ramsey C (2009) Bayesian analysis of radiocarbon dates. Radiocarbon 51(1):337–360

Carpelan C (1979) Om asbestkeramikens historia i Fennoskandien. Finskt Museum 85:5–25

Cerbing M (2016) Utgrävningarna på Sandsøy och Grytøya, Harstad kommune, 2014 (Report). Tromsø University Museum, UiT The Arctic University of Norway, Tromsø

Diinhoff S (2005) Den vestnorske agrarbosætning. Fra sen stenalder til folkevandringstid. Arkeologiske resultater fra et tiår med fladeafdækninger på Vestlandet. AmS-Varia 43:75–85

Fjærvoll K (1961) Korndyrkinga i Hålogaland i gammal tid: 1500–1600 åra. Hålogaland Historielag, Svorkmo

Fremstad E (1997) Vegetasjonstyper i Norge [Norwegian vegetation types]. NINA Temahefte 12:1–279 (in Norwegian with English summary). NTNU, Trondheim

Fægri K, Iversen J (1989) *Textbook of Pollen Analysis*. 3. Revised edition by Fægri K, Kaland PE, Krzywinski K. John Wiley & Sons, 328pp

Hansson A-M (2003) Plant remains from Borg. In: Munch GS, Johansen OS, Roesdahl E (eds) Borg in Lofoten: A chieftain's farm in north Norway. Lofotr-Vikingmuseet på Borg, Tapir Academic Press, Trondheim. ISBN: 978825191825

Heinerud J, Larsson T (2013) Bronsåldershus vid Umeälvens mynning. Populär Arkeologi (4):24–25

Hjelle KL, Solem T, Halvorsen LS, Åstveit LI (2012) Human impact and landscape utilization from the Mesolithic to medieval time traced by high spatial resolution pollen analysis and numerical methods. J Archaeol Sci 39:1368–1379

Holmblad P (2010) Coastal communities on the move: House and polity interaction in southern Ostrobothnia 1500 BC–AD 1. Archaeology and Environment 26. Umeå University, Umeå

Hultgreen T, Johansen OS, Lie R (1985) Stiurhelleren i Rana: Dokumentasjon av korn, husdyr og sild i yngre steinalder. Viking 48:83–102

Høeg HI (2000) Pollenanalytiske undersøkelser i Finnmark, Nord-Norge [Pollenanalytical surveys in Finnmark, northern Norway]. AmS-Varia 37:53–97

Jensen C (1991) Vegetation history and anthropogenic development of a 'leikvang-farm' in Vesterålen: A palynological investigation. Cand. Scient. thesis, University of Tromsø, Tromsø (in Norwegian with English summary and captions)

Jensen CE (2012) Tidlig jordbruk i det nordligste Norge: Botaniske vitnesbyrd. In: Kaul F, Sørensen L (eds) Agrarsamfundenes ekspansion i nord. National Museum of Denmark, Copenhagen, pp 169–183

Jensen C, Ahlqvist J (2016) Botaniske analyser av stratifiserte kulturlag på Nordsand - Bjarkøyforbindelsen, g.nr. 138, b.nr. 1, 14, Sandsøy i Harstad k., Troms (Report 2016/5). Arkeologisk Museum, University of Stavanger, Stavanger

Jensen C, Arntzen JE (2016) A Late Bronze Age sheep farm north of the Arctic Circle? In: Iversen F, Petersson H (eds) Agrarian living in the North BC 1800–1000 AD: Studies in rural settlements and farming in Norway. Portal, Kristiansand, Norway, pp 173-202

Jensen C, Westling S (2013) Ekren, gnr. 55, bnr. 2 Hadsel k, Nordland: Botaniske analyser (Report 7/2013). Arkeologisk Museum, University of Stavanger, Stavanger

Johansen OS (1976–1980) Dagbøker, rapporter og korrespondanse fra de arkeologiske utgravningene på Hofsøy, Tranøy kommune 1976–1980. Various reports, diaries and correspondence. Tromsø University Museum, UiT The Arctic University of Norway, Tromsø

Johansen OS (1990) Synspunkter på jernalderens jordbrukersamfunn i Nord-Norge. Stensilserie B 29:1–64

Josefsson T, Ramquist PH, Hörnberg G (2014) The history of early cereal cultivation in northernmost Fennoscandia as indicated by palynological research. Veg Hist Archaeobot 23:821–840

Jørgensen R (1986) The early metal age in Nordland and Troms. Acta Borealia 3(2):61–87

Jørgensen R (1989) En boplass fra eldre metallalder i Nordland. In: Bertelsen R, Reymert PK, Utne A (eds) Framskritt for fortida i nord: I Povl Simonsens fotefar. Tromsø Museums Skrifter 22:135–145. Tromsø University Museum, UiT The Arctic University of Norway, Tromsø

Jørgensen R, Olsen B (1988) Asbestkeramisk grupper i Nord-Norge 2100 f.Kr.–100 e.Kr. Tromura (Tromsø Museums rapportserie) Kulturhistorie 13. Tromsø University Museum, UiT The Arctic University of Norway, Tromsø

Kaland PE (1986) The origin and management of Norwegian coastal heaths as reflected by pollen analysis. In: Behre K-E (ed) Anthropogenic indicators in pollen diagrams. A.A. Balkema, Rotterdam, pp 19–36

Kamerling IM, Schofield JE, Edwards KJ, Aronsson K-Å (2017) High-resolution palynology reveals the land use history of a Sami renvall in northern Sweden. Veg Hist Archaeobot 26:369–388 https://doi.org/10.1007/s00334-016-0596-5

Lahtiperä P (1980) Osteologisk analys fra materialet på Hofsøy, Tranøy kommune, Troms. J. S. 635. (Research report). Museum of Zoology, University of Bergen, Bergen

Lid J, Lid DT (2005) Norsk flora 7th edn, Elven R (ed). Det Norske Samlaget, Oslo

Lindqvist A-K, Granholm N (2016) Stolpbyggda hus vid Umeälvens mynning: Den yngre bronsåldern i ny skepnad. Arkeologi i Norr 15:37–62

Moen A (1999) National atlas of Norway: Vegetation. Norwegian Mapping Authority, Hønefoss

Moltu T (1988) Engeløya i Steigen. Studier over den forhistoriske busetnaden tufta på arkeologi og vegetasjonshistorie [Engeløya in Steigen: Studies of the prehistoric settlement based on archaeology and vegetation history]. Dissertation, University of Tromsø

Moore PD, Webb JA, Collinson ME (1991) Pollen analysis. Blackwell Scientific Publications.Oxford.

Munch GS, Johansen OS, Roesdahl E (eds) (2003) Borg in Lofoten: A chieftain's farm in north Norway. Lofotr-Vikingmuseet på Borg, Tapir Academic Press, Trondheim

Mökkönen T (2010) Kivikautinen maanviljely Suomessa. Suomen Museo 116:5–35

Nilssen EJ (2010) Staḑir-Gårder. En pollenanalytisk undersøkelse av syv gårder med endelsen -stad (-staðir) i Vesterålen, Lofoten og Nord-Trøndelag. MUNIN, University of Tromsø, Tromsø. http://www.ub.uit.no/munin/

Odgaard BV (1994) The Holocene vegetation history of northern west Jutland, Denmark. Oper Bot 123:1–171

Olsen B (1993) Bosetning og samfunn i Finnmarks forhistorie [Prehistoric settlement and society in Finnmark]. Universitetsforlaget, Oslo. https://www.nb.no/items/URN:NBN:no-nb_digibok_2013060306066

Östman S (2014) Miljöarkeologiska analyser av anläggningar från den arkeologiska utgrävningen inför dragningen av Västra länken i Umeå 2013: Klabbölevägen, Umeå socken Raä 366, Umeå kommun, Västerbottens län (Report 2014-018). Miljöarkeologiska laboratoriet, Institutionen för idé- och Samhällsstudier, Umeå University, Umeå

Pilø L (1989) Early soapstone vessels in Norway from the Late Bronze Age to the Early Roman Iron Age. Acta Archaeologica 60:87–100

Prøsch-Danielsen L, Simonsen A (2000) Palaeoecological investigations towards the reconstruction of the history of forest clearances and coastal heathlands in south-western Norway. Veg Hist Archaeobot 9:189–204

Reimer PJ, Bard E, Bayliss A, Beck JW, Blackwell PG, Bronk Ramsey C, Grootes PM, Guilderson TP, Haflidason H, Hajdas I, Hatt C, Heaton TJ, Hoffmann DL, Hogg AG, Hughen KA, Kaiser KF, Kromer B, Manning SW, Niu M, Reimer RW, Richards DA, Scott EM, Southon JR, Staff RA, Turney CSM, van der Plicht J (2013) IntCal13 and marine13 radiocarbon age calibration curves 0–50.000 years cal BP. Radiocarbon 55(4):1869–1887

Rønne P (2011) Highlights from the northernmost Bronze Age societies in Norway. In: Mahler DL, Andersen C (eds) Farming on the edge: Cultural landscapes of the North (Short papers from the network meeting in Lerwick, Shetland, September 7th–10th 2010). Northern Worlds. The National Museum of Denmark, Copenhagen: 58–69. ISBN: 978-87-7602-158-0

Sjögren P (2010) Pollenanalys. In: Arntzen JE, Sommerseth I (eds) Den første gården i Nord-Norge: Jordbruksbosetting fra bronsealder til jernalder på Kveøy. TROMURA 39. Tromsø University Museum, UiT The Arctic University of Norway, Tromsø

Sjögren P, Arntzen J (2013) Agricultural practices in Arctic Norway during the first millennium B.C. Veg Hist Archaeobot 22:1–15

Soltvedt EC (1984) Makrofossilanalyse Borg 1984: Intern arbeidsrapport (Report). University of Tromsø

Soltvedt, EC (2000) Carbonized cereal from three Late Neolithic and two Early Bronze Age sites in western Norway. Environ Archaeol 5:49–62

Tveraabak LU (2004) Lowland *Calluna* heath vegetation along the coast of north Trøndelag and Nordland, Norway: Present state, development and changes during the last 4–5000 years. Dissertation, University of Tromsø

Tveraabak LU, Alm T (1997) En pollenanalytisk undersøkelse av Sandvika på Kvaløya (Tromsø kommune): 1–20 (Unpublished report in the topographical archive). Tromsø University Museum, UiT The Arctic University of Norway, Tromsø

Utne A (1973) En veidekulturs-boplass i Lofoten. Storbåthallaren ved Nappstraumen. Thesis, University of Tromsø

Valen CR (2007) Jordbruksimpulser i neolitikum og bronsealder I Nord-Norge? En revisjon av det arkeologiske gjenstandsmaterialet og de naturvitenskapelige undersøkelsene. Master's thesis, University of Tromsø

Vankilde H, Rahbek U, Rasmussen KL (1996) Radiocarbon dating and the chronology of Bronze Age southern Scandinavia. Acta Archaeologica 67:183–198

Vorren K-D (2005) Farm development at the Arctic cereal limit in northern Norway: Continuity and discontinuity. Veg Hist Archaeobot 14:161–170

Vorren K-D (1986) The impact of early agriculture on the vegetation of northern Norway: A discussion of anthropogenic indicators in biostratigraphical data. In: Behre K-E (ed) Anthropogenic indicators in pollen diagrams. A.A. Balkema, Rotterdam, pp 1–18

Vorren K-D, Nilssen E, Mørkved B (1990) Age and agricultural history of the 'staðir' farms of north and central Norway. Norwegian J Geography 44:79–102

Vorren K-D, Blaauw M, Wastegård S, van der Plicht J, Jensen C (2007) High-resolution stratigraphy of the northernmost concentric raised bog in Europe: Sellevollmyra, Andøya, northern Norway. Boreas 36:253–277

Wallin J-E (2009) Bebyggelsen i Laihela, Österbottens kustland under bronsålder och förromersk järnålder: En studie av vegetationsförändringar och odling baserad på pollenanalys (Report from Pollenlaboratoriet i Umeå Miljöarkeologiska laboratoriet). Institutionen för idé- och Samhällsstudier, Umeå University, Umeå

Indications of rye (*Secale cereale*) cultivation from 7th century south-western Norway

Sara Westling & Christin Eldegard Jensen

Author addresses:
Arkeologisk museum, University of Stavanger,
Peder Klows gate 30A, 4036 Stavanger, Norway

E-mail: sara.westling@uis.no

Abstract

An assemblage of charred rye (*Secale cereale*) grains has been found in a posthole of a longhouse at the settlement site of Hove-Sørbø, located in Sandnes, Rogaland, south-western Norway. The rye has been directly radiocarbon dated to AD 618–683 and is the earliest find of a large collection of rye grains in Norway. Plant macrofossil and pollen data are presented from the site, and these are discussed in the light of other relevant finds from northern Europe. The introduction of rye into south-western Norway has not hitherto been the subject of specific studies due to the limited empirical data available from the region. This article contributes to our knowledge of agricultural developments. Possible local cultivation or imports of a new crop species, perhaps as a strategy for coping with climatic stress and/or social crisis, are discussed.

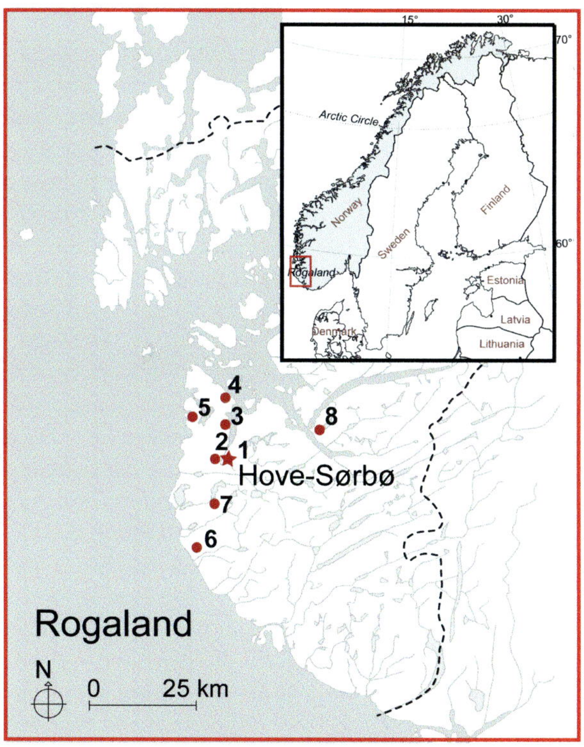

Fig. 1. Map of the county of Rogaland showing the location of the main site of Hove-Sørbø. Sites with finds of rye (*Secale cereale*) grains or pollen earlier than the Viking Age are also marked: 1 Hove-Sørbø; 2 Sørbøtunet; 3 Gausel; 4 Tastarustå; 5 Tjora; 6 Audemotland-Kvia; 7 Kvåle; 8 Forsandmoen. Insert: Map of northern Europe showing the location of Rogaland.

Introduction

Finds of rye (*Secale cereale*) are rarely recorded from prehistoric contexts in Norway, and this cereal species is regarded as a late arrival in Scandinavia (Ahokas 2009; Grabowski 2011; Viklund 1998). Although sporadic charred grains and pollen of rye are found earlier, it is not until the Middle Ages (which in Norway span the period AD 1050–1537) that there is substantial archaeobotanical evidence of rye cultivation in Norway (cf. Hjelle 2007; Sandvik 2018). Barley is the dominant cereal in prehistoric contexts in south-western Norway, followed by oat (*Avena sativa*) and wheat (*Triticum* sp.) (Fredh et al. 2018; Prøsch-Danielsen and Soltvedt 2011). In the light of this, the discovery of an assemblage of more than a thousand charred rye grains, directly radiocarbon dated to AD 618–683, in a posthole associated with a three-aisled house structure at the site of Hove-Sørbø in Rogaland (Fig. 1), is remarkable. In addition to this large collection of charred grains, a few individual rye grains were recorded from other house structures at Hove-Sørbø, and rye pollen was found in buried field and soil layers within the prehistoric farm site.

The dating of the rye grains to the 7th century AD connects them to the transitional time of the late Migration (AD 400–550) and the Merovingian (AD 550–800) periods. Widespread farm abandonment and significant changes evident in the archaeological record for the 6th and 7th centuries in south-western

Norway, as well as in other parts of Scandinavia, have led to an interpretation of this time as a period of stress or crisis (Axboe 1999; Fredriksen et al. 2014; Myhre 1991; Pedersen and Widgren 1998; Price and Gräslund 2015). Social, political and demographic factors have been proposed as contributing to the changes that occurred in Scandinavia during the 6th and 7th centuries, and many researchers emphasise the complexity of events during this period (Iversen 2013; Myhre 2002; Näsman 2012; Solberg 2000; Widgren 2013).

Two dramatic events that occurred during the 6th century have also been discussed as possible contributory factors to the evident transformation. One is a climate deterioration, following "the dust veil event" of AD 536, when a mysterious cloud obscured the sun for as long as 18 months (Arjava 2005). Its cause has yet to be confirmed (cf. Rigby et al. 2004), but recent research suggests that volcanic eruptions were probably involved (Sigl et al. 2015; Toohey et al. 2016). A fall in temperature, caused by the dust veil, has been recorded in tree-ring chronologies from northern Europe (Büntgen et al. 2016), and crop failure and famine are described in contemporary sources from the Mediterranean region (Arjava 2005; Barnish 1992). The cold period that followed may have lasted for several decades (Büntgen et al. 2016; Larsen et al. 2008; Toohey et al. 2016) and could have had a considerable effect on conditions for agriculture in Norway (Stamnes 2016).

A few years after the dust veil event, in AD 542, the Justinian Plague struck Europe (Harbeck et al. 2013; Wagner et al. 2014). This first recorded incidence of bubonic plague killed millions of people in a series of outbreaks (Little 2007; Stathakopoulos 2007). A possible connection between plague outbreaks and climatic variation has been suggested, and the Justinian Plague may have been triggered by the dust veil event (Rosen 2006; Stenseth et al. 2006). How far north the plague reached has not yet been established, but there is evidence that the bacterium spread north of the Alps (Harbeck et al. 2013). An outbreak in Scandinavia has been discussed as one possible factor behind the changes that occurred in south-western Norway during the 6th and 7th centuries (Iversen 2016; Myhre 2002; Rødsrud 2016), but there is still a lack of evidence in support of this hypothesis. Regardless of the factors involved, new circumstances or a period of stress could have made it desirable, or even necessary, to adapt agriculture and adopt a new crop. Rye may have been an appropriate choice in this respect.

The ecology and history of rye in the Nordic countries

The ecological benefits of rye, compared with those of barley, wheat and oat, are its ability to be productive on soils of a lower nutrient content and its greater tolerance of low temperatures during the growing season (Behre 1992). Engelmark (1992) highlights the advantages of expanding arable fields without having to increase livestock numbers to produce extra manure during periods of demographic pressure, and Behre (1992) notes that the flowering of rye is actually stimulated by a period of low temperature (vernalisation). Autumn sowing of rye (winter rye) is the most common practice in Norway, as the longer growing season results in larger crop yields and earlier ripening of the grains, but spring sowing, followed by livestock grazing after harvest, has also been practised in modern times, particularly in eastern Norway. Both forms of management would therefore be applicable during prehistoric times. Like oat, rye thrives on soil with a low pH, but rye is sensitive to poor drainage and prefers dry, sandy soils. It is, however, vulnerable to deep snow cover and fungal attacks, particularly by ergot (*Claviceps purpurea*) and rusts (*Puccinia* spp.). Combined with its tolerance of low winter temperatures, the physiological qualities of rye make it profitable to grow in the more continental areas of northern Europe. Due to its high tolerance of acid soils, rye was the preferred cereal in the slash-and-burn cultivation of spruce forests in Finland (Ahokas 2009; Alenius 2007; Soininen 1974), and this practice was introduced later to Sweden and eastern Norway with the immigration of Finns in the late Middle Ages (Bjørnstad 2012). Rye is hardly cultivated at all in south-western Norway today (Stabbetorp 2018).

According to Behre (1992), rye cultivation was established on the sandy soils of north-western Germany during the Roman Iron Age, evidently in two- and three-field rotation systems. Behre (1992) suggests that the spread of rye into northern Europe followed the expansion of the Roman Empire. His hypothesis may be supported by recent investigations in Lithuania, where a rye grain from a Roman hillfort has been radiocarbon dated to AD 85–250 (Grikpedis and Motuzaite-Matuzeviciute 2016). In Denmark, rye grains have been recorded from the Late Bronze Age (Robinson 1993; 2004). However, these early finds are thought to represent the presence of rye as a weed. Finds of a magnitude and context that may represent cultivation first occur in the Roman Iron Age (Robinson et al. 2009). According to Regnell (2002), rye cultivation in Denmark may have reached a level corresponding to that of the Viking Age from the Roman Iron Age onwards. In an investigation of an iron-smelting site at Drengsted, in Denmark, rye was found mixed with spring weeds and barley with summer weeds (Mikkelsen 1999; Mikkelsen and Nørbach 2003), leading to the hypothesis that winter rye and spring barley were cultivated there. According to Grabowski (2011), rye was introduced into Sweden during the Roman Iron Age, but winter rye did not become an integrated part of the crop rotation system until the Merovingian period or the Viking Age. The greater climatic and edaphic tolerance of rye is

Table 1. Radiocarbon dates from archaeological structures and soil layers at Hove-Sørbø. unspec. = unspecified.

Laboratory code	Material	Context	^{14}C age BP	Cal. age AD/BC, 2 sigma
UBA-23360	Charred grain, *Secale cereale*	House 36, posthole	1376±30	AD 608–683
Beta-397517	Charred grain, *Hordeum vulgare* var. *vulgare*	House 36, posthole	1300±30	AD 660–770
UBA-24838	Charred grain, *Hordeum vulgare* var. *vulgare*	House 19, posthole	1283±21	AD 671–729, 736–769
Beta-397518	Charred grain, *Hordeum vulgare* var. *vulgare*	House 19, posthole	1250±30	AD 690–750, 760–775
UBA-24847	Charcoal, deciduous tree (not *Quercus*)	House 51, hearth	1252±38	AD 672–780, 787–876
UBA-23382	Charcoal, *Betula* sp.	House 51, hearth	1099±26	AD 889–995, 1008–1011
UBA-25605	Charcoal, deciduous tree (cf. *Sorbus*, not *Quercus*)	House 51, hearth	1228±25	AD 692–747, 762–881
Beta-397519	Charred grain, *Secale cereale*	House 51, posthole	170±30	AD 1665–1685, 1730–1780, 1795–1810, 1925–post 1950
UBA-23384	Charcoal, *Alnus*	House 59, hearth	1411±27	AD 597–662
UBA-24849	Charred grain, *Hordeum*	House 59, lower floor layer	1721±22	AD 252–306, 311–386
UBA-24850	Charred grain, Cerealia	House 59, posthole (door)	1628±24	AD 353–366, 380–437, 445–472, 486–534
UBA-25084	Charcoal, *Betula*	House 59, upper floor layer	1613±27	AD 392–478, 482–536
UBA-24856	Charred grain, cf. *Hordeum*	Field 2, cultural layer	1175±29	AD 770–901, 920–954, 957–961
UBA-24857	Charred grain, *Hordeum vulgare,* unspec.	Field 2, cultural layer	1183±22	AD 772–893, 933–936
UBA-24858	Charred grain, *Hordeum vulgare* var. *vulgare*	Field 3, cultural layer	1592±23	AD 413–537
UBA-24859	Charred grain, *Hordeum vulgare* var. *vulgare*	Field 3, cultural layer	1523±27	AD 429–495, 507–520, 527–604
UBA-24860	Charred grain, *Hordeum vulgare* var. *vulgare*	Field 3, cultural layer	1601±22	AD 404–477, 482–536,
UBA-24861	Charred grain, *Hordeum vulgare* var. *vulgare*	Field 3, cultural layer	1600±22	AD 405–478, 482–536
UBA-24862	Charred grain, *Hordeum vulgare* var. *vulgare*	Field 3, cultural layer	1692±26	AD 258–296, 321–410

commonly used to explain a coherence of rye cultivation and settlement history in Finland (Solantie 1988). Viklund (1998) questions this theory, pointing out that the earliest finds of rye in Sweden are from the agriculturally favourable region of Scania. Rye cultivation related to slash-and-burn practices is, however, documented in south-eastern Finland from AD 300–600 onwards (Alenius 2007).

Site description

Among the sites in Rogaland with records of prehistoric rye, Hove-Sørbø stands out due to the magnitude of the find here. The site lies in the southern part of the municipality of Sandnes, in the county of Rogaland, south-western Norway (Fig. 1). This region has a highly oceanic climate and a vegetation characterised by open heathland and coastal species, dependent upon a high moisture level in the air; the vegetation has no continental features (Moen 1999). The landscape is undulating, with the prehistoric settlements located on high ground covered by a well-drained till substrate. The local bedrock is gneiss, but phyllite and mica schists also occur (Birkeland 1981), providing a good nutrient level for plant growth.

An archaeological excavation of the site was carried out in 2011–2012 by the Arkeologisk museum (Museum of Archaeology), University of Stavanger. The excavated area, which was divided into five sub-

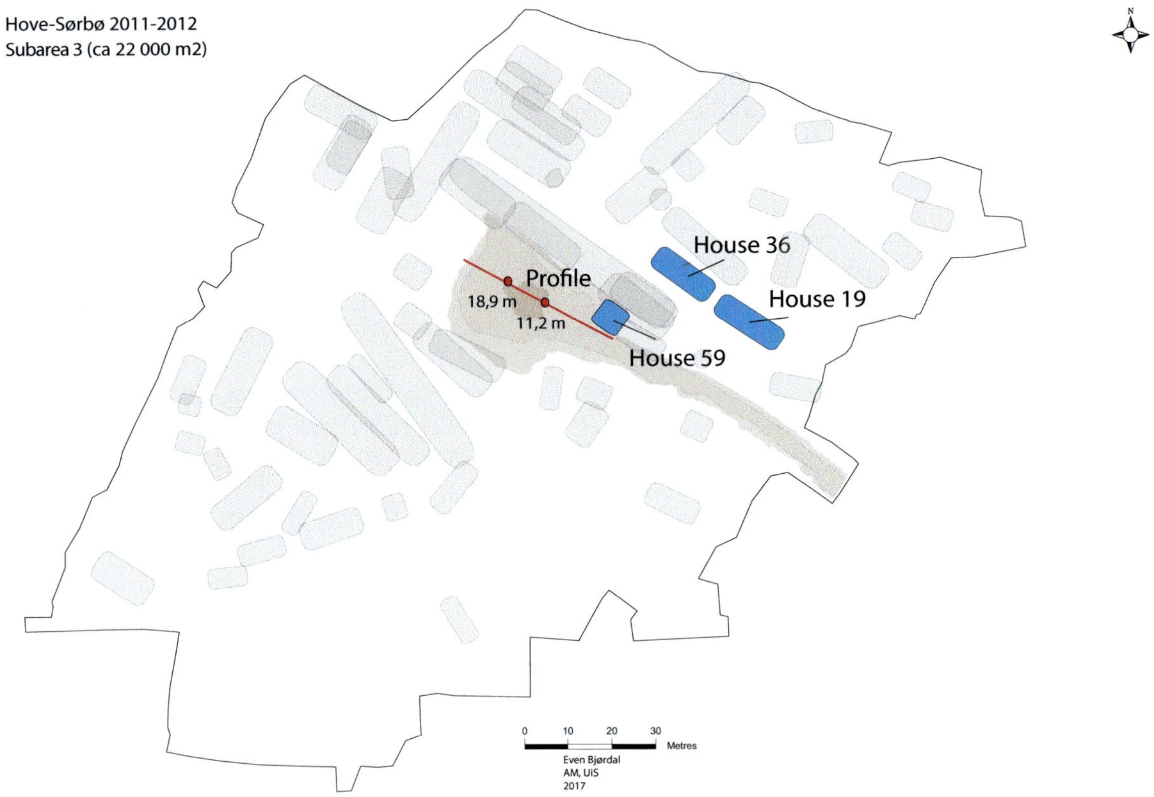

Fig. 2. Map of subarea 3 at Hove-Sørbø. Houses and soil profiles with finds of rye (*Secale cereale*) are highlighted

areas, covered about 33,100 m^2 in total. A total of 47 confirmed, 33 probable and 21 possible structures were uncovered. Radiocarbon dates from the site extend from the Late Neolithic to the early Middle Ages (Table 1). This article deals mainly with material found in subareas 1, 2 and 3, where sub-fossil remains of rye were recorded.

Most of the structures found during the excavation were located in subarea 3, including house 36, which contained the large find of charred rye grains (Fig. 2). House 36 (Figs. 2–3) is a three-aisled building located in the eastern corner of the subarea. It has been radiocarbon dated to the Merovingian period (Table 1). The house was at least 13 m, and possibly up to 17 m, in length and 3–4.5 m wide and had curved walls. It appears to have been divided into functional units, with a central room containing the hearth. Towards the southern part of the building are postholes indicating the existence of a dividing wall. The structure had been disturbed by modern ploughing; therefore no floor layer survived and the southern limit of the house could not be conclusively established.

Materials and methods

During the excavation, the topsoil was mechanically stripped off to reveal the remains of structures and other archaeological features. Hearths and postholes associated with roof-bearing posts and doorposts, as well as stratigraphic layers in soil profiles, were given highest priority when it came to sampling for archaeobotanical analyses. A total of 253 samples were analysed, and the average sample volume was 3.9 litres.

Plant macrofossil analysis

The samples were processed by flotation, with the flots being retained on a 0.5 mm mesh sieve. Charred diaspores and other plant macro-remains were examined using a stereo microscope with 7.5× to 112.5× magnification and identified with the aid of the reference collection held at the museum of archaeology at the University of Stavanger, as well as relevant literature (Anderberg 1994; Berggren 1969, 1981; Cappers et al. 2006; Jacomet 2006; Korsmo et al. 1981; Mossberg et al. 1992).

A metric analysis was undertaken on two large finds of, respectively, rye and hulled barley from house 36, to obtain an impression of the grain size. A total of 50 grains from each sample were randomly selected using a sample splitter. The length, width and thickness of the grains was measured using a stereo microscope, and the average, median and variance were calculated. The resulting data will enable comparison with other finds in the future.

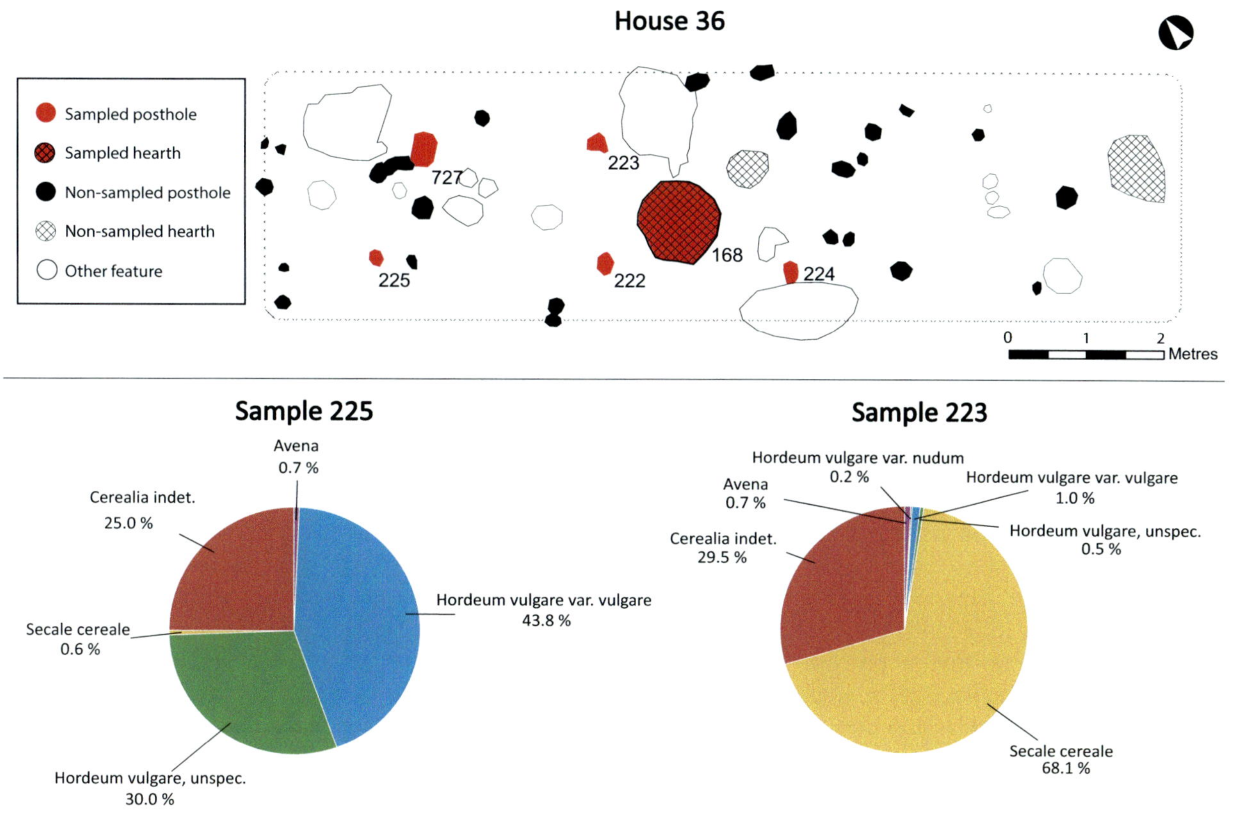

Fig. 3. Plan of house 36. The grain composition in samples 225 and 223 is shown in the pie charts. The total number of grains is 1823 in sample 225 and 1741 in sample 223.

Table 2. Localities with finds of plant macrofossils or pollen of rye (Secale cereale) in the county of Rogaland.

Locality	Geographical coordinates (lat./long.)*	Bedrock; superficial deposit	Archaeological context	Type of plant remains	Age	Reference
Hove-Sørbø	58° 49'59" N 05° 43'59" E	Gneiss, phyllite; till	Soil profiles	Charred seeds and pollen (grains)	MVP	Westling & Jensen (this volume)
Sørbøtunet	58° 49'60" N 05° 43'24" E	Gneiss, phyllite; till	Lynchet†	Pollen		Sandvik 1999 (macro-remains); Jensen, C. unpublished data (pollen)
Gausel	58°54'41" N 05°44'58" E	Gneiss, granite, meta-supracrustals; till (thick)	House (posthole)	Charred seeds	MVP	Børsheim & Soltvedt 2002
Kvåle	58°43'31" N 05°40'14" E	Gneiss, granite, meta-supracrustals; till (thick)	Lynchet†	Pollen		Soltvedt et al. 2007
Tastarustå	58° 59' 12" N 05° 41' 70" E	Phyllite; till (thick)	Soil profiles (lynchet?) †	Pollen		Soltvedt & Enevold 2009
Tjora‡	58° 54' 52" N 05° 35' 59" E	Gneiss, granite, phyllite; till, aeolian sand	Soil profile†	Pollen		Soltvedt & Jensen 2011; Westling & Jensen (this volume)
Audemotland-Kvia‡	58°39'16" N 05°37'39" E	Gneiss, granite	†	Pollen		Prøsch-Danielsen unpublished data
Forsandmoen†	58°51'45" N 06°8'18" E	Granitic gneiss; fluvial deposits	Soil profiles†	Pollen		Prøsch-Danielsen 1988; Prøsch-Danielsen & Simonsen 1988

*www.norgeskart.no
†Not a radiocarbon-dated context
‡No samples taken for plant macrofossil analysis
MVP = Merovingian and Viking Period

Pollen analysis

A total of 42 pollen samples, each with a volume of 1 cm^3, were collected in plastic tubes from soil profiles within subareas 1, 2 and 3. Of these, 28 samples, randomly selected from the different soil layers, were analysed. The samples were subjected to chemical pre-treatment with 37.5% hydrofluoric acid (HF) to dissolve minerogenic particles, with 10% potassium hydroxide (KOH) to remove humic material, and with standard acetolysis (Fægri and Iversen 1989) to remove cellulose and concentrate the content of pollen, spores and charcoal dust. The identification of pollen types was undertaken with reference to Moore and colleagues (1991), Beug (2004) and Fægri and Iversen (1989). The software Tilia, version 2.0.41 (Grimm 2012), was used in the production of the pollen diagrams.

To identify possible diagnostic features in the pollen flora related to the local cultivation of rye, a dataset comprised of analyses from seven sites in Rogaland with records of rye pollen from stratigraphic soil layers was established (Table 2). A detrended correspondence analysis (DCA) of all herbaceous plants (square root transformed), excluding cereal pollen taxa, rendered a short gradient length (1.6 SD), recommending the use of a linear model (Leps and Smilauer 2003; ter Braak and Prentice 1988). A principal component analysis (PCA) was therefore performed, using cereal pollen taxa as supplementary variables.

Results

Plant macro-remains from the house samples

Six samples were collected and analysed from house 36. Sample 223 (Fig. 3) contained 1182 charred grains of rye, as well as smaller numbers of other cereals and weeds. Much of the material was poorly preserved and fragmented, and 29.5 % of the grains could only be identified as unspecified cereals, i.e. Cerealia indeterminable. In the remaining material, rye constituted 68.1 % of the cereal grains identified to species (Figs. 3–4, Table 3).

Sample 225 (Fig. 3), from another posthole in house 36, contained 779 grains of hulled barley, which were dated to the same period as the rye. Preservation was poor, and 25 % of the grains could not be identified to species. The identified material in the sample was dominated by hulled barley (*Hordeum vulgare* var. *vulgare*), which constituted 43.8 % of the grains. A further 30 % of the grains were recorded as barley (*Hordeum vulgare*) but could not be identified further. Since no grains of naked barley (*Hordeum vulgare* var. *nudum)* were recorded, however, it is highly likely these represent poorly preserved grains of hulled barley (Figs. 3 and 5, Table 3).

Rye was documented in three other buildings at the site, houses 19, 59 (Fig. 2) and 51 (subarea 5). One grain of rye was identified from house 19. The grain was too small to be directly dated, but the house was radiocarbon dated to the Merovingian period. One cereal grain of rye was also found in house 51, but radiocarbon dating revealed this to be modern (Table 1), and it was therefore interpreted as contamination. Two grains were found in house 59, but due to their small size no direct date could be obtained. The house was interpreted as having had numerous phases, extending in date from the Roman Iron Age (AD 1–400) to the Merovingian period.

The results of the metric analysis (Table 4) show that the average length of the rye grains in sample 223 was 4.85 mm, the average width was 1.98 mm and the average thickness was 1.78 mm. In sample 225, the average length of the hulled barley grains was 5.74 mm, the average width was 3.08 mm and the average thickness was 2.38 mm. A calculation of the cumulative average measurements shows that measuring 50 grains is enough to give a representative value.

Siliceous scoria droplets are formed by the melting of siliceous material in the soil during exposure to high temperatures and are associated with buildings destroyed by fire (Thy et al. 2015). Such droplets were documented in five out of six analysed samples from house 36 (Table 3), and three samples contained more than 10 of these. In house 19, only one sample contained siliceous scoria droplets, and in this sample, fewer than 10 of these were recorded. Large amounts of charcoal could also indicate a house fire. The average charcoal content in house 36 was 12 ml per litre (Table 3). In house 19, the average charcoal content was 5 ml per litre.

Pollen and plant macro-remains from soil profiles

The field investigations of the soil profiles revealed three or perhaps four possible ancient farming horizons, based on the texture and colour of the sediments. In the various soil profiles, a compact moraine substrate was transformed into poorly sorted, humic, silty sand with gravel and visible charcoal remains (Table 5). Possible traces of mechanical soil management were observed, but later biological activity (bioturbation) has probably disturbed any clear ard marks. The pollen grains show signs of a high degree of corrosion and mechanical abrasion, typical of unstable aerobic environments, such as those prevailing in cultivated fields. Some radiocarbon dates (Table 1) from within the same profile overlap (Figs. 6–8). This may be a result of cultivation activity in the field or soil being moved upwards by bioturbation. The results of the botanical analyses confirmed the assumption of farming practices through the presence and abundance of diagnostic pollen and plant macro-remains that indicate animal grazing activity and cereal cultivation (Figs. 6–8). Although chronostratigraphical control is generally poor in managed

Fig. 4. Rye (*Secale cereale*) grains from Hove-Sørbø, sample 223.

Fig. 5. Hulled barley (*Hordeum vulgare* var. *vulgare*) grains from Hove-Sørbø, sample 225.

field systems, the botanical content of discrete field layers may still adequately reflect the local flora during the period of activity.

The pollen assemblages evident in the three analysed profiles are very much in accordance with each other in regard to the main environmental features. They reflect a semi-open to open cultivated landscape with grassland and possibly heather (*Calluna*) heathland and mires. Care should, however, be taken in interpreting the pollen of dwarf shrubs and mire herbs as evidence for the presence of this vegetation close to the fields, as transported peat and turf may have been used to enrich the soil in the fields. The composition of the tree pollen flora indicates the presence of deciduous forest, mainly comprised of alder (*Alnus*), hazel (*Corylus*) and birch (*Betula*), but with the additional presence of lime (*Tilia*), oak (*Quercus*), elm (*Ulmus*), ash (*Fraxinus*), willow (*Salix*) and juniper (*Juniperus*). Rye pollen was recorded in the profiles from all three subareas and was always found together with pollen of barley and unidentified cereals (Figs. 6–8). The charred cereal grains were identified as hulled barley or barley species (*Hordeum vulgare*) and oat. No secure identification of other cereals was possible for the pollen or the plant macro-remains in the soil profiles, although single possible wheat pollen grains were recorded from the same field layers as the rye pollen. The weed pollen flora includes nitrophilous taxa, such as goosefoot (Chenopodiaceae) and nettles (*Urtica*), but also taxa that are not dependent on manuring, such as corn spurrey (*Spergula arvensis),* knotweed *(Persicaria)* and common knot-

Table 3. Results of the plant macrofossil analysis of samples from house 36. fragm. = fragment; unspec. = unspecified; x = present; xx = abundantly present.

Sample information						
Sample number	168	222	223	224	225	727
Structure type	Hearth	Posthole	Posthole	Posthole	Posthole	Pit
Sample volume (dm³)	4	3,5	4	3,8	4,5	4
Carbonised material content						
Avena fatua					1	
Avena sativa			13		1	
Avena sp.	2			3	10	
Avena, glume base					1	
Hordeum vulgare var. *nudum*			3			
Hordeum vulgare var. *vulgare*	1	3	18	5	779	
Hordeum vulgare var. *vulgare*, glume fragm.					20	
Hordeum vulgare, unspec.		1	8	5	546	
Secale cereale		8	1182	9	11	
Secale cereale, glume fragm.			3			
Cerealia indet.	2	5	513	12	451	
Cerealia indet., spikelet fork			1	1		
Cerealia indet., rachis fragm.					4	
Carex distigmaticaea		1				
Carex tristigmaticae						1
Chenopodium album			3	4		
Corylus avellana, shell fragm.						1
Galeopsis sp.	1		4	2	6	
Juncus sp.			1			
Persicaria lapathifolia			18		6	
Persicaria sp.	3	5		2	3	
Plantago lanceolata			2		1	
Poaceae sp.	4				1	
Rumex acetosella	21					
Spergula arvensis		2	4		1	
Stellaria media	1	2	1	2	1	
Trifolium sp.						1
Silicious scoria droplets	x	xx	xx	xx	x	
Charcoal volume (ml)	100	25	50	50	25	30

Table 4. Average, median and variance of measurements on hulled barley and rye grains from samples 225 and 223.

	Hulled barley			**Rye**		
	Length (mm)	Width (mm)	Thickness (mm)	Length (mm)	Width (mm)	Thickness (mm)
Average	5.74	3.08	2.38	4.85	1.98	1.78
Median	5.77	3.14	2.35	4.85	1.96	1.80
Variance	0.51	0.24	0.19	0.61	0.09	0.08

weed (*Polygonum aviculare*). Typical grazing indicators, in addition to grasses, are sorrel (*Rumex acetosa*) and buttercup (*Ranuculus acris*), ribwort plantain (*Plantago lanceolata*) and clover (*Trifolium* spp.).
The soil profile in the sectioned area in the middle of subarea 3 (Fig. 8) stands out by having a wider range of plant taxa in its plant macrofossil and pollen flora than is evident in the field layers sampled in the soil profiles in subareas 1 and 2. The greater biodiversity here may be related to the presence of a multipurpose area or a pathway or crossroads.

Charred barley grains from two of the soil profiles have been radiocarbon dated, one from subarea 2 and one from subarea 3 (Table 1). The radiocarbon dates from the soil profile at 11.2 m from the eastern end in the sectioned area of subarea 3, assigned layer 2 to the Late Roman Iron Age, layers 3 and 5 to the Migration period, and layer 4 to the Migration/Merovingian period. Rye pollen was found in layers 2, 3 and 5. In soil profile 1, in subarea 2, layers 2 and 3 gave overlapping radiocarbon dates at the transition between the Merovingian period and the Viking Age.

Table 5. Lithological description of the soil profiles at Hove-Sørbø.

Subarea/profile	Soil layer	Depth (cm)	Lithological description
1/2	4A	0–26	Very silty, sandy, humic mixed topsoil. Homogenous, compact, poorly sorted. Moderately fine sand. Very dark, greyish brown. Modern roots.
1/2	3 A	26–50	Very silty sand, moderately fine, compact, homogenous, with a few stones. Dark brown with charcoal. Modern roots. Sharp upper boundary.
1/2	2 A	50–60	Highly silty sand, moderately fine, compact. Slightly humic. Dark yellowish brown. Mixed zone with grey, brown and orange spots. Possible traces of bioturbation and ploughing. Gradual upper and lower boundary.
1/2	1C	60–>70	Silty sand, moderately fine. Slightly humic. Dark yellowish brown, grey, brown and orange spots, less than in the layer above. Compact material. Possible traces of bioturbation.
2/1	4 A	0–25	Very silty sand, humic rich. Moderately coarse, mixed topsoil, slightly loose material. Very dark brown.
2/1	3 A	25–58	Very silty sand, humic rich. Poorly sorted, mixed topsoil. Compact material. Very dark brown. Modern roots.
2/1	2 A	58–79	Slightly silty, moderately coarse sand, moderately humic. Compact material. Very dark greyish brown with yellow sand spots. Upper sharp boundary.
2/1	1 C	79–94	Very silty, moderately coarse sand. Yellowish brown with yellow, orange and reddish spots. Compact material. Possible traces of bioturbation.
3/1	6 A	0–17	Mixed topsoil, silty humic sand with small stones and modern roots. Brownish grey.
3/1	5 A	17–47	Humic sand with some 5–10 cm diameter stones near the lower boundary. Brownish. Modern roots.
3/1	4 A	47–62	Humic sand with some gravel and small stones. Greyish brown with some charcoal fragments.
3/1	3 A	62–77	Moderately coarse sand, moderately humic. Dark brownish grey with charcoal.
3/1	2 A	77–95	Fine sand, humic. Greyish brown with some charcoal fragments.

The results from the PCA ordination indicate a strong gradient in the dataset along the first axis (Fig. 9), separating samples containing weed species occurring in arable fields, to the left, from samples containing species that thrive in heath and grassland habitats, to the right, thereby separating off the majority of the samples from Forsandmoen. The gradient of secondary importance (along axis 2) reflects grazing impact, with particularly high scores for grasses (Poaceae) and ribwort plantain. The samples from the soil profiles in subareas 1 and 2 at Hove-Sørbø and from Tjora correlate positively with the latter group, while the samples from the section area of subarea 3 cluster with the samples from the ancient earth terraces (lynchets) at Sørbøtunet and one of the soil profiles at Tastarustå.

Discussion

Was the rye cultivated locally?

The large finds of grain in house 36 could represent stored cereals that became charred during a catastrophic fire. Similar grain concentrations in houses, between the hearth and the nearest gable wall, have been found in several Danish contexts, and this part

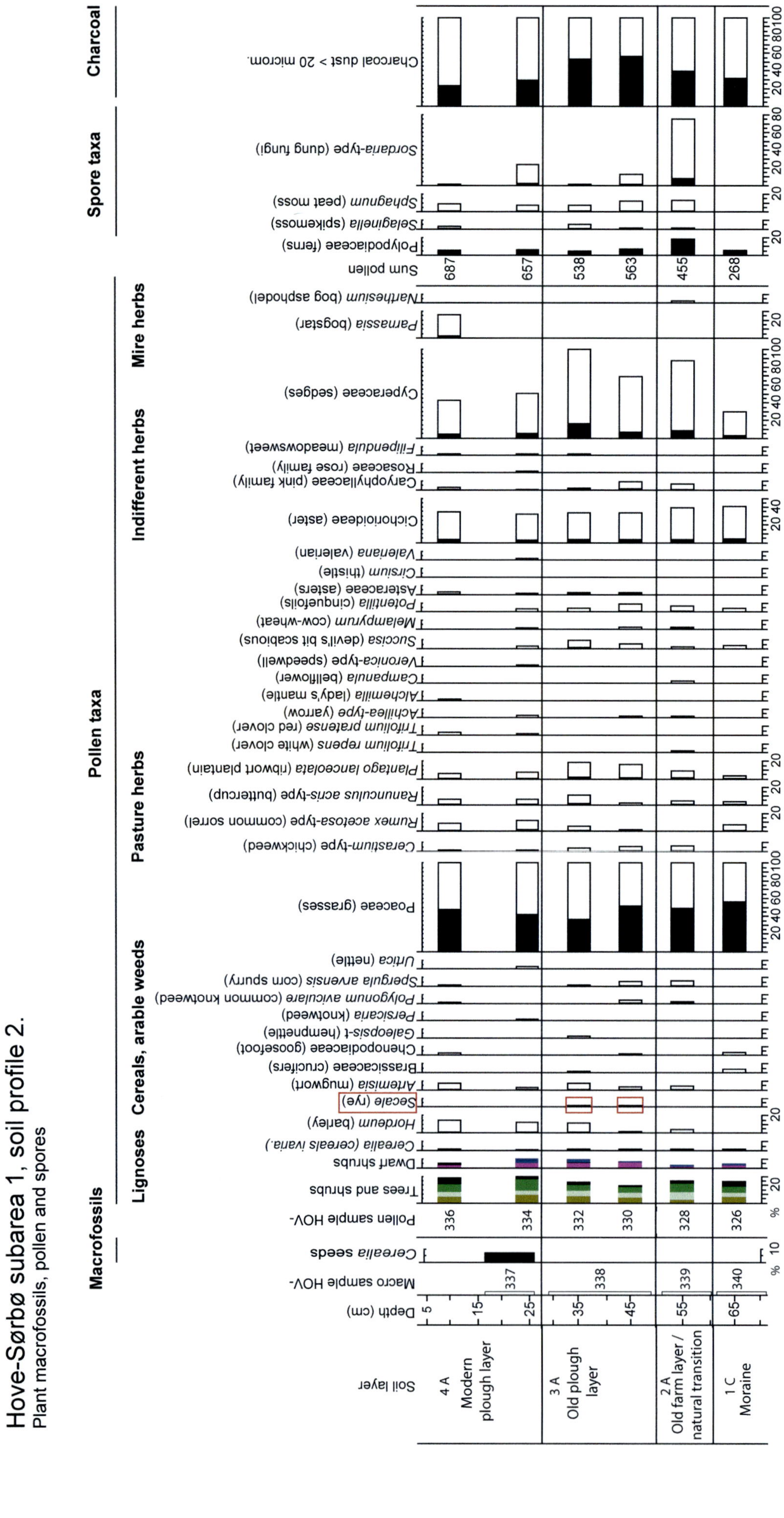

Fig. 6. Plant macrofossil, pollen and spore diagram (percentages) from soil profile 2 in subarea 1 at Hove-Sørbø.

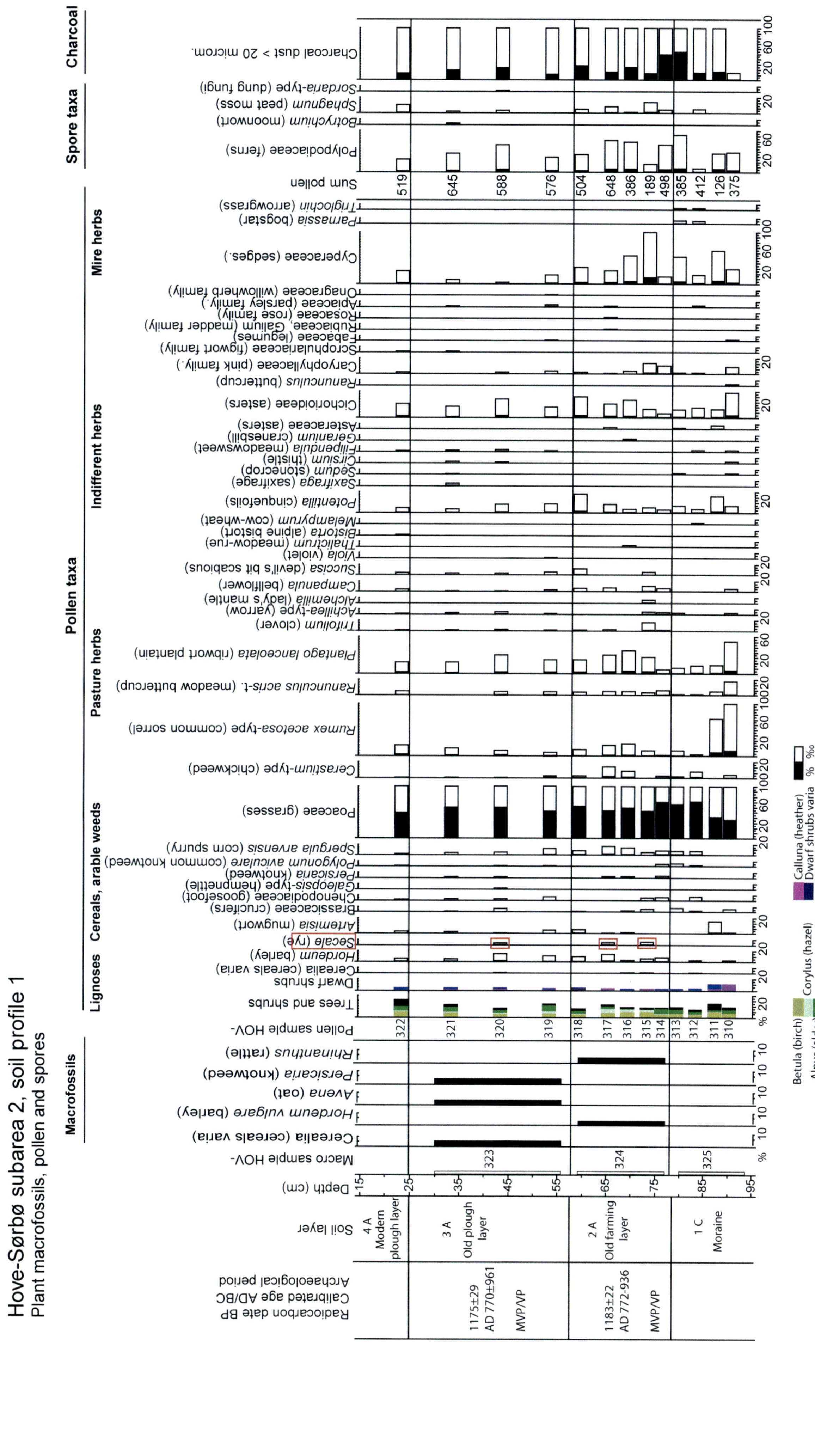

Fig. 7. Plant macrofossil, pollen and spore diagram (percentages) from soil profile 1 in subarea 2 at Hove-Sørbø.

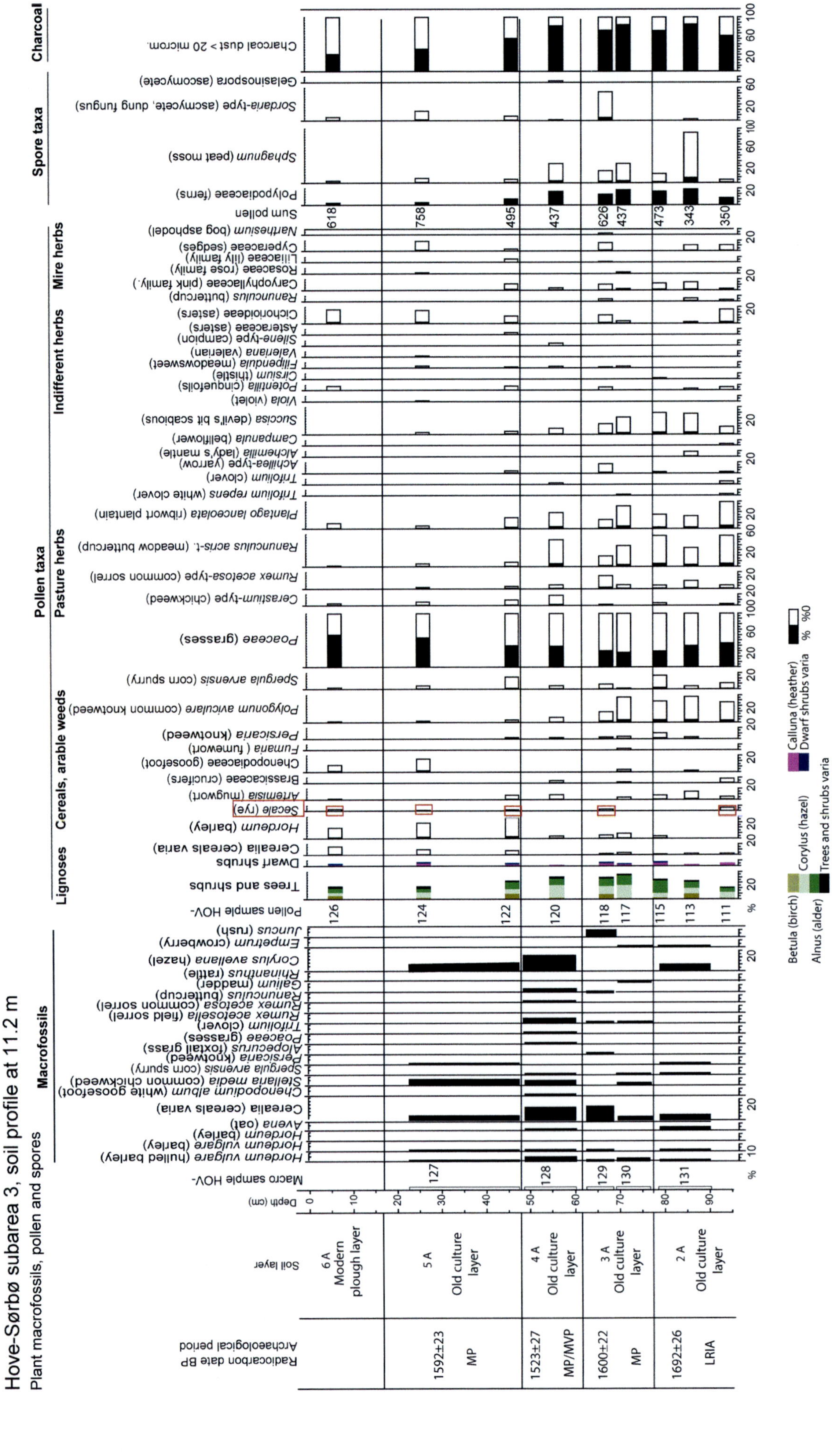

Fig. 8. Plant macrofossil, pollen and spore diagram (percentages) from soil profile at 11.2 m on transect in subarea 3 at Hove-Sørbø.

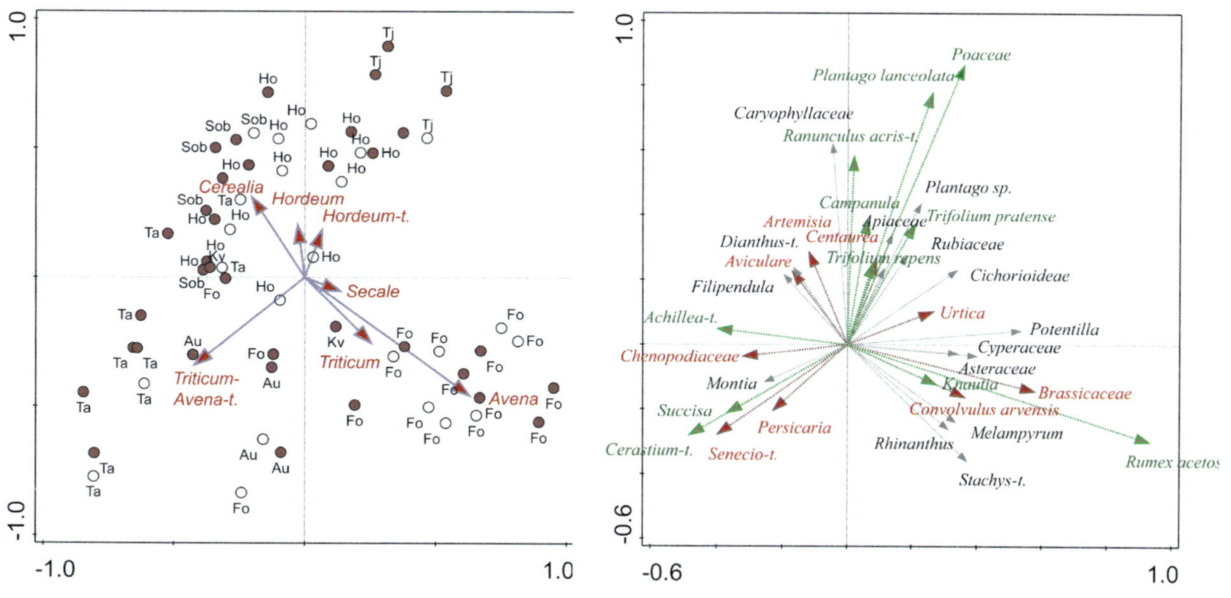

Fig. 9. Results of the principal component analysis (PCA) of herb pollen taxa from soil layers in Rogaland with records of *Secale* pollen. Cerealia pollen taxa are treated as supplementary variables. The eigenvalues of axes 1–4 are 0.2955, 0.206, 0.1085 and 0.0663. The explained cumulative variation of the primary and secondary axes (displayed in the figure) is 50 % and 67 % in total, respectively. -t = -type. Sample plot with supplementary variables (indicated by arrows) (left). Red dots indicate presence of *Secale*. Au Audemotland-Kvia; Fo Forsand; Ho Hove-Sørbø; Kv Kvåle; Ta Tastarustå; Tj Tjora; Sob Sørbøtunet. Species plot (right). Red arrow indicates arable weeds, green arrow indicates apophytic herbs of pasture, black arrow indicates ecologically indifferent herb taxa.

of the building has often been interpreted as a storage area associated with the kitchen or cooking facilities (Viklund 1998). Containers or sacks of grain may have been stored on the floor or hanging from the ceiling (Henriksen and Robinson 1996), but since no floor layers are preserved in the houses at Hove-Sørbø, interpretation is difficult. Charcoal was found in much greater quantities in house 36 than in the adjacent house 19, together with relatively large numbers of siliceous scoria droplets. One possible explanation is that the differences are due to the houses having had different functions; interpretation of the large grain finds in house 36 as ritual deposits has been put forward as a possibility (Regnell 1997). However, we believe that the most likely explanation, due to the position in the building and the quantities of charcoal and siliceous scoria droplets present, is that a grain store in the dwelling was exposed to fire.

The rye grains in sample 223 and the hulled barley grains in sample 225 are both associated with a small proportion of weed seeds and chaff. They give the impression of having been processed, i.e. threshed and cleaned, and were probably intended as seed corn or for human consumption. The weed species present are very similar in both samples, despite the different crops involved. Mikkelsen and Nørbach (2003) have studied iron smelting furnaces in south-western Jutland, Denmark. They found that the rye and barley grains recovered from this site were associated with different weed floras, and they conclude that the two crops were most likely grown separately. They suggest that if they had been grown as spring crops on the same field, rye following barley, the weed species would not differ substantially. If the crops found at Hove-Sørbø were locally grown, this could explain the similarity in weed composition between samples 223 and 225 from house 36. A maslin crop, where two crops are grown together on the same field, seems an unlikely explanation, given the great dominance of one of the species in each sample. The small proportion of barley in the rye-dominated sample and the small proportion of rye in the barley-dominated sample likely represent either stray plants or contamination that occurred during crop processing.

Common corn cockle (*Agrostemma githago*) and corn flower (*Centaurea cyanus*) are generally seen as indicators of winter rye cultivation, but these may be a later addition to the weed flora, being derived from rye imported to Scandinavia in the Viking Age and the Middle Ages (Hjelle 2007; Mikkelsen 1999; Mikkelsen and Nørbach 2003). There are also examples of regional differences: Wild radish (*Raphanus raphanistrum*) has been considered an indicator of winter rye in Sweden (Viklund 1998), but a clear connection is lacking in the Danish material (Mikkelsen and Nørbach 2003). The absence of common corn cockle, cornflower, wild radish or any other non-local or uncommon weeds, together with the similarity of the weed composition between the two samples, tends to indicate that the rye was locally grown.

A comparison of grain sizes can give an idea of the conditions under which the crops were grown. The size and shape of grains are affected by climate, soil conditions and degree of manuring. The grain size within archaeological samples relates to the stage of maturation at harvest, the stage in crop processing and the size-sorted fraction represented by the find (whether seed corn, grain for human consumption, animal fodder, etc.). Unripe grains may be more easily preserved (Henriksen and Robinson 1996; Viklund 1998), and they would normally be smaller and possibly narrower (Helbæk 1955). Grain size is, however, also affected by the charring process, and charred grains cannot be compared metrically with uncharred material. There are no other large finds of rye from Rogaland earlier than the Middle Ages (cf. Sandvik 2018), and consequently there is no local material for direct comparison in the metric analysis. Hulled barley is a more common find in Rogaland, but metric analyses have not been part of the standard documentation process. The measurements presented in this article are, therefore, mainly provided for future comparison and research.

The charred plant remains recovered from the soil profiles are generally thought to be derived from household ash spread over the field area as a means of soil improvement, but they may also be the result of stubble burning in autumn. The species composition of the remains from the soil profiles represents domestically used and cultivated species. However, due to taphonomic processes, no conclusions can be drawn from the numbers of seeds, or about species that are missing from the assemblages. Pollen may derive from multiple sources, e.g. from flowering plants growing in and around the field and from animal dung used as manure or from animals that periodically grazed the field, as well as from transported peat or turf used for soil enrichment. This mixed signal is evident in the pollen assemblages recorded from the soil profiles. The soil has been reworked over time, with the original farming traces becoming blurred in the process. The indication of grazing impact appears to harmonise with the impact of agricultural cultivation in the pollen record. Several options are possible with regard to management scenarios that can potentially explain this, such as rotation cycles with cultivation and fallow periods involving animal grazing, i.e. seasonal grazing during the spring and autumn, or manuring with animal dung.

Rye pollen has been observed in field layers dated to the late Roman Iron Age, the Migration period and the Migration period–Merovingian period transition in subarea 3 (Fig. 8), and to the Merovingian period–Viking Age transition in subarea 2 (Fig. 7). There is no exact chronological overlap with the collection of rye grains dated to the early Merovingian period found in house 36. We should, however, bear in mind the chronological limitations associated with investigations of this kind. Only some of the records of rye from structures and field layers have been subjected to radiocarbon dating, and there is only one radiocarbon date for each of the field layers. Given the taphonomic challenges associated with ancient fields, the representativeness of these dates relative to the period of activity is not completely clear and must therefore be subject to some doubt.

Compared with natural organic sediments, field layers are low in abundance of micro- and macrofossils. Rye, contrary to other cereals, such as barley and wheat, is cross-pollinated. It has open flowers which produce a large amount of pollen, and it employs wind as its main dispersal agent. Pollen values greater than only 1 or 2 % are therefore to be expected if rye grew on-site. Roemer (1932) demonstrated, however, that the dispersal range for rye pollen is very limited: less than 500 m in an open field. Given the smaller size of prehistoric fields, and given the less open landscape, the presence of rye may be difficult to document unless the samples are taken within the actual prehistoric field. It is therefore possible that rye was cultivated on neighbouring fields that were not investigated at Hove-Sørbø, or that it was present in some quantity as a weed in barley fields. A comparison of Hove-Sørbø with other sites in Rogaland with records of rye pollen from prehistoric field layers (Fig. 9) clarifies the positive correlation with barley. The PCA ordination also demonstrates that some sites with records of rye pollen are characterised by oat and wheat rather than barley. Moreover, the Hove-Sørbø pollen samples are positively correlated with herbs representing agricultural land, in particular pasture, while samples from the sites characterised by oat and wheat are positively correlated with herbs representing natural heaths. These results may demonstrate the ecological resilience of rye, i.e. its ability to grow and flower at sites with varying local climatic and edaphic conditions.

Was the introduction of rye a reaction to stress or crisis?

As its name suggests, the Migration period was a transformative time across all of Europe. Migrating people, new trading networks and cultural influences are all possible explanations for why and how rye was introduced into south-western Norway. As an established and affluent farm, Hove-Sørbø would have had trading contacts extending to the south and east, where rye was a common crop in the 7th century AD (Grabowski 2011; Robinson et al. 2009). In a time of crisis or stress, rye could well have been imported, either for direct consumption or as seed corn. Rye tolerates low winter temperatures, giving it an advantage over the already established crops in a colder climate (Behre 1992). Climate data are inconclusive with respect to the effects of the 6th century dust veil on temperatures in northern Europe, but most researchers suggest an average temperature decrease of 1–4 degrees Celsius (Briffa et al. 1990; Sigl et al. 2015; Toohey et al. 2016). A study undertaken by Stamnes (2016) in Nord-Trøndelag, a region located 500 km north of Rogaland, includes calculations of how temperature changes during the dust veil event would have affected agriculture. The study concludes that a 1 degree drop in average temperature would have severely limited the areas where cereal crops could be successfully grown, while a 3.5 degree decrease would have made cereal cultivation impossible in the entire region. This research, though not directly applicable to south-western Norway, suggests that cereal cultivation may also have been impeded farther south.

In the case of a lingering cold after the dust veil had lifted, as has been suggested by several researchers (Büntgen et al. 2016; Larsen et al. 2008; Toohey et al. 2016), a hardier crop, such as rye, could have increased crop yields. Rye also demands fewer soil nutrients than, for example, barley and wheat (Behre 1992; Engelmark 1992), and if the population fell following an epidemic or a period of starvation, growing rye could have been advantageous, since fewer resources would have been required to manure the fields. A lack of manure, due to cold summers yielding insufficient fodder, would also have reinforced the advantageousness of this choice.

Conclusions

The large assemblage of charred rye grains presented in this article provides the earliest indication to date of rye cultivation in Norway. Rye was in domestic use at Hove-Sørbø during the 7th century AD, and the weed composition of the sample suggests that the crop was cultivated locally. Rye pollen is present in ancient field layers that have been radiocarbon dated to earlier and later than the rye grains, covering the time span from the Roman Iron Age to the Viking Age. The lack of contemporaneous records of rye pollen from the fields, matching the charred rye grains, complicates, but does not invalidate, the conclusion about the rye being the result of local cultivation.

Dramatic events in the 6th and 7th centuries could provide an explanation for the introduction of rye at this time. In the case of a deterioration in climate or a decline in population, the hardiness and low nutritional demands of rye would have been favourable qualities. The charred rye grains from Hove-Sørbø, though unique to date, may be followed by other, similar finds, indicating more widespread cultivation, or they may remain the only record, representing measures taken perhaps during a crisis, which were abandoned again when good times returned.

Acknowledgements

We are grateful to the Arkeologisk museum (archaeology museum) at the University of Stavanger for supplementary financial support for the development-led cultural heritage project. This publication is a contribution to the University of Stavanger research programmes *Prehistoric farming at the north European fringe – Interdisciplinary aspects* and *BEAM – Biological Environmental and Archaeological interdisciplinary research on life-course, Material and materiality in human depositions*, which financed some of the radiocarbon dates and the professional language revision. We appreciate the constructive suggestions for improvement made by two anonymous referees with reference to an earlier version of the manuscript, and we thank Even Bjørdal for valuable discussions regarding the archaeological structures at the Hove-Sørbø site.

References

Ahokas H (2009) The exported winter rye form evolved in co-cultivation with spring-sown seasonal crops in Finland where the slash-and-burn and riihi traditions selected against the rachis brittleness of the weedy rye. Interdisciplinary Biol, Agricult, Linguistics Antiquities (4): 1–153

Alenius T (2007) Environmental change and anthropogenic impact on lake sediments during the Holocene in the Finnish-Karelian inland area. Dissertation University of Helsinki

Arjava A (2005) The mystery cloud of the year 536 CE in the Mediterranean sources. Dumbarton Oaks Papers 59:73–94

Anderberg AL (1994) Atlas of seeds and small fruits of northwest-European plant species: Part 4. Resedaceae-Umbelliferaea. Swedish Museum of Natural History, Stockholm

Axboe M (1999) The year 536 and the Scandinavian gold hoards. Medieval Archaeol 43:186–188

Barnish SJB (ed & tr) (1992) The variae of Magnus Aurelius Cassiodorus senator. Liverpool University Press, Liverpool

Behre KE (1992) The history of rye cultivation in Europe. Veg Hist Archaeobot 1:141–156

Berggren G (1969) Atlas of seeds and small fruits of northwest-European plants species. Part 2. Cyperaceae. Swedish Natural Science Research Council, Stockholm

Berggren G (1981) Atlas of seeds and small fruits of northwest-European plants species. Part 3. Salicaceae-Cruciferae. Swedish Natural Science Research Council, Stockholm

Birkeland T (1981) The geology of Jæren and adjacent districts: a contribution to the Caledonian Nappe Tectonics of Rogaland, southwest Norway. Norwegian J Geology 61:213–235

Beug HJ (2004) Leitfaden der Pollenbestimmung für Mitteleuropa und angrenzende Gebiete. Pfeil, München

Bjørnstad Å (2012) Vårt daglige brød: Kornets kulturhistorie. Vidarforlaget, Oslo

Briffa KR, Bartholin TS, Eckstein D, Jones PD, Karlén W, Schweingruber FH, Zetterberg PJN (1990) A 1,400-year tree-ring record of summer temperatures in Fennoscandia. Nature 346:434–439

Büntgen U, Myglan VS, Charpentier Ljungqvist F, McCormich M, Di Cosmo N, Sigl M, Kirdyanov AV (2016) Cooling and societal change during the Late Antique Little Ice Age from 536 to around 660 AD. Nat Geosci 9:231–236

Børsheim RL, Soltvedt EC (2002) Gauselutgravningene 1997–2000. AmS-Varia 39. Arkeologisk museum, University of Stavanger, Stavanger

Cappers RTJ, Bekker RM, Jans JEA (2006) Digitale zadenatlas van Nederland (Digital seed atlas of the Netherlands). Barkhuis Publishing & Groningen University Library, Groningen

Engelmark R (1992) A review of the farming economy in south Scania based on botanical evidence. In: Larsson L et al (eds) The archaeology of the culture landscape: field work and research in a south Swedish rural region. Acta Archaeol Lundensia Ser in 4o 19:369–374

Fredh D, Prøsch-Danielsen L, Jensen CE (2018) A synthesis of pollen composition in prehistoric cultivation layers in south-western Norway. Environ Archaeol. https://doi.org/10.1080/14614103.2018.1536499

Fredriksen PD, Kristoffersen ES, Zimmerman U (2014) Innovation and collapse: bucket-shaped pottery and metalwork in the terminal Migration Period. Nor Archaeol Rev 47:119–140

Fægri K, Iversen J (1989) Textbook of pollen analysis, 4th edn. Wiley, New York

Grabowski R (2011) Changes in cereal cultivation during the Iron Age in southern Sweden: a compilation and interpretation of the archaeobotanical material. Veget Hist Archaeobot 20(5):479–494

Grikpėdis M, Motuzaite-Matuzevičiūtė G (2016) The beginnings of rye (*Secale cereale*) cultivation in the East Baltics. Veg Hist Archaeobot. https://doi.org/10.1007/s00334-016-0587-6

Grimm EC (2012) TILIA 1.7.16 Software. Springfield, IL: Research and Collection Center, Illinois State Museum

Harbeck M, Seifert L, Hänsch S, Wagner DM, Birdsell D, Parise KL, Wiechmann I, Scholz HC (2013) *Yersinia pestis* DNA from skeletal remains from the 6th century AD reveals insight into Justinian plague. PLoS Pathog. https://doi.org/10.1371/journal.ppat.1003349

Helbæk H (1955) The botany of the Vallhagar Iron Age field. In: Stenberger M (ed) Vallhagar: a Migration Period settlement on Gotland/Sweden. Part II. Munksgaard, Copenhagen, pp 653–699

Henriksen PS, Robinson, D (1996) Early Iron Age agriculture: archaeobotanical evidence from an underground granary at Overbygård in northern Jutland, Denmark. Veg Hist Archaeobot 5:1–11

Hjelle KL (2007) Foreign trade and local production: plant remains from medieval times in Norway. In: Karg S (ed) Medieval food traditions in northern Europe. National Museum of Denmark, Copenhagen, pp 161–180

Iversen F (2013) Big bang, lordship or inheritance? Changes in the settlement structure on the threshold of the Merovingian Period, south-eastern Norway. In: Klapste J (ed) Hierarchies in rural settlements. Brepols, Turnhout, pp 1–18

Iversen F (2016) Estate division: social cohesion in the aftermath of AD 536–7. In: Iversen F, Petersson H (eds) The agrarian life of the north 2000 BC–AD 1000: studies in rural settlement and farming in Norway. Cappelen Damm Akademisk/Nordic Open Access Scholarly Publishing, Oslo, pp 41–75

Jacomet S (2006) Identification of plant remains from archaeological sites, 2nd edn. Archaeobotanical Lab IPAS, Basel University, Basel

Korsmo E, Vidme T, Fykse H (1981) Korsmos ogräsplancher. LTs, Stockholm

Larsen LB, Vinther BM, Briffa KR, Melvin TM, Clausen HB, Jones PD ... Grudd H (2008) New ice core evidence for a volcanic cause of the AD 536 dust veil. Geophys Res Lett 35(4). https://doi.org/10.1029/2007GL032450

Leps J, Smilauer P (2003) Multivariate analysis of ecological data using CANOCO. Cambridge University Press, Cambridge

Little LK (2007) Life and afterlife of the first plague pandemic. In: Little LK (ed) Plague and the end of antiquity: the pandemic of 541–750. Cambridge University Press, Cambridge, pp 3–32

Mikkelsen PH (1999) Da vinterrugen kom til Sønderjylland. Særtryk af Sønderjyske Museer. Nordslesvigske Museer, ny række, bind 1, pp 35–41

Mikkelsen PH, Nørbach LC (2003) Drengsted: Bebyggelse, jernproduktion og agerbrug i yngre og ældre germansk jernalder. Jysk Arkæologisk Selskab, Moesgård

Moen A (1999) National atlas of Norway: vegetation. Norwegian Mapping Authority, Hønefoss

Moore PD, Webb JA, Collinson ME (1991) Pollen analysis. Blackwell Scientific, Oxford

Mossberg B, Stenberg L, Ericsson S (1992) Den nordiska floran. Wahlström & Widstrand, Stockholm

Myhre B (1991) Bosetning og politisk organisasjon i Vest-Norge før vikingtid. In: Joensen, JP, Johansen R, Kløvstad J (ed) Nordantlantiske foredrag. Seminar om nordatlantisk kulturforskning i Nordens hus på Færøerne: 27–30 August 1990. Norðurlandahúsið í Føroyum, Tórshavn

Myhre B (2002) Jorda blir levevei: 4000 f.Kr.–1350 e.Kr. Norges Landbrukshistorie 1. Samlaget, Oslo

Näsman U (2012) Comments on "An Iron Age shock doctrine: The 536–37 event as a trigger of large-scale social change in the Mälaren valley area" by Daniel Löwenborg. [referee comment later published by Uppsala Universitet]

Pedersen EA, Widgren M (1998) Järnålder, 500 f.Kr.–1000 e. Kr. In: Welinder S, Pedersen EA, Widgren M (eds) Jordbrukets första femtusen år: Natur och kultur. LTs, Uppsala, pp 239–454

Price N, Gräslund B (2015) Excavating the Fimbulwinter? Archaeology, geomythology and the climate event(s) of AD 536. In: Riede F (ed) Past vulnerability: volcanic eruptions and human vulnerability in traditional societies past and present. Aarhus University Press, Aarhus, pp 109–132

Prøsch-Danielsen L (1988) Principal components analysis of pollen types from prehistoric agricultural settlements at Forsandmoen, south-west Norway: correlations between cereals, herbs (weeds) and grasses. AmS-Skrifter 12:63–71

Prøsch-Danielsen L, Simonsen A (1988) Principal components analysis of pollen, charcoal and soil phosphate data as a tool in prehistoric land-use investigations at Forsandmoen, south-west Norway: correlations between cereals, herbs (weeds) and grasses. Nor Archaeol Rev 21(2):85–102

Prøsch-Danielsen L, Soltvedt E-C (2011) From saddle to rotary hand querns in south-western Norway and the corresponding crop plant assemblages. Acta Archaeologica 82:129–162

Regnell M (1997) Växtoffer: En förbisedd fyndkategori i Huslämningar. In: Karsten P (ed) Carpe Scaniam: Axplock ur Skånes förflutna. Riksantikvarieämbetet, Arkeologiska undersökningar, Skrifter 22, Riksantikvarieämbetet, Stockholm, pp 102–110

Regnell M (2002) Skånska järnålder skördar: Växtmakrofossil-analyser och odlingshistoriska iakttagelser från tolv boplatser. In: Carlie A (ed) Skånska regioner: Tusen år av kultur och samhälle i förändring. Riksantikvarieämbetet, Arkeologiska Undersökningar, Skrifter 40, Riksantikvarieämbetet, Stockholm, pp 25–50

Rigby E, Symonds M, Ward-Thompson D (2004) A comet impact in AD 536? Astron Geophys 45(1):1.23–1.26

Robinson DE (1993) Dyrkede planter fra Danmarks forhistorie: arkæologiske udgravninger i Danmark 1993. Det Arkæologiske Nævn: 20–39

Robinson DE (2003) Neolithic and Bronze Age agriculture in southern Scandinavia – Recent archaeobotanical evidence from Denmark. Environ Archaeol 8:145–165

Robinson DE, Hambro Mikkelsen P, Malmros C (2009) Agerbrug, driftsformer og planteressourcer i jernalder og vikingetid (500 f.Kr.–1100 e.Kr.). In: Odgaard B, Rydén Rømer J (eds) Danske landbrugslandskaber gennem 2000 år: Fra digevoldinger til støtteordninger. Aarhus Universitetsforlag, pp. 117–142

Roemer T (1932) Über die Reichweite des Pollens beim Roggen. Zeitschrift für Züchtung (Reihe A: Pflanzenzüchtung) 17:14–35

Rosen W (2006) Justinian's flea: plague, empire and the birth of Europe. Pimlico, London

Rødsrud CL (2016) Why did pottery production cease in Norway during the transition to the late Iron Age? In: Iversen F, Petersson H (eds) The agrarian life of the north 2000 BC–AD 1000: studies in rural settlement and farming in Norway. Cappelen Damm Akademisk/Nordic Open Access Scholarly Publishing, Oslo, pp 77–92

Sandvik P (1999) Analysar av plantemakrofossilar i jordprøver frå Sørbøtunet, Skei gnr. 43 og Sørbø gnr. 45, bnr. 1 og 2–4, Sandnes kommune, Rogaland. AmS Oppdragsrapport 1999/2, Arkeologisk museum, University of Stavanger, Stavanger

Sandvik PU (2018) Tusenårsstedet Stavanger torg: Naturvitskaplege undersøkingar. AM Oppdragsrapport 2018/17, Arkeologisk museum, University of Stavanger

Sigl M, Winstrup M, McConnell J, Welten K, Plunkett G, Ludlow F ... Dahl-Jensen D (2015) Timing and climate forcing of volcanic eruptions for the past 2,500 years. Nature 523: 543–549

Soininen AM (1974) Vanha maataloutemme: Maatalous ja maatalousväestö Suomessa perinnäisen maatalouden loppukaudella 1720-luvulta 1870-luvulle (in Finnish with English summary). Historiallisia tutkimuksia 96. Suomen Historiallinen Seura, Helsinki

Solantie R (1988) Climatic conditions for the cultivation of rye with reference to the history of settlement in Finland. Fennoscandia Archaeologica 5:3–20

Solberg B (2000) Jernalderen i Norge: Ca. 500 f. Kr.–1030 e. Kr. Cappelen Akademiske Forlag, Oslo

Soltvedt EC, Enevold R (2009) Naturvitenskapelige undersøkelser i spor etter jernalderbebyggelsen på Tastarustå. Øvre Tasta gnr. 28, bnr. 54, 63, 11 og 26, Stavanger kommune. AM Oppdragsrapport 2009/29, Arkeologisk museum, University of Stavanger, Stavanger

Soltvedt EC, Jensen CE (2011) Makrosubfossil: Og pollenanalyser fra forhistoriske åkerlag, hustomter og graver på Tjora. Tjora gnr. 10 bnr. 5, 17 og 19, Sola kommune. AM Oppdragsrapport 2011/24, Arkeologisk museum, University of Stavanger, Stavanger

Soltvedt EC, Løken T, Prøsch-Danielsen L, Børsheim RL, Oma K (2007) Bøndene på Kvålehodlene. Boplass-, jordbruks- og landskapsutvikling gjennom 6000 år på Jæren, SV Norge. AmS-Varia 47. Arkeologisk museum, University of Stavanger, Stavanger

Stabbetorp H (2018) Dyrkingsomfang og avling i kornproduksjonen. In: Strand E (ed) Jord- og Plantekultur 2018. Forsøk i korn, olje- og proteinvekster, engfrøavl og potet 2017. NIBIO Bok 4(1) Norwegian Institute of Bioeconomy Research (NIBIO), Ås, Norway, pp 14–25

Stathakopoulos D (2007) Crime and punishment: The plague in the Byzantine Empire 541–749. In: Little LK (ed) Plague and the end of antiquity: the pandemic of 541–750. Cambridge University Press, Cambridge

Stamnes AA (2016) Effects of temperature change on iron age cereal production and settlement patterns in mid-Norway. In: Iversen F, Petersson H (eds) The agrarian life of the north 2000 BC–AD 1000: studies in rural settlement and farming in Norway. Cappelen Damm Akademisk/Nordic Open Access Scholarly Publishing, Oslo, pp 27–39.

Stenseth NC, Samia NI, Viljugrein H, Kausrud KL, Begon M, Davis S ... Ageyev VS (2006) Plague dynamics are driven by climate variation. PNAS 103(35):13110–13115

ter Braak CJF, Prentice IC (1988) A theory of gradient analysis. Adv Ecol Res 18:271–313

Toohey M, Krüger K, Sigl M, Stordal F, Svensen H (2016) Climatic and societal impacts of a volcanic double event at the dawn of the Middle Ages. Climatic Change 136(3–4):401–412

Thy P, Willcox G, Barfod GH, Fuller DQ (2015) Anthropogenic origin of siliceous scoria droplets from Pleistocene and Holocene archaeological sites in northern Syria. J Archaeol Sci 54:193–209

Viklund K (1998) Cereals, weeds and crop processing in iron age Sweden: methodological and interpretative aspects of archaeobotanical evidence. Archaeology and Environment 14. Department of Archaeology, Umeå University, Umeå

Wagner DM, Klunk J, Harbeck M, Devault A, Waglechner N, Sahl JW ... Lumibao C (2014) *Yersinia pestis* and the Plague of Justinian 541–543 AD: a genomic analysis. The Lancet Infec Dis 14(4):319–326

Welinder S, Pedersen EA, Widgren M (2004) Jordbrukets första femtusen år: 4000 f.Kr.–1000 e.Kr. Natur och Kultur/LTs företag, Uppsala

Widgren M (2013) Climate and causation in the Swedish Iron Age: learning from the present to understand the past. Geografisk Tidsskrift–Danish Journal of Geography 112(2):126–134

The development of plant use and cultivation in the Sukhona basin, north-west Russian taiga zone

Wiebke Kirleis[1], Magdalena Wieckowska-Lüth[1], Henny Piezonka[1], Nadezhda Nedomolkina[2], Sebastian Lorenz[3], Vanessa Elberfeld[1], Jens Schneeweiß[4]

Author addresses:
[1] Institute for Prehistoric and Protohistoric Archaeology, Christian Albrechts University Kiel, Germany
[2] Vologda State Museum-Preserve of History, Architecture and Decorative Arts, Vologda, Russia
[3] Institute for Geography and Geology, Greifswald University, Germany
[4] Centre for Baltic and Scandinavian Archaeology, Schleswig, Germany

E-mail: wiebke.kirleis@ufg.uni-kiel.de

Abstract

The Sukhona basin in the north-west Russian taiga zone is an aquatic ecosystem providing excellent conditions to explore limnic resources from prehistory to Modern times. Concerning the use of food plants, there is potential for the gathering of wild fruits and seeds. Modern crop growing is restricted to local subsistence and is based on a limited number of species (such as rye, barley, oat and potato) that can cope with the short vegetation period and poor soil conditions. Up until now, central questions concerning plant use and cultivation have remained open: When were plant domesticates introduced into the area? Which further edible plants have supplemented the diet in various periods of pre- and protohistory? To answer these questions, archaeobotanical on-site and near-site investigations were carried out on the multi-period archaeological site of Veksa, oblast' Vologda, within the frame of a research project at the Kiel Graduate School 'Human Development in Landscapes' and Greifswald University in collaboration with Vologda State Museum, Russia, which was funded by the German Research Foundation (DFG). The site of Veksa is characterised by a unique sequence of archaeological layers comprising all periods from the Early Neolithic (Ceramic Mesolithic in European terminology) of the 6th millennium BCE through to the medieval period, including preserved post-built structures of the Late Neolithic. The results from pollen analyses of a near-site sediment profile and an on-site archaeological stratigraphy provide evidence for cereal cultivation most probably not before the medieval period, thus suggesting a late transition to agriculture in this area. Macrobotanical finds of charred water lily seeds indicate that the prehistoric fishers and hunters at Veksa supplemented their animal-based diet with wild plant resources. Domesticates occur in the Early Historic cultural layers, having been radiocarbon dated to the 5th–6th century CE. As the evidence consists of only three single charred grains from hulled barley and free-threshing wheat, cultivation cannot be proven for this period, but cereals were present and consumed by that time in the Russian north.

Introduction

The timing and nature of the onset of agriculture in the north-western taiga zone is still an unsolved research question. There is no archaeological evidence for the assumption that the Final Neolithic Fatyanovo group triggered the introduction of crop growing and herding in the northern forest zone, as suggested e.g. by Bashenkin (1997). The only available prehistoric evidence of domestic animal occurrence are horse teeth from the Bronze Age layers of Veksa, in the Sukhona region. It is only for medieval times that finds of agricultural tools are reported (Nikitinski 1992; Nedomolkina 2007), representing the earliest reliable, albeit indirect, indication of farming in this region.

Farther south-west, initial hints of animal husbandry in the west Russian Dvina-Lovat' region date to the 3rd millennium BCE. Bones of domestic animals from pile dwelling settlements contribute 14 per cent to the overall spectrum of faunal remains. These are attributed to the Zhizhitsian and the North Byelorussian cultures, the latter of which shows influences from the European Corded Ware horizon (Arslanov et al. 2009, 118). Evidence of domesticated animals other than dog from earlier prehistoric phases in this region at best encompasses small amounts of specimens that may be linked to imported animals rather than to local husbandry. At the Early Neolithic settlement of Serteya X (7th millennium BCE), dog is the only domestic animal present, and cereal cultiva-

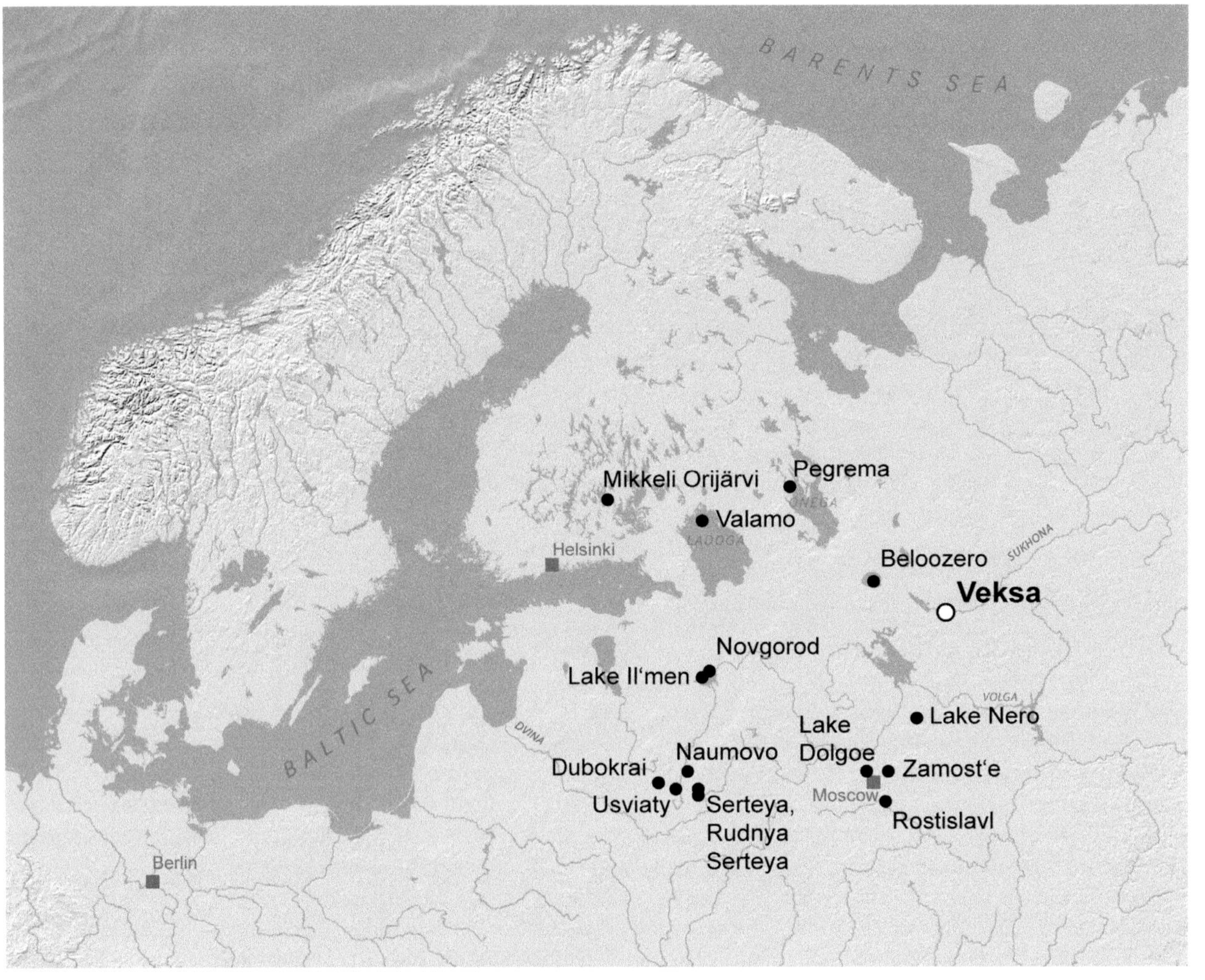

Figure 1. Map showing sites mentioned in the text (map: Henny Piezonka).

tion is not shown in the pollen diagram from nearby Rudnya-Serteya (Dolukhanov et al. 1989). From five Middle Neolithic sites in the area dating to the late 5th and the 4th millennium BCE, only 1 per cent of the mammal bones have been attributed to domestic animals. Bones of dog, horse and cattle stem from the three sites of Usviaty 'B', Serteya II and Serteya XI. Late Neolithic material of the 3rd millennium BCE from Usviaty IV 'A', Naumovo 'A' and Usviaty 'alpha' contains a larger portion of domesticates, reaching an overall average of 10 per cent within the faunal assemblage (Mazurkevich et al. 2009).

With respect to crop growing, the 4.6 m deep pollen sequence of Serteya II suggests that the sporadic onset of the cereal pollen curve is connected to the late phase of pile dwellings of the Late Neolithic Usviaty culture (3rd millennium BCE). The following Early Bronze Age Zhizhitsian culture, which continued the pile dwelling tradition, shows a marked increase in the cereal pollen ratios, indicating the increasing importance of agriculture towards the end of the 3rd millennium BCE. However, the time-depth model for the sequence remains vague. Wooden agricultural implements were recovered from Usviaty IV (hoe) and Dubokrai I (plough) (Arslanov et al. 2009, 119; Mazurkevich and Dolbunova 2011, 161ff, fig 13.9). A pollen diagram derived from a sediment sequence from Lake Myshetskoe-Dolgoe, in the Moscow region (Kremenetski et al. 2000), shows a continuous record for cereal pollen for the past 2000 years, whereas a sequence from Lake Nero, in the Rostov region, about 200 km north-east from Moscow, does not show evidence for rye growing before 1500 CE (Wohlfarth et al. 2006).

Farther north, at Lake Il'men, the oldest pollen signals for cereals date to the late 2nd millennium BCE (Dolukhanov et al. 2009, 134). Recent data from a sequence of the eastern shore of Lake Il'men indicate agricultural activities for 1250–1100 BCE (Yeremeyev et al. 2007). From older investigations in the Riberik mire, near Novgorod, indications for agriculture are inferred for stratigraphic levels dating to 3370–2200, 2300–1700 and 600–150 BCE (Königsson et al. 1997). Based on these results, it has been suggested that the transition to agriculture was driven by a general fall in the water-level of the Ladoga-Volkhov-Il'men hydrological system, which provided wide, low-lying, fertile soils, thus stimulating a rapid expansion of agriculture in the late 2nd millennium BCE (Dolukhanov et al. 2009).

For the Sukhona region, farther to the north-east (Fig. 1), so far, no data have been published on the beginning of agriculture, and thus the question of the chronological and cultural setting of this process has remained open. In this chapter, we present macrobotanical and palynological evidence from Veksa, shedding new light on the environmental setting, economic developments and cultural background of plant use and cultivation in prehistoric and historic times.

Environmental setting

The archaeological complex of Veksa is situated within the wide-stretched basin of the Late Weichselian proglacial Lake Sukhona, 400 km north of Moscow (cf. Kvasov 1979). Nowadays this basin comprises the River Sukhona, the River Vologda and many smaller waterways, including the small River Veksa, which originates from the shallow Lake Molotov. Meanders of the latter two rivers came closer to one another, until River Vologda tangentially eroded the banks of the River Veksa and created its current confluence. The area of Lake Molotov has been severely hydromeliorated since the 1980s and is now temporarily desiccated.

The study area belongs to the taiga biome, consisting of western boreal spruce forests, partly with pine, locally with birch, alder or mixed forests (Federal Agency for Nature Conservation (BfN): EuroVegMap 2.0). Typical soils for the hydromorphic areas of the Sukhona basin are gleysols, stagnosols and histosols, while in the better-drained, surrounding morainic areas albeluvisols, podzoluvisols and cambisols are present (Stolbovoi 2000, 17–18; Shahgedanova 2003, 111–112; Zech et al. 2014, 10–11). In general, the soils show tendencies of lacking nutrients, of acidic conditions and of stagnant water, which favours forests and meadows, but is less suitable for arable fields (Zech et al. 2014, 16–21). The Sukhona region is characterised by a humid boreal climate (Dfb sensu Köppen-Geiger classification). Today, the mean July and January temperatures reach peaks of +18 and −18°C, respectively (Gey et al. 2001; Tarasov et al. 1999), while the average annual temperature of the area is 2.4°C. The annual precipitation averages 574 mm (Vologda station, climate-data.org). During summer, high temperatures and dense vegetation cause strong evaporation with low water levels, while snow and a late thaw induce high water levels in springtime. The level surface of the Sukhona basin and its fine-grained sediment inventory strongly influence the recent hydrography, with low flow velocities and the suspension of enormous loads in the rivers. The humid boreal climate determines large annual water level fluctuations of 3–4 m that, together with the alluvial sedimentation, induce strong channel dynamics and have created a dense network of active rivers, backwaters and shallow lakes.

The multi-period archaeological site of Veksa, oblast' Vologda: A diachronic reference site for the north-east European forest zone

The multi-period site of Veksa is a key location with regard to the cultural development of north-eastern Europe. The cultural layers comprise all periods from the Early Neolithic (Ceramic Mesolithic in European terminology) of the 6th millennium BCE through to the medieval period, including an extensive concentration of wooden piles of the Late Neolithic and remains of a large Iron Age settlement (Nedomolkina 2000; Nedomolkina 2004; Nedomolkina 2007; Nedomolkina, Piezonka 2016). Veksa thus represents a crucial site in the European taiga zone for the diachronic assessment of human–environment interactions, including the introduction of agriculture (Piezonka et al. 2020).

Recent excavations

The archaeological remains extend along the northern bank of the River Vologda on both sides of the small tributary the Veksa, with the sites of Veksa 1 and 3 (Fig. 2). The cultural layers indicate seasonal riverside settlements and have been excavated since the 1990s (Nedomolkina 2000). During several seasons of Russian-German fieldwork between 2011 and 2016, systematic sampling for archaeobotanical on-site investigations was carried out, supplemented by extensive drilling to assess the geomorphological situation and reconstruct the vegetation history (Lorenz et al. 2014; Nedomolkina et al. 2015b). Two archaeological trenches provide information on the past plant economy (Veksa 1, trench 2016; Veksa 3, trench 1). A third trench (Veksa 3, trench 2) on the modern-day riverbank in the area of the c. 1800 wooden piles from around 3000 cal BCE (Nedomolkina et al. 2015a; Piezonka 2015) only allows reconstruction of the natural shoreline environment, not of the past plant economy.

Anthropogenic indicators in the pollen records from Veksa

One on-site and one near-site pollen record from Veksa contribute to our understanding of the development of plant cultivation in the north-west Russian taiga zone (Fig. 2).

Material and methods

The near-site sediment core VKS 16, of about 4 m, was obtained from an ancient drainage channel of Lake Molotov at a distance of about 200 m from the excavation trenches, using a percussion corer with

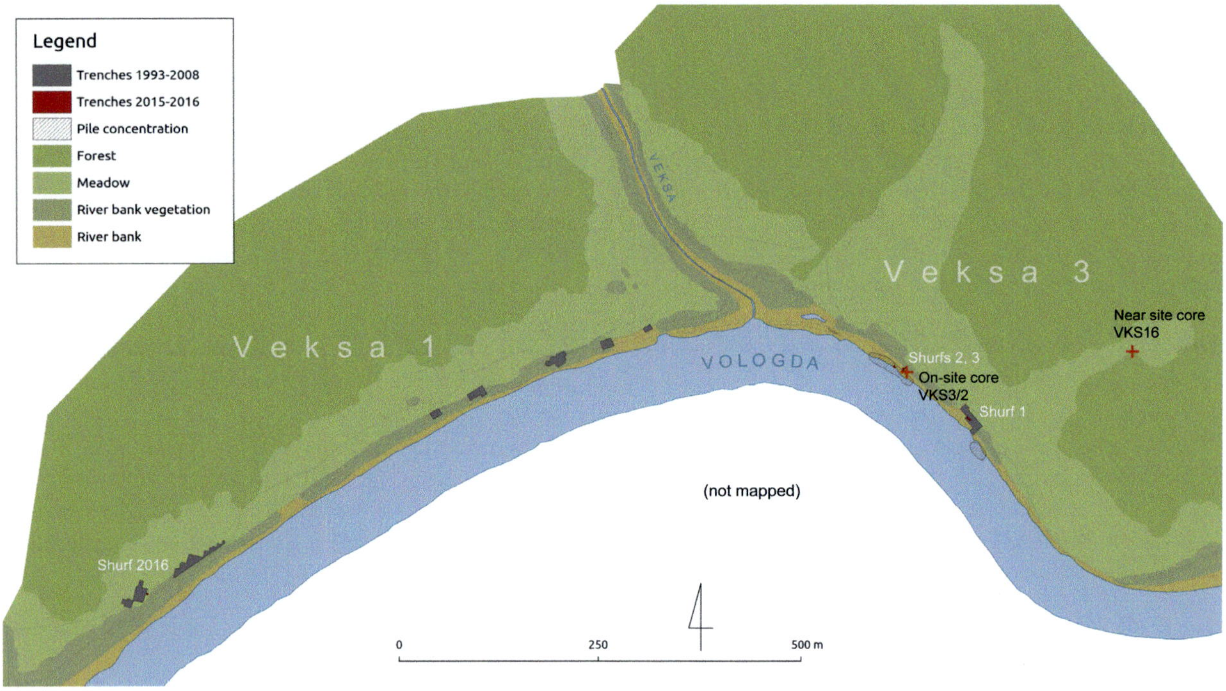

Figure 2. Map of Veksa, Vologda province, Russia, showing excavation trenches ('shurfs') and near-site pollen core VKS 16 on the northern bank of the Vologda River (illustration: Christoph Engel, Nadezhda Nedomolkina, Henny Piezonka).

Table 1. Radiocarbon dates from the VKS16 core (calibration: OxCal, Bronk Ramsey 2013; IntCal13, Reimer et al. 2013). Poz = Poznań; C = carbon.

Lab code and number	Depth (cm)	Material dated	^{14}C age BP	Calibrated age, 2σ range	Notes
Poz-82632	16–19	pollen	1075 ± 35	893–1020 CE	
Poz-82633	68–71	micro-charcoal	18 390 ± 160	20 605–19 909 BCE	small total organic content: 0.5 mg C
Poz-82634	172	wood	5920 ± 40	4904–4709 BCE	
Poz-82635	257	wood	6240 ± 40	5311–5066 BCE	
Poz-82636	318–324	micro-charcoal	18 790 ± 170	21 127–20 396 BCE	small total organic content: 0.18 mg C
Poz-82637	392–398	micro-charcoal	14 720 ± 80	16 186–15 724 BCE	

piston and liner. Accurate age determination of the fossilised drainage channel is difficult due to repeated alluvial influence during spring floods. Nevertheless, this location is the only hydromorphic and therefore wet depression that provides sediments with suitable, albeit poor, pollen preservation. The sediment of the core consisted exclusively of alluvial loam. In order to trace pollen signals of human activity in the course of the Holocene, sediment samples were taken at 16–20 cm intervals and, in the uppermost section, at 4 cm intervals. Poor preservation conditions and the large amount of time required for analysis explain the low temporal resolution of the VKS 16 core, which is based on 29 pollen spectra.

To provide a chronological frame for the near-site pollen record, four samples consisting of pollen and micro-charcoal concentrations from bulk samples were used for radiocarbon dating. Two samples included wood remains. The AMS radiocarbon dating was conducted at the Poznań Radiocarbon Laboratory (Table 1). Calibration to the calendar timescale was carried out using the OxCal v.4.2.3 calibration software (Bronk Ramsey 2013) and the IntCal13 calibration dataset (Reimer et al. 2013). The calibration results used for the age-depth model are given as mean values of the calibrated ages within the 2σ range. Three samples provided ages that were significantly too old; two of these are affected by very low values of total carbon content. Thus, the chronology of VKS16 (Fig. 3) is based on three AMS radiocarbon dates and is complemented by a biostratigraphical classification according to Khotinskiy (1984) and Kuosmanen et al. (2016).

The assignment to archaeological periods follows the information on the site stratigraphy with its specific cultural horizons (Nedomolkina 2000; Piezonka et al. 2017). The sedimentation rates are non-linear. Thus, the Neolithic is well represented, with 14 samples, while only four samples date to the Metal Ages, three to the medieval period and two to Early Modern times. Based on major changes in the arboreal pollen (AP) proportions, four local pollen assemblage zones, A–D,

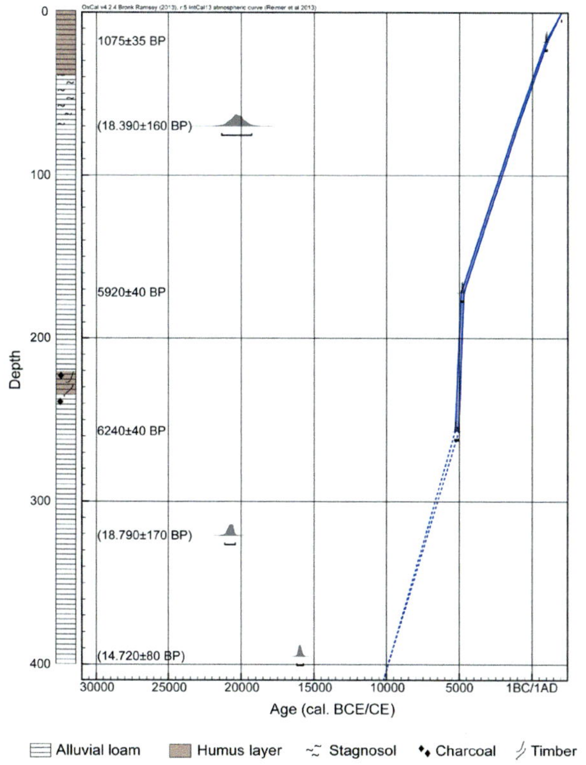

Figure 3. Age–depth model for the VKS16 core (OxCal, Bronk Ramsey 2013; IntCal13, Reimer et al. 2013) (illustration: Magdalena Wieckowska-Lüth)

were distinguished for VKS16 (Fig. 4) and further subdivided into subzones based on minor changes. VKS16-A is dated pollen-stratigraphically to the Boreal period due to the presence of the thermophilous taxa *Quercus*, *Ulmus*, *Tilia*, *Corylus* and *Alnus*. The following expansion of *Tilia* and *Ulmus*, along with the two ^{14}C dates, of 5311–5066 BCE (257 cm) and 4904–4709 BCE (172 cm), characterises VKS16-B as representing the Atlantic period. The spread of *Picea* during VKS16-C marks the Subboreal, whereas the declines in *Ulmus*, *Quercus*, *Tilia* and *Corylus*, together with the third ^{14}C date, of AD 893–1020 (16–19 cm), identify VKS16-D as the Subatlantic period. Due to this tentative chronology, only estimates for archaeological periods were able to be specified in the pollen diagram of the VKS16 profile.

From the excavation profile of Veksa 3, trench 2 (VKS3/2), a monolith was obtained using overlapping sample boxes (Fig. 5). Apart from the bottom section, which consisted of clayey palaeo-lake deposits, its sediment was composed of alluvial loam. In order to detect changes in on-site conditions during occupation phases, 15 sediment samples were taken for palynological analyses at distinctive stratigraphical transitions: two samples from layer 7, one from layer 6, one from the transition between layers 6 and 5b, three from layer 5b, five from layer 5a, two from layer 4, and one from layer 2. Five local pollen assemblage zones, A–E, were distinguished based on the different deposition layers (Fig. 6). Zones VKS3/2-A–B belong to the archaeological horizons of the late 4th and early to mid-3rd millennium BCE. The following zone, VKS3/2-C, represents the Bronze Age layers, and the uppermost zone, VKS3/2-D–E, reflects deposits of the recent past.

All samples were prepared for pollen analysis according to the standard techniques outlined by Moore et al. (1991). Pollen counts were carried out under 400× or, in ambiguous cases, 1000× magnification. For the identification of Cerealia-type pollen, phase-contrast microscopy was applied. On average, 350 arboreal pollen grains per sample were counted, due to low pollen concentrations. The nomenclature for pollen types follows Beug (2004) and that of spores, Moore et al. (1991). The calculation of pollen percentages is based on the total terrestrial pollen sum. Diagrams were produced with the programs C2 version 1.7.7 (Juggins 2014) and CountPol (Feeser, unpublished).

Results: Local woodland development and anthropogenic impact

During the Boreal period (VKS16-A), the forest around the site was quite open and consisted mainly of the conifers *Picea* and *Pinus,* with herbs, grasses, and dwarf-shrubs in the undergrowth, before *Betula* gained increasing importance (VKS16-A2). This light woodland on the higher terraces was admixed with other deciduous taxa, especially *Quercus* and *Corylus*, whereas in the lowlands, *Alnus* stands dominated, together with *Salix* and Cyperaceae. At the transition to the following Atlantic period, the woodland became increasingly dense (VKS16-B), having a negative impact on the light-demanding *Corylus*, in parallel with grasses, herbs and sedges. Nonetheless, *Sorbus/Rubus*-type, *Viburnum opulus*-type, *Sambucus nigra*-type and *Rosa* still indicate small-scale openings in the woodland that may be linked to human activity. A new type of forest developed during the Early Neolithic, around 5600 BCE. The appearance of *Ulmus* in the near-site pollen record indicates the spread of elm in moist areas, probably adjacent to the drainage channel. In contrast, local stands of *Tilia* were probably restricted to higher-lying dry sites. Nevertheless, *Betula* still dominates by far the surrounding forest inventory, suggesting frequent succession stages within the lowland woodland. Probably during the Middle Neolithic, around 4000 BCE, the forest changes again. Now, a gradual reduction in the proportions of *Betula* occurs, whereas *Corylus* increases (VKS16-B3), indicating light woodland on the higher terraces and less forest succession stages in the lowlands. During the course of the Middle Neolithic, there is a slight rise in the open-land vegetation (VKS16-B3). In particular, pollen from *Artemisia* is found continuously. Although there is no direct indication for it, changes in the vegetation cover in this period may have been driven by anthropogenic impact.

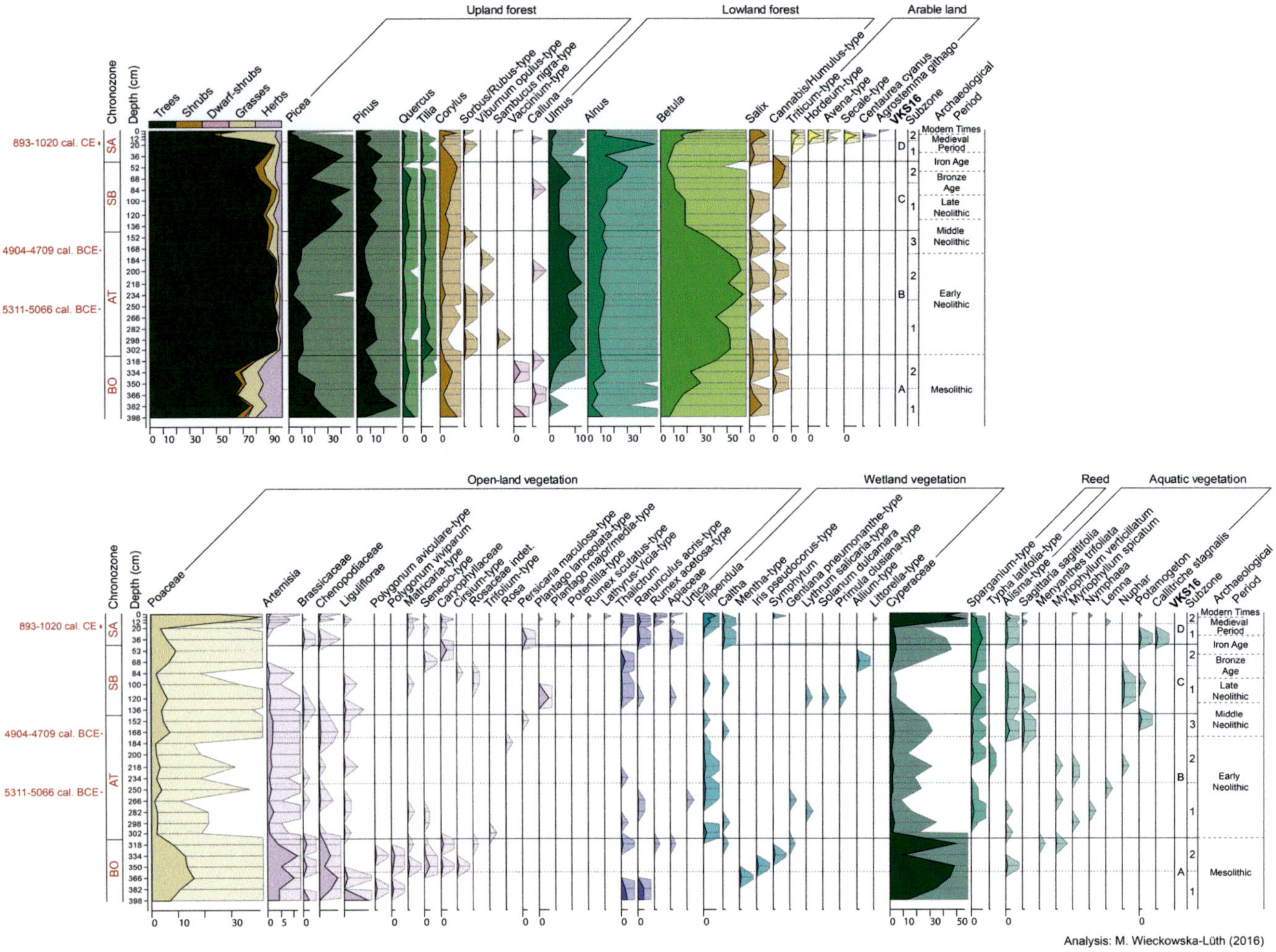

Figure 4. Simplified percentage diagram for Veksa 16 (VKS16), showing selected pollen taxa. Percentages are based on the sum of terrestrial pollen (AP+NAP) excluding wetland and aquatic pollen types. Curves are exaggerated by a factor 10 (marked as shaded). The red bars mark the position of radiocarbon dates (from bottom to top, calibrated ages; see Table 1): 5311–5066 BCE, 4904–4709 BCE and 893–1020 CE. Estimates for archaeological periods according to Russian terminology are shown in dashed lines (analysis and illustration: Magdalena Wieckowska-Lüth).

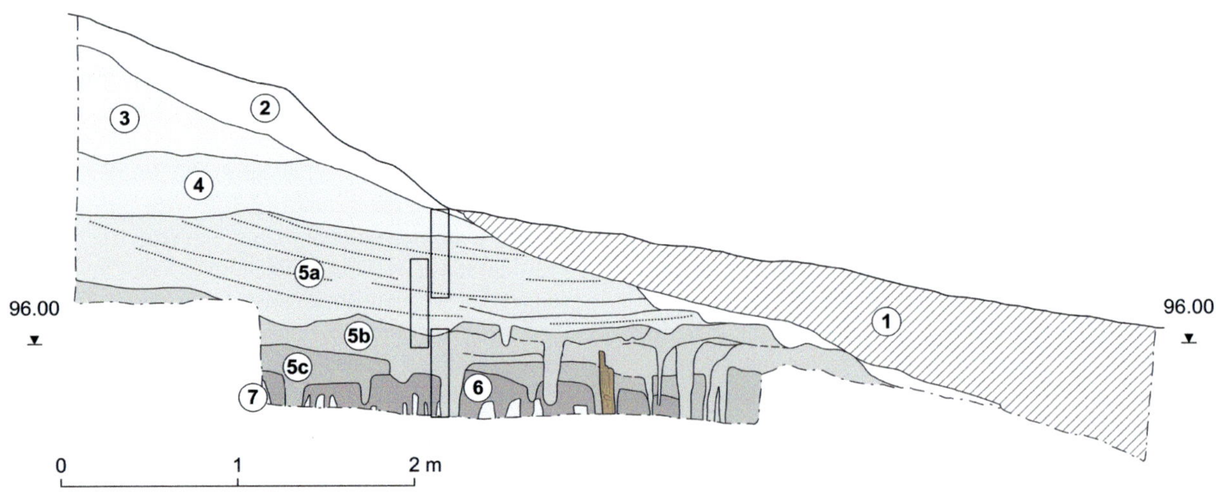

Figure 5. Veksa 3, trench 2. Stratigraphy of eastern section wall. 1 layer 1, spoil heap of river dredging works in 2014; 2 layer 2, topsoil before dredging works in 2014; 3 Bronze Age layer 3 (c. 1500–1000 cal BCE); 4 Bronze Age layer 4 (c. 1500–1000 cal BCE); 5a Early Bronze Age layer 5a (c. 2500–2000 cal BCE); 5b, c Late Neolithic layers, later phase of pile settings (c. 3000–2750 cal BCE) with wooden post in situ; 6 Late Neolithic layer, early phase of pile settings (around 3000 cal BCE); 7 paleo-lake clay sediments layer. The boxes indicate the locations of metal box soil samples. The heights are relative to the local, on-site system (illustration: Christoph Engel, Henny Piezonka).

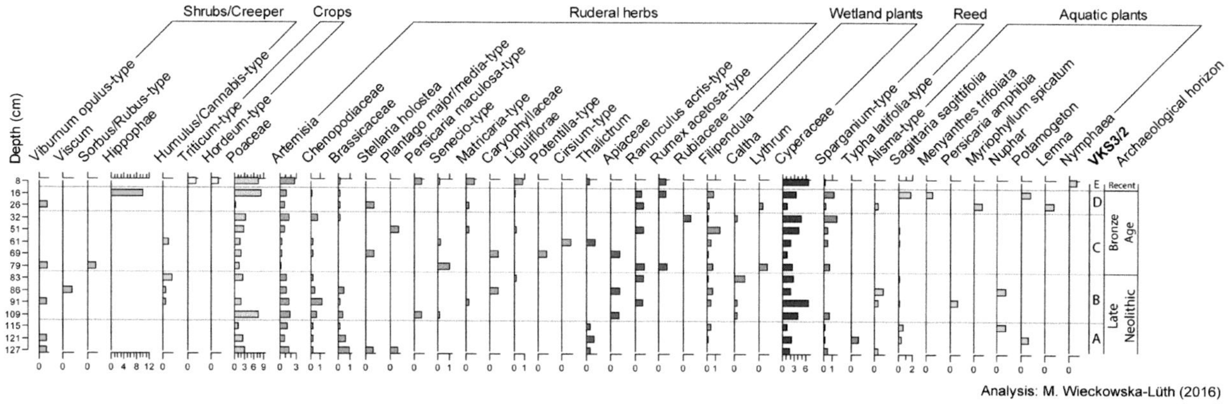

Figure 6. Simplified percentage diagram for Veksa 3, trench 2 (VKS3/2), showing the bars of selected pollen taxa. Percentages are based on the sum of terrestrial pollen (AP+NAP) excluding wetland and aquatic pollen types (analysis and illustration: Magdalena Wieckowska-Lüth).

During the following stage of the Subboreal, the decline in *Ulmus* suggests a reduction in moist locations around the study site. At the same time, the conifers gain in significance in the surroundings, along with *Quercus, Corylus* (VKS16-C) and *Viburnum opulus*–type (VKS3/2-A–B). In the course of this period, the amount of *Alnus* increases distinctly (VKS16-C2), suggesting the colonisation of the dried-up channel with alder stands. The synchronous rise in *Humulus/ Cannabis*-type demonstrates its increasing local presence. The on-site occurrence of this taxon is also verified in the Late Neolithic horizons, along with that of *Viscum* (VKS3/2-A–B). There is little change within the open-land vegetation (VKS16-C1). Probably during the Late Neolithic, the *Plantago lanceolata*–type—a taxon characteristic of open grazed land, fallow land, and trampled vegetation (Düll and Kutzelnigg, 2005)—occurs temporarily in the near-site pollen record, in relatively high proportions Further ruderals in the on-site pollen record, including *Plantago major/media*–type, *Persicaria maculosa*–type and *Stellaria holostea* (VKS3/2-A–B), emerge contemporaneously with the pile constructions and thus may hint at human activity in the area. Since the beginning of the Subatlantic period, the deciduous trees *Ulmus, Quercus,* and *Tilia* become rare (VKS16-D), pointing to extensive woodland clearance, in particular on the higher terraces. In the lowlands, an *Alnus* carr forest develops (VKS16-D1) that is, however, later largely replaced by *Salix* shrubbery (VKS16-D2). Synchronously, increasing proportions of the light-demanding *Betula, Corylus* and *Sorbus/ Rubus*-type may be indicative of increased human activities also on the lower terraces. In the uppermost part of the near-site pollen record, a considerable change in the landscape is also shown by rising quantities of Poaceae and an increasing diversity of ruderal herbs of higher locations (VKS16-D). Of particular importance is the occurrence of cereal pollen grains, pointing to the presence of arable fields close to the site, probably from the early medieval period onwards. Initial evidence of *Triticum*-type is followed a little later by pollen of other cereals, including *Hordeum*-type, *Avena*-type and *Secale* (VKS16-D2 and VKS3/2-E). The concurrent presence of pollen from *Centaurea cyanus* and *Agrostemma githago* supports the indication for wheat and rye.

Macrobotanical investigations into plant use at the sites of Veksa 3 and Veksa 1

The macrobotanical investigations carried out at Veksa 3 focus on the Early Neolithic (EN) and the Late Iron Age to Early Modern periods. The charred remains from Veksa 1 comprise only seeds and fruits from the 1st millennium CE. The taxonomic nomenclature for the macrobotanical remains follows Erhardt et al. (2002).

The charred plant remains from Veksa 3

Altogether, 96 samples from the site of Veksa 3, trench 1, were processed by flotation with meshes of 300 µm width to extract the charred plant remains (Table 2). A total of 42 out of 96 soil samples contained charred seeds, fruits and vegetative remains. The low find densities reach in average of 3 per litre. Altogether, 313 charred plant remains were identified, 245 (78%) of which originate from the EN and the Early to Middle Neolithic (MN) transition layer, covering the second half of the 6th–beginning of the 5th millennium BCE (Table 2). The MN and the Eneolithic layers did not contain any charred plant remains.

With 140 specimens, making up 57 per cent, more than half of the charred seeds could be identified as *Nuphar lutea* (Fig. 7). Further seeds belong to *Persicaria lapathifolia*, *P. maculosa* and to Cyperaceae, *Poa annua*, Rosaceae, Poaceae p.p. and Chenopodiaceae. In addition, there are single finds e.g. from raspberry and cranberry bush. Ubiquity values show the great importance of water lily, followed by curly-top knotweed and lady's thumb. Absolute dates have

Figure 7. Early Neolithic charred seeds of yellow water lily (*Nuphar lutea*) from Veksa 3, trench 1 (photos: Agnes Heitmann, Wiebke Kirleis).

Table 2. The charred plant remains from the Early to Middle Neolithic layers from Veksa 3, trench 1. Nomenclature according to Erhardt et al. (2002).

	Number	Proportion	Ubiquity	
Overall sum	245	100%	37	**Overall sum**
Aquatic plants				**Aquatic plants**
Nuphar lutea	140	57.1%	54.1%	Yellow water lily
Potamogeton sp.	1	0.4%	2.7%	Pond weed
Fruits				**Fruits**
cf. *Rubus idaeus*	1	0.4%	2.7%	Raspberry
Viburnum opulus	1	0.4%	2.7%	Cranberry bush
Herbs				**Herbs**
Persicaria lapathifolia	32	13.1%	27.0%	Curlytop knotweed
Persicaria maculosa	12	4.9%	21.6%	Lady's thumb
Poa annua	2	0.8%	5.4%	Annual bluegrass
cf. *Bolboschoenus*	1	0.4%	2.7%	Bulrush
Chenopodiaceae	5	2.0%	10.8%	Goosefoot family
Cyperaceae	12	4.9%	13.5%	Sedge family
Poaceae	5	2.0%	10.8%	Grass family
Rosaceae	4	1.6%	2.7%	Rose family
Polygonaceae	1	0.4%	2.7%	Knotweed family
Miscellaneous				**Miscellaneous**
Varia	3	1.2%	8.1%	Varia
Indet.	19	7.8%	29.7%	Indet.
Vegetative remains	6	2.5%	16.2%	Vegetative remains

been retrieved for *Nuphar* and *Polygonum lapatifolium/persicaria* from the EN layer and for *Polygonum lapatifolium/persicaria* from the EN–MN transition layer. In addition, modern seeds of *Nuphar lutea* from the River Vologda at Vologda city, c. 20 km west of Veksa, were dated to gain information on possible reservoir effects for this aquatic plant (Table 3). All archaeological samples date to the last quarter of the 6th millennium BCE. For the EN layer, the dates are in accordance with the chronology suggested by other ^{14}C dates and typology, although the small age offset of the *Nuphar* sample may point to a slight influence of a freshwater reservoir effect (Piezonka et al. 2017: 131–132; see also Kulkova et al. 2015). The sample from the EN–MN transition layer returned a date several hundred years older than that suggested by charcoal dates and typology (Piezonka et al. 2017: 131–132). It is possible that the dated seeds may have been relocated from the older layers, as postholes associated reach into the layers below.

Further archaeobotanical charred remains in the uppermost layer date to Early Modern times. Among these are *Malus* sp., Cyperaceae, Poaceae and Chenopodiaceae. Only a single find of cf. *Setaria* most probably is of Late Medieval origin (see raw data in Supplementary table 1, which is available online only, see Acknowledgements).

In general, the charred plant remains relate to the vegetation of disturbed places within a wetland area. They mainly belong to the ruderal and riverine flora or are aquatic plants. As the material is charred and originates from the cultural layers and specific ar-

Table 3. Radiocarbon dates on botanical macroremains from archaeological contexts from Veksa 1 and Veksa 3 and on modern seed of *Nuphar lutea* from the Vologda River in Vologda city. pMC percentage of modern carbon (calibration: Poz-84485, Poz-84486, Poz-97773, Poz-98846, Poz-92587: OxCal, Bronk Ramsey 2013; IntCal13, Reimer et al. 2013; Poz-92589: Bomb13NH1 Curve (Bomb13NH1.14c) Post-bomb atmospheric NH1 curve, Hua et al. 2013). pMC = percent modern carbon; Poz = Poznan; C = carbon.

Lab code and number	Context	Material dated	^{14}C age BP	Calibrated age, 2σ range	Notes
Poz-92589	Vologda, River Vologda	seed of *Nuphar lutea* plant growing in summer 2016	−59 ± 27	1953–1959 CE	pMC: 100.74 ± 0.34
Poz-97773	Veksa 1, trench 2016, layer 4	charred barley grain of *Hordeum vulgare*	1545 ± 35	423–589 CE	
Poz-98846	Veksa 1, trench 2016, layer 5, feature 2	charred wheat grain of *Triticum* sp.	1555 ± 35	418–579 CE	0.9 mg C
Poz-84486	Veksa 3, trench 1, layer 9, object 446, sample 96a	charred seeds of *Polygonum lapatifolium/persicaria*	6110 ± 40	5208–4942 BCE	
Poz-92587	Veksa 3, trench 1, layer 7, object 43, sample 46	charred seeds of *Polygonum lapatifolium/persicaria*	6170 ± 40	5208–5062 BCE	
Poz-84485	Veksa 3, trench 1, layer 9, object 446, sample 96b	charred seeds of *Nuphar lutea*	6240 ± 40	5311–5066 BCE	

Table 4. Charred plant remains dating to the 1st millennium CE from Veksa 1, trench 2016.

	Number	Proportion	
Overall sum	18	100%	**Overall sum**
Cereals			**Cereals**
Triticum sp.	1	5.56%	Wheat
Hordeum vulgare	2	11.11%	Barley
Weeds			**Weeds**
Galium sp.	1	5.56%	Cleavers
Chenopodiaceae	4	22.22%	Goosefoot family
Cyperaceae	2	11.11%	Sedge family
Poaceae	4	22.22%	Grass family
Miscellaneous			**Miscellaneous**
Indet.	3	16.67%	Indet.
Vegetative remains	1	5.56%	Vegetative remains

chaeological features, the deposition of these plant remains was linked with human activity. Several of the seeds and fruits are rich in starch and are edible, and they thus can be interpreted as originating from economic plants. Most prominent among these gathered seeds are those from yellow water lily (Fig. 7). Further evidence for plants possibly exploited by people includes seeds and fruits from cranberry bush, knotweed and the goosefoot family.

The charred plant remains from Veksa 1

Nine soil samples from the site Veksa 1, trench 2016, were processed by use of the flotation method with a mesh aperture of 300 µm width in order to retrieve the charred plant remains. Only four of these nine samples contained a total of 18 poorly preserved charred plant remains (Table 4).

One sample contained one half grain of charred *Triticum* sp. (Fig. 8) accompanied by two seeds of Chenopodiaceae and a tiny Poaceae. Another sample provided two charred grains of hulled barley (Fig. 9) with cracked surface structure that were accompanied by *Galium* sp. and nutlets of Cyperaceae and Poaceae. Further weed remains stem from the two neighbouring samples. According to the radiocarbon dates (Table 3), the cereal grains date to the 5th–6th century CE, which is older than expected based on the palynological results, but which does fit into the broad timeframe of the 1st millennium CE, as suggested based on pottery and metal finds.

Discussion: Past plant use in the north-west Russian taiga zone

The results from the archaeobotanical investigations at Veksa provide twofold insight into past plant use in the Upper Sukhona basin. First, the use of local wild plant resources, and in particular of yellow water lily, is attested since Early Neolithic times (Table 2).

Second, the use of cereals, which apparently are a late arrival in the 1st millennium CE, are indicated here for the first time by macrobotanical evidence.
Before looking into the details, a critical assessment of the data is necessary to discuss the limitations of the Veksa evidence. The near-site sediment sequence VKS 16 is affected by overall unfavourable preservation conditions for pollen and spores. Thus, pollen counts reach only 350 arboreal pollen grains/sample, and the temporal resolution of the sequence is hardly satisfying. From an archaeological point of view, the sites of Veksa 1 and Veksa 3 provide rich and diverse spatial, stratigraphic and material data on the Early Neolithic, the Metal Ages and the medieval period. However, archaeobotanical on-site studies on the plant economy have been more or less restricted to the Neolithic period and the 5th century CE onwards. This is due to preservation issues and the limited representation of certain archaeological horizons in the sampled trenches. On this basis, information on the plant economy and the onset of plant cultivation remains selective.

Figure 8. Charred wheat grain fragment with embryo, lateral and frontal views of the embryo, from Veksa 1, trench 2016, directly dated to the 5th–6th century CE (Poz-98846, see Table 4) (photos: Agnes Heitmann, Wiebke Kirleis).

The use of local wild plant resources

The pollen data from the near-site pollen profile VKS16 and the on-site profile VKS3/2 indicate a late onset of plant cultivation in the 1st millennium CE. Macrobotanical evidence for crop plants is present only from the 5th–6th century CE onwards. This said, the palynological and archaeobotanical investigations have shown that a wide range of potentially edible plants has been present in the vicinity of Veksa throughout time—plants that could have been gathered by fishers and hunters. For example, fruits of *Sambucus nigra, Sorbus*, *Rosa* and *Rubus* were abundant in the local woodland during the Early Neolithic, whereas in the Late Neolithic, the surrounding vegetation offered increased amounts of nuts of *Corylus* and *Quercus*, as well as numerous edible herbs. In addition, tubers and seeds from aquatic plants can be considered as dietary components, based on evidence from other archaeological sites (Berihuete 2018; Kubiak-Martens 2002; Warren et al. 2014; Wolters 2016; O'Neill and Rana 2016).

Nuphar as food plant: Examples from ethnography and archaeobotany

From ethnographical studies e.g. on Native North Americans, it is known that the starch-rich seeds and rhizomes of *Nuphar* species are a welcome dietary contribution. Harvesting is carried out by canoe in late summer or early autumn. The rhizomes are prepared similarly to other vegetables by boiling or roasting. Rhizome slices and seeds are torrefied, ground to powder and used to thicken soups and other dishes. Roasted seeds produce a variant of popcorn

Figure 9 Charred barley grains, dorsal and ventral views, from Veksa 1, trench 2016, directly dated to the 5th–6th century CE (Poz-97773, see Table 4) (photos: Agnes Heitmann, Wiebke Kirleis).

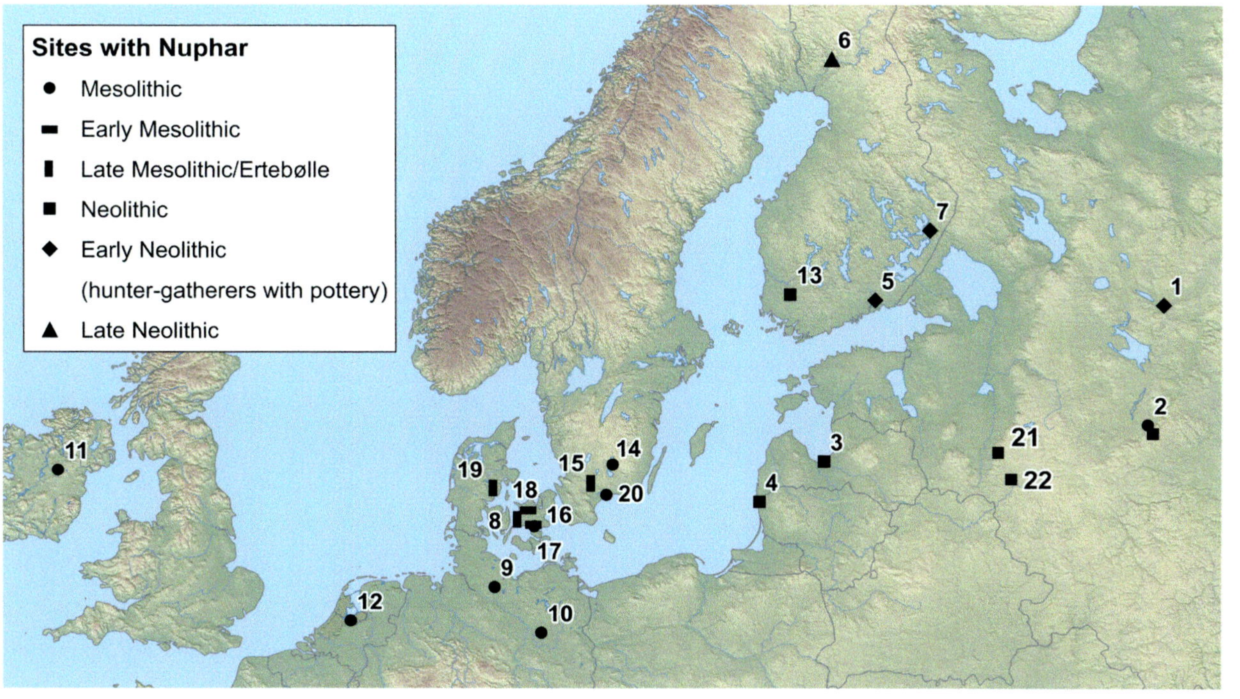

Figure 10. Map showing archaeological sites with macroremains of *Nuphar lutea* or, if specified in Table 5, *N. pumila*. Archaeological periods according to local terminology. Numbering of the sites follows that in Table 5 (map: Christoph Engel).

(Mears and Hillman 2007, 123, 126; Moerman 2009[8], 358–359).

Archaeobotanical investigations in Europe comprise several Mesolithic to Neolithic sites with evidence for *Nuphar* seeds, from Ireland to central Russia (Fig. 10, Table 5). Out of these, at least five show substantial amounts of more than one hundred to several thousands of charred and/or waterlogged seeds. In particular, carbonised seeds that originate from cultural layers and features interpreted as hearths indicate the status of these plant parts as food component. On a transect spanning from northern to central Europe, the northernmost site containing *Nuphar*, with 127 charred seeds, is the Late Neolithic site of Rovaniemi Koskenniska, in Finland, 3400–1500 BCE (Vanhanen 2014). Further Late Neolithic to Bronze Age sites in Finland and the Baltic show evidence for single charred seeds (Bērziņš et al. 2016; Grikpėdis and Motuzaite Matuzeviciute 2017; Lavento 1992; Pesonen 1993; Pesonen 1994; Vanhanen and Pesonen, 2016). The north-west Russian site of Veksa that is presented here yielded 140 charred seeds of yellow water lily dating to the 6th–5th millennium BCE. In central Russia, an extensive collection of charred and waterlogged seeds (>2000) was found at the Late Mesolithic–Middle Neolithic site of Zamostje 2, in the Moscow region (Berihuete-Azorín and Lozovskaya 2014; Berihuete 2018). From northern Germany, high numbers of waterlogged (5389) and charred (60) water lily seeds came to light at the Mesolithic site of Friesack 4, dating to 9100–5500 BCE (Wolters 2016). For southern Scandinavia, six Early–Late Mesolithic sites show evidence for waterlogged *Nuphar* seeds here listed in a semi-quantitative way (Regnell 1998). These are supplemented by finds from Halsskov, in Denmark (Kubiak-Martens 2002), and from Norje Sunnansund, in southern Sweden (Per Lagerås, pers. comm.). From early Swifterbant surface hearths of the Dutch site of Hoge Vaart (4900–4500 BCE), 182 charred seeds possibly indicate food processing (Brinkkemper et al. 1999). In Ireland, there are many sites that indicate the use of *Nuphar* as foodstuff. For example, the Mesolithic site of Lough Kinale provided several thousands of waterlogged seeds that have been interpreted as either storage finds or processing remains from food preparation (Warren et al. 2014).

Based on this evidence, it can be concluded that *Nuphar* seeds have been a common component in the diet of pre-agricultural communities throughout north-central and north-eastern Europe. This food tradition continued, at least in some of the regions, into the agricultural Neolithic, showing that knowledge on useful plants was bequeathed over centuries and millennia. Macrobotanical evidence for the use of *Nuphar* tubers as food has so far not been recorded in this area, which may be connected to problems of preservation.

The introduction of crops in the Sukhona basin inferred from the first palynological and macrobotanical data from Veksa

The new archaeobotanical investigations at Veksa provide the first evidence for the onset of crop cultivation in the north-western Russian taiga zone. The

Table 5. Archaeological sites with evidence of *Nuphar lutea* or, if specified, *N. pumila*. w = waterlogged; c = charred; Meso = Mesolithic; Neo = Neolithic.

No. on map	Site	Country	Archaeological period (local terminology)	Absolute dating	Number of seeds *N. lutea/N. pumila*	Archaeological context of the seeds	Sources
1	Veksa 3	Russia	Early Neolithic (fisher-hunter-gatherers)	6th–early 5th mill. cal BCE	140 c	cultural layers 9–7, single features	this chapter
2a	Zamostje 2	Russia	Late Mesolithic, Early and Middle Neolithic (fisher-hunter-gatherers)	start of 7th to middle of the 5th mill. cal BCE	Meso: 960 w/c Neo: 1338 w/c	macrobotanical remains from non-systematic samples	Berihuete-Azorín and Lozovskaya 2014
2b	Zamostje 2	Russia	Late Mesolithic, Early and Middle Neolithic (fisher-hunter-gatherers)	start of 7th to middle of the 5th mill. cal BCE	Late Meso: 48 w/c Final Meso: 16 w/c Meso/Neo: 1 w/c Neo: 25 w/c	systematic botanical samples from test pit 2 (2013)	Berihuete 2018
3	Priedaine Riga	Latvia	Late Neolithic (fisher-hunter-gatherers)	c. 3700–3500 cal BCE	3 w 1 c		Bērziņš et al. 2016
4	Šventoji 6	Lithuania	Late Neolithic (fisher-hunter-gatherers)	3092–2916 cal BCE (direct date on *Nuphar*)	n, c		Grikpėdis and Motuzaite Matuzeviciute 2017
5	Anjalankoski Ankkapurha	Finland	Early Neolithic (and Bronze Age)	5200–4000 cal BCE (1500–500 cal BCE)	2 c		Vanhanen and Pesonen 2016
6	Rovaniemi Koskenniska	Finland	Late Neolithic (and Bronze Age)	3400–1500 cal BCE (1500–500 cal BCE)	127 c		Vanhanen 2014
7	Rääkkylä Pörrinmökki	Finland	Early Neolithic–Bronze Age	5200–500 cal BCE	4 c		Lavento 1992; Pesonen 1993; Pesonen 1994
8	Halsskov	Denmark	Late Mesolithic (Ertebølle)	5400–4800 cal BCE	*N. pumila*, 10 c	3 samples from pit h	Kubiak-Martens 2002
9	Duvensee, Wohnplatz 6	Schleswig-Holstein, Germany	Early Mesolithic	7100–6900 BCE	1 c	hearth	Holst 2010
10	Friesack 4	Brandenburg, Germany	Early–Late Mesolithic	9100–5500 cal BCE	5389 w 60 c	cultural layers (*Schichtkomplexe*)	Wolters 2016
11	Lough Kinale	Ireland (→ many more sites)	Mesolithic		several thousand, w		Warren et al. 2014
12	Hoge Vaart	The Netherlands	Late Mesolithic, Early Neolithic (Swifterbant)	5200–5000 cal BCE 4900–4500 cal BCE	182 c	Late Mesolithic fire pits and Early Swifterbant surface hearths	Brinkkemper et al. 1999
13	Järvensuo	Varsinais-Suomi, Finland	Neolithic		11–50, w		Regnell 1998
14	Trummen	Smaaland, Sweden	Mesolithic		11–50, w		Regnell 1998
15	Bökeberg III	Skaane, Sweden	Late Mesolithic		>50, w		Regnell 1998
16	Holmegaard	Sjaelland, Denmark	Mesolithic		>50, w		Regnell 1998
17	Holmegaard mose	Sjaelland, Denmark	Early Mesolithic		1–10, w		Regnell 1998
18	Ulkestrup	Sjaelland, Denmark	Early Mesolithic		>50, w		Regnell 1998
19	Ringkloster	Jylland, Denmark	Late Mesolithic (Ertebølle)		43 w		Andersen 1975; Regnell 1998
20	Norje Sunnansund	Skaane, Sweden	Early Mesolithic (Early Maglemose)	7600–6600 cal BCE	n, c		Lagerås, pers. comm.
21	Naumovo	Russia	Neolithic		n, w/c		Berihuete 2018
22	Serteya I, II	Russia	Neolithic		n, w/c		Berihuete 2018

palynological results indicate a late onset of cereal cultivation in the vicinity of Veksa, towards the end of the 1st millennium CE. This is in accordance with the fact that cereal pollen from self-pollinated wheat and barley in general rarely occurs in high numbers earlier than the early medieval period, continuing from then on into Modern times. The cereal-type pollen hints to a local onset of crop cultivation with wheat species, soon supplemented by barley, oat and rye (Fig. 4, VKS16-D2, and Fig. 6, VKS3/2-E). The cereal pollen is accompanied by pollen of the weedy species *Centaurea cyanus* and *Agrostemma githago.*. These herbs are strongly associated with arable fields and as both are winter annuals (the cornflower germinates in summer, too) they may also demonstrate the cultivation of winter cereals. The high proportion of grass pollen in the uppermost layers of the sedimentation sequence may indicate that shifting cultivation was carried out, as it is assumed e.g. for early medieval Novgorod (Alsleben 2001, 365–366), based on comparison with agricultural traditions of the early 20th century in the region north of Lake Ladoga, in Karelia. A small number of charred cereal grains, one wheat and two barley grains, found at Veksa 1 support this data. They stem from the Late Iron Age to early medieval cultural layers and have been directly radiocarbon dated to the 5th–6th century CE. While these single grains do indicate the presence and thus the use of these species at this site, an actual local cultivation cannot be deduced from this small assemblage. Iron Age evidence for crop growing is proven farther south in the Upper Volga region for the D'yakovo culture (7th century BCE–5th century CE) (Vishnevski 1994; Lebedeva 2005; Makarov et al. 1998; Makarov et al. 2001). At the fortified site of Rostislavl, by the Oka River, numerous grains of charred cereals and threshing remains indicate crop growing, including of emmer, free-threshing wheat, barley and broomcorn millet. Single finds of rye and oat may still be interpreted as weed species, while linseed, pea and hemp clearly supplemented the assemblage of cultivars (Lebedeva 2005, 163). Makarov et al. (1998; 2001) compiled archaeobotanical data from the Beloozero and the Upper Sheksna regions, indicating crop cultivation since the end of the 9th century CE at the northern periphery of the Kievan Rus'. Based on the lack of cultivar evidence farther north-east, a north-eastern border for crop growing is postulated in the Beloozero region in the pre-Mongolian period, before 1300 CE (Makarov et al. 1998: 84). With the new archaeobotanical data from Veksa at hand, we now have evidence for the presence of cereals in this part of northern Russia in the 5th–6th century CE. This may hint to possible exchange between the substantial Iron Age settlement at Veksa (Nedomolkina 2000) and the D'yakovo sites farther south.

Summary: Hunting, fishing and plant use at Veksa over time

The archaeobotanical evidence from Veksa attests the use of wild plants by the local hunter-fisher communities from the 6th millennium BCE onwards and the comparatively late introduction of domestic plants (cereals), in the 1st millennium CE. For much of the Holocene, up until the medieval period, hunting and fishing constituted major components within the subsistence economy. Veksa is located in an environment that, ever since the drying up of the Sukhona palaeolake, has been well suited for hunting and fishing, especially in summer, when low water levels enable seasonal settlement on the riverbank. Osteological data as well as archaeological finds and features show a wide spectrum of hunting and fishing activities over the millennia, with beaver, elk and deer dominating in the Early and Middle Neolithic (Nedomolkina et al. 2015a). Analyses of stable isotopes in pottery food crusts indicate increased processing of aquatic foods over the course of the Neolithic, a pattern also observed at other central and northern Russian sites (Piezonka et al. 2016). The Late Neolithic to Eneolithic complexes of the late 4th–3rd millennium BCE at the riverbank site of Veksa 3 indicate intensive fishing, with six wooden fish traps and hundreds of vertical posts probably representing stationary fishing constructions (Nedomolkina et al. 2015a; Piezonka et al. in press). Based on this evidence, and taking into account the environmental particularities of the site, it is likely that Veksa was foremost a site for hunting and fishing for most of its history. Cultivation of crops may have been possible in the immediate vicinity of the Veksa settlements in dry years, although better-suited areas, with drier and more fertile morainic soils, lie beyond the margins of the former palaeo-lake to the north.

To evaluate the data from Veksa on a wider scale with respect to the beginning of farming in north-western Russia, it is necessary to separately regard the results of archaeozoological on-site investigations showing evidence for domestic animals, of palynological studies on the occurrence of cereal pollen in sediment profiles, and of macrobotanical finds from archaeological contexts. Early evidence for animal husbandry comes from Middle and Late Neolithic lakeshore settlements in the west Russian Dvina-Lovat' region (Arslanov et al. 2009; Mazurkevich et al. 2009). Animal husbandry as a minor component of the subsistence economy is thereby proven for the Neolithic in more southerly parts of the forest zone. Palynological investigations at Serteya II and in the Novgorod region (Arslanov et al. 2009; Dolukhanov et al. 2009; Mazurkevich and Dolbunova 2011) hint to an onset of cereal cultivation already in the 3rd millennium BCE. But, similar to this study, the temporal resolution of the pollen data is low, and cereal pollen can only be identified as cereal-type pollen (Beug 2004), the different types also including pollen from wild grasses. A convincing identification as either *Triticum, Hordeum* or *Avena* pollen would require detailed measurements and would be best supported by macrobotanical evidence. Thus, the indications for cereal cultivation in the 3rd millennium BCE at Serteya II and also in the Novgorod region (Arslanov et al. 2009; Dolukhanov et al. 2009; Mazurkevich and Dolbunova 2011) remain vague and for the time being must be regarded as not altogether convincing. We suggest these sequences be re-investigated to evaluate the reliability of the early dates for cereal cultivation. Profound macrobotanical data for the establishment of crop growing in north-western Russia are available only for the 5th–9th century CE and relate to the appearance of Slavonic groups and the constitution of the Ancient Rus. In particular rye is regarded as a Slavonic contribution to the crop spectrum (Alsleben 2012).

In the neighbouring Eastern Baltic region and Fennoscandia, initial evidence for crop plants is a hotly debated issue (e.g. Piličiauskas et al. 2016; Lahtinen and Rowley-Conwy 2013). Pollen data from the Estonian alvar areas shows continuous cereal curves that indicate the adoption of arable farming from the Bronze Age onwards. Throughout other parts of Estonia, though, the adoption of farming took place in the Iron Age (from c. 500 BCE onwards) and went along with the implementation of iron tools, ploughing and manuring (Poska et al. 2004). For Lithuania, published macrobotanical evidence for cereals said to date to the Neolithic did not withstand re-evaluation. The earliest AMS date on a barley grain revealed a Bronze Age date (1392–1123 BCE). Further, it remains an open question whether the total of eight grains indicate cultivation or import (Grikpėdis and Moutezaite Matuzeviciute 2017). In Finland and Russian Karelia, arguments for Late Neolithic cereal cultivation rely on single cereal-type pollen grains in co-occurrence with *Cannabis*-type or *Tilia* pollen, the latter being interpreted to indicate woodland management (e.g. Alenius et al. 2013; Alenius et al. 2017). Being based solely on single microfossil occurrences, these interpretations on crop growing are problematic (Lahtinen and Rowley-Conwy 2013; Lahtinen et al. 2017). However, based on a critical assessment of 112 pollen records from Finland and Russian Karelia with a focus on the start of continuous cereal pollen curves with reliable radiocarbon dates, integrated with two independent proxies—the archaeological record and the pollen data—three main phases of advance of farming could be identified. The first, modest phase indicating the onset of cereal cultivation dates to the Early Bronze Age, at approximately 1500 BCE. A second advance took place in the Late Bronze Age, from c. 900 BCE onwards, while the broad adoption occurred in the 1st millennium CE (Lahtinen et al. 2017). Macrobotanical evidence for crop species in Russian Karelia stems from pollen sequences at Lake Onega and is extremely rare. The earliest crop find is one single seed of flax (*Linum usitatissimum*) in the sequence from Igumeeninlampi, on Valamo Island, dating to the Late Iron Age. Four small capsule fragments of flax from 280–240 cm depths in the pollen profile Pegrema S possibly originate from using the shallow shores of Lake Onega for flax retting after 1460 CE (Vuorela et al. 2001a; Vuorela et al. 2001b). Reliable pollen data for cereal growing based on high percentage values were recently shown for a sequence in northern Sweden and date as late as from the 14th century CE onwards (Hörnberg et al. 2014).

Generally, single cereal pollen grains demand integration with further proxies. However, macrobotanical investigations that could substantiate the use and—if mass finds are available—cultivation of crops are sparse in the northern regions. The few published macrobotanical studies prove that crop growing hardly took place before 500 CE (Vanhanen 2010; for earlier finds, see Vanhanen, this volume). The site of Mikkeli Orijärvi, in eastern Finland, is an outstanding example for the successful application of a multiproxy approach. Here, both the microfossil and the macrobotanical record prove the late onset of cereal cultivation, in medieval times (Alenius et al. 2007; Vanhanen 2010). These findings fit very well with our combined micro- and macrobotanical data from Veksa in the north-west Russian taiga zone, where we have been able to show that the beginning of actual cereal cultivation in the Sukhona basin apparently did not start before early medieval times and must thus be seen as a late arrival compared with more southerly and westerly regions.

Conclusions

Plant use and the development of plant cultivation in the north-western Russian taiga zone have been investigated at the multiperiod archaeological sites of Veksa 1 and 3, oblast' Vologda, by a multiproxy approach combining archaeological, sedimentological and archaeobotanical analyses. It shows intensive use of local plant resources in the Stone Age and a late introduction of crops, in the 1st millennium CE.

It is in particular the combination of pollen analysis and macrobotanical studies that has been undertaken for the first time in this region. This approach, although confronted with partly difficult preservation conditions for organic matter, has succeeded in generating new data with respect to the exploitation of wild plants by local communities and the beginning of cereal cultivation in the forest zone. Our data clearly show that the region has a long-lasting tradition in the use of wild plants, with special emphasis in the Neolithic period on the starch-rich seeds of yellow water lily. The onset of cereal cultivation is more difficult to pinpoint. Macrofossils show that cereal grains have been present in the area since the 5th–6th century CE, possibly indicating trade connections to Late Iron Age communities of the D'yakovo culture in the Upper Volga region farther south that had already integrated cultivated plants into their economy. Based on the palynological data at hand, it seems that actual cereal cultivation in the vicinity of Veksa did not set in before early medieval times.

Altogether the archaeological, archaeozoological and palaeo-environmental evidence suggests that Veksa has been a focus for hunting and fishing activities from the beginning of settlement in the 6th millennium BCE up until Early Historic times. The Upper Sukhona basin, being a low-lying, wet area prone to periodical flooding, was well suited for seasonal foraging activities, and the botanical evidence of the collection and preparation of aquatic plants such as *Nuphar* for food already in the most ancient layers of the 6th millennium BCE underlines the prevalence of this type of economy. The late occurrence of cereals at Veksa, around the middle of the 1st millennium CE, and a possibly even later onset of cereal cultivation could be connected to this specific environmental setting of the site. At the same time, it may also be part of a larger-scale pattern of arable farming practices only being adopted very late in this part of north-eastern Europe.

Acknowledgements

Financial support was granted to Henny Piezonka by the German Research Foundation (DFG grant no. PI 1120/2-1) and by the Eurasia Department of the German Archaeological Institute. The study was further supported within the frame of CRC 1266, *Scales of transformation* (DFG, Project Ref. 2901391021—SFB 1266) in subproject F3 *Dynamics of plant economy,* PI Wiebke Kirleis). We are particularly grateful to Ljudmila Shumilovskikh, Göttingen, who supported the archaeobotanical on-site investigations. Sincere thanks are due to Christoph Engel, Ilia Heit and Marina Tydo, all from Berlin, for providing their archaeological and technical expertise during the excavations at Veksa and the subsequent post-excavation work. The investigations would not have been possible without our great Russian–German excavation team and the kind support of the Vologda State Museum. We express our warm thanks to all of them.

Electronic supplementary material for the paper is accessible at the data exchange platform of the Johanna Mestorf Academy at Kiel University (https://www.jma.uni-kiel.de/en/research-projects/data-exchange-platform).

References

Alenius T, Mökkönen T, Holmqvist E, Ojala A (2017) Neolithic land use in the northern Boreal zone: High-resolution multi-proxy analyses from Lake Huhdasjärvi, south-eastern Finland. Veg Hist Archaeobot 26:469–486

Alenius T, Mökkönen T, Lahelma A (2013) Early Farming in the Northern Boreal Zone: Reassessing the History of Land Use in Southeastern Finland through High-Resolution Pollen Analysis. Geoarchaeology 28:1–24

Alenius T, Mikkola E, Ojala AEK (2007) History of agriculture in Mikkeli Orijärvi, eastern Finland as reflected by palynological and archaeological data. Veg Hist Archaeobot 17:171–183

Alsleben A (2012) The plant economy of northern Russia. In: Brisbane MA, Makarov N, Nosov E (eds) The archaeology of medieval Novgorod in context: A study in centre/periphery relations. Oxbow Books, Oxford, pp 321–350

Alsleben, A (2001) Angebot und Nachfrage: Frühmittelalterliche Nahrungswirtschaft im Umland von Novgorod. In: Müller-Wille M, Janin VL, Nosov EN, Rybina EA (eds) Novgorod: Das mittelalterliche Zentrum und sein Umland. Wachholtz, Neumünster, pp 359–368

Andersen S (1975) Ringkloster, en jysk indlandsboplads med Ertebøllekultur. Ringkloster. An inland Ertebølle settlement in Jutland. Kuml 27, 1973–74, 11–108.

Arslanov KA, Dolukhanov PM, Savelieva LA, Dzinoridze EN, Kuzmin GF, Denisenko VP (2009) The Holocene environments in north-western and central Russia. In: Dolukhanov PM, Sarson GR, Shukurov AM (eds) The East European plain on the eve of agriculture. BAR Int. Ser. 1964. BAR Publishing, Oxford, pp 109–121

Bashenkin AN (1997) Vologodskaya oblast' v drevnosti i srednevekovye. Vologda: Istoriko-kraevedcheski almanakh 2:5–35.

Berihuete-Azorín M, Lozovskaya O (2014) Evolution of plant use at the wetland site Zamostje 2, Russia: First results. In: Russian Academy of Science, Institute for the History of Material Culture, The State Heritage Museum (eds) Palaeo-environment and models of adaptations of lake settlements in the Mesolithic and Neolithic of the forest zone of eastern Europe (Materials of the International conference, 19–21 May 2014, St. Petersburg). IIMK RAN, St. Petersburg, pp 74–79.

Berihuete M (2018) First results of the archaeobotanical study of the Test pit 2 profile column. In: Lozovskaya O (ed) Site Zamostje 2 and landscape evolution in the Volga-Oka region during the Holocene. IIMK RAN, St. Petersburg, pp 41–48

Bērziņš V, Ceriņa A, Kalniņš M, Lõugas L, Lübke H, Meadows J (2016) Priedaine: A Neolithic site at the head of the Gulf of Riga. Archaeologia Baltica 23:12–37

Beug HJ (2004) Leitfaden der Pollenbestimmung für Mitteleuropa und Angrenzende Gebiete. Pfeil, München

Brinkkemper O, Hogestijn WJ, Peeters H, Visser D, Whitton C (1999) The Early Neolithic site at Hoge Vaart, Almere, the Netherlands, with particular reference to non-diffusion of crop plants, and the significance of site function and sample location. Veg Hist Archaeobot 8:79–86

Bronk Ramsey C (2013) OxCal 4.2 manual. https://c14.arch.ox.ac.uk/oxcal.html

Dolukhanov PM, Gey NA, Miklyayev AM, Mazurkiewicz AN (1989) Rudnya-Serteya, a stratified dwelling site in the upper Duna basin (a multidisciplinary research). Fennoscandia Archaeologica 6:23–27

Dolukhanov PM, Subetto DA, Arslanov KA, Davydova NN, Zaitseva GI, Dijnoridze EN, Kuznetsov DD, Ludikova AV, Sapelko TV, Savelieva LA (2009) The Holocene history of the Baltic Sea, and the Ladoga Lake. In: Dolukhanov PM, Sarson

GR, Shukurov AM (eds) The East European plain on the eve of agriculture, BAR Int. Ser. 1964. BAR Publishing, Oxford, pp 123–138

Düll R, Kutzelnigg H (2005) Taschenlexikon der Pflanzen Deutschlands: Ein Botanisch-ökologischer Exkursionsführer. Quelle & Meyer, Wiebelsheim

Erhardt W, Götz E, Bödeker N, Seybold S (2002) Zander, Handwörterbuch der Pflanzennamen. Ulmer, Stuttgart

Federal Agency for Nature Conservation (2004) Interactive CD-ROM map of the natural vegetation of Europe (EuroVegMap 2.0)

Feeser I (unpublished) CountPol - counting program for pollen analyses

Gey V, Saarnisto M, Lunkka JP, Demidov I (2001) Mikulino and Valdai palaeoenvironments in the Vologda area, NW Russia. Global and Planetary Change 31:347–366

Grikpėdis M, Motuzaite Matuzeviciute G (2017) A review of the earliest evidence of agriculture in Lithuania and the earliest direct AMS date on cereal. Eur J Archaeol 21(2):1–16

Hörnberg G, Josefsson T, Liedgren L (2014) Revealing the cultivation history of northernmost Sweden: Evidence from pollen records. The Holocene 24:318–326

Holst D (2010) Hazelnut economy of early Holocene hunter-gatherers: A case study from Mesolithic Duvensee, northern Germany. J Archaeol Sci 37:2871–2880

Hua Q, Barbetti M, Rakowski AZ (2013) Atmospheric radiocarbon for the period 1950–2010. Radiocarbon 55(4):2059–2072

Juggins S (2014) C2 user guide: Software for ecological and palaeoecological data analysis and visualisation. University of Newcastle, Newcastle upon Tyne

Khotinskiy NA (1984) Holocene vegetation history. In: Velichko AA (ed) Late quaternary environments of the Soviet Union. University of Minnesota Press, Minneapolis

Königsson LK, Possnert G, Hammar T (1997) Economical and cultural changes in the landscape development at Novgorod. Tor 29:353–387

Kremenetski KV, Borisova OK, Zelikson EM (2000) The Late Glacial and Holocene history of vegetation in the Moscow region. Paleontol J 34(Suppl. 1):S67–S74

Kubiak-Martens L (2002) New evidence for the use of root foods in pre-agrarian subsistence recovered from the Late Mesolithic site at Halsskov, Denmark. Veg Hist Archaeobot 11:23–32

Kulkova M, Mazurkevich A, Dolbunova E, Regert M, Mazuy A, Nesterov E, Sinai M (2015) Late Neolithic subsistence strategy and reservoir effects in 14C dating of artefacts at the pile-dwelling site Serteya II (NW Russia). Radiocarbon 57(4):611–623

Kuosmanen N, Seppä H, Reitalu T, Alenius T, Bradshaw RHW, Clear JL, Filimonova L, Kuznetsov O, Zaretskaya N (2016) Long-term forest composition and its drivers in taiga forest in NW Russia. Veg Hist Archaeobot 25:221–236

Kvasov DD (1979) The Late Quaternary history of large lakes and inland seas of eastern Europe. Annales Academiae Scientiarum Fennicae, Series A (Helsinki) 127:71

Lahtinen M, Rowley-Conwy P (2013) Early farming in Finland: Was there cultivation before the Iron Age (500 BC)? Eur J Archaeol 16:660–684

Lahtinen M, Oinonen M, Tallavaara M, Walker JWP, Rowley-Conwy P (2017) The advance of cultivation at its northern European limit: Process or event? The Holocene 27(3):427–438

Lavento M (1992) Rääkkylä Jaamankangas Pörrinmökki 1991 (Excavation report in the archives of the National Board of Antiquities, Helsinki)

Lebedeva EY (2005) Kul'turnye rasteniya Rostislavlya: arkheobotanicheskie materialy iz d'yakovskogo gorodishcha i drevnerusskogo goroda. In: Chernykh EN, Zavyalov VI (eds) Archaeology and methods of natural sciences. IA RAN, Moscow, pp 159–180

Lorenz C, Nedomolkina NG, Piezonka H (2014) Geomorfologicheski analiz paleolandshafta v bassejne Verkhei Sukhony: Na primere poselenija Veksa 3. In: Lozovski VM, Lozovskaya OV (eds) Prirodnaya sreda i modeli adaptacii ozernykh poseleni v mezolite i neolite lesnoi zony Vostochnoi Evropy (Materialy mezhdunarodnoi nauchnoi konferencii, Saint Petersburg, 19–21 maya 2014 g) IIMK RAN, St. Petersburg, pp 11–14

Makarov NA, Zakharov SD, Kir'yanova NA (1998) Paleobotanicheskie materialy iz Beloozera: k istorii stanovleniya zemledeliya na severnykh okrainakh Drevnei Rusi. Rossiiskaya Arkheologiya 1:175–186

Makarov NA, Zakharov SD, Kir'yanova NA (2001) Paleobotanicheskie materialy iz Beloozera: k istorii stanovleniya zemledeliya na severnykh okrainakh Drevnei Rusi. In: Makarov NA, Zakharov SD, Buzhilova AP (eds) Medieval settlement in Beloozero region. Yazyk Russkoi Kul'tury, Moscow, pp. 112–123

Mazurkevich A, Dolukhanov P, Shukurov A, Zaitseva G (2009) Late Stone–Early Bronze Age sites in the western Dvina-Lovat area. In: Dolukhanov PM, Sarson GR, Shukurov AM (eds) The East European plain on the eve of agriculture. BAR Int. Ser. 1964. BAR Publishing, Oxford, pp 145–158

Mazurkevich A, Dolbunova E (2011) Underwater investigations in northwest Russia: Lacustrine archaeology of Neolithic pile dwellings. In: Benjamin J, Bonsall C, Pickard C, Fischer A (eds) Submerged prehistory. Oxbow Books, Oxford, pp 158–72

Mears R, Hillman G (2007) Wild Food. Hodder and Stoughton, London

Moerman, DE (2009[8]) Native American ethnobotany. Timber Press, Portland, London

Moore PD, Webb JA, Collison ME (1991) Pollen analysis. Blackwell Scientific Publications, Oxford

Nedomolkina NG (2000) Mnogosloinoe poselenie Veksa. Tverskoi Arkheologicheski Sbornik 4(1):277–283

Nedomolkina NG (2004) Neoliticheskie kompleksy poselenii Veksa i Veksa III basseina Verkhnei Sukhony i ikh khronologiya. In: Timofeev VI, Zajtseva GI (eds) Problemy khronologii i etnokul'turnykh vzaimodeistvii v neolite Evrazii. IIMK RAN, St. Petersburg, pp 265–279

Nedomolkina NG (2007) Neolit Verkhnei Sukhony. Avtoreferat dissertacii na soiskanie uchenoi stepeni k.i.n. Institute of the History of Material Culture of the Russian Academy of Sciences, St. Petersburg (unpublished dissertation summary)

Nedomolkina NG, Piezonka H, Lorenz S, Schmölke U (2015a) Novye arkheologicheskie, osteologicheskie i geomorfologicheskie issledovaniya na komplekse mnogosloinykh poseleni Veksa v basseine Verkhnoi Sukhony. Tverskii Arkheologicheski Sbornik 10(1):75–84

Nedomolkina NG, Piezonka H, Meadows J, Craig O, Lorenz S (2015b) Neoliticheskie kompleksy poseleniya Veksa v basseine Vekhnei Sukhony, Severo-Zapadnaya Rossiya: Novye estestvennonauchnye issledovaniya. In: Lozovski VM, Lozovskaya OV, Vybornov AA (eds) Neoliticheskie kul'tury vostochnoi Evropy: Khronologiya, paleoekologiya, tradicii. IMMK RAN, St. Petersburg, pp 151–158

Nedomolkina NG, Piezonka H (2016) Radiocarbon chronology of the upper Sukhona region in Early and Middle Neolithic (sites Veksa I and Veksa III). In: Zaitseva GV, Lozovskaya OV, Vybornov AA, Mazurkevich AN (eds) Radiouglerodnaya khronologiya epokhi neolita Vostochnoj Evropy VII–III tysyacheletiya do n.e. Svitok, Smolensk, pp 425–433

Nikitinski IF (1992) K voprosu o rannem zemledelii na Vage i Sukhone. Arkheologiya i istoriya Pskova i Pskovskoi zemli 9:109–111

O'Neill AR, Rana SK (2016) An ethnobotanical analysis of parasitic plants (Parijibi) in the Nepal Himalaya. J Ethnobiol Ethnomedicine 12:14

Pesonen P (1993) Rääkkylä Jaamankangas Pörrinmökki 1992 (Excavation report in the archives of National Board of Antiquities, Helsinki)

Pesonen P (1994) Rääkkylä (Jaamankangas) Täitimänniemi Pörrinmökki 1993 (Excavation report in the archives of National Board of Antiquities, Helsinki)

Piezonka H (2015) Jäger, Fischer, Töpfer. Wildbeuter mit früher Keramik in Nordosteuropa im 6. und 5. Jahrtausend v. Chr. Archäologie in Eurasien 30. Habelt, Bonn

Piezonka H, Meadows J, Hartz S, Kostyleva E, Nedomolkina N, Ivanishcheva M, Kozorukova N, Terberger T (2016) Stone Age pottery chronology in the northeast European forest zone: New AMS and EA-IRMS results on foodcrusts. Radiocarbon 58(2):267–289

Piezonka H, Nedomolkina N, Ivanishcheva M, Kosorukova N, Kul'kova M, Meadows J (2017) The Early and Middle Neolithic in NW Russia: Radiocarbon chronologies from the Sukhona and Onega regions. Documenta Praehistorica 44:122–151

Piezonka H, Nedomolkina NG, Heußner KU, Kirleis W, Lorenz S, Wieckowska-Lüth M, Elberfeld V (2020) Pile dwellers in the Sukhona basin? Wooden structures of the 4th and 3rd millennium BCE at Veksa, Northern Russia. In: Hafner A, Dolbunova E, Mazurkevich A, Pranckenaite E, Hinz M (eds) Settling waterscapes in Europe The archaeology of Neolithic and Bronze Age pile-dwellings. Open Series in Prehistoric Archaeology 1. Propylaeum, Bern and Heidelberg, pp 59–94

Piličiauskas G, Kisielienė D, Piličiauskienė G (2016) Deconstructing the concept of Subneolithic farming in the southeastern Baltic. Veg Hist Archaeobot 26:183–193

Poska A, Saarse L, Veski S (2004) Reflections of pre- and early-agrarian human impact in the pollen diagrams of Estonia. Palaeogeogr Palaeocl 209:37–50

Regnell M (1998) Archaeobotanical finds from the Stone Age of the Nordic countries: A catalogue of plant remains from archaeological contexts. Lundqua Report 36. Department of Quaternary Geology, Lund University, Lund

Reimer P, Bard E, Bayliss A, Beck J, Blackwell P, Ramsey C, Van der Plicht J (2013) IntCal13 and Marine13 radiocarbon age calibration curves 0–50,000 years cal BP. Radiocarbon 5:1869–1887

Shahgedanova, M (2003) The physical geography of northern Eurasia. Oxford University Press, Oxford

Stolbovoi, V (2000) Soils of Russia: Correlated with the revised legend of the FAO soil map of the world and world reference base for soil resources. Research report RR-00-13. International Institute for Applied Systems Analysis, Austria

Tarasov PE, Peyron O, Guiot J, Brewer S, Volokova VS, Bezusko LG, Dorofeyuk NI, Kvavadze EV, Osipova IM, Panova NK (1999) Last glacial maximum climate of the former Soviet Union and Mongolia reconstructed from pollen and plant macrofossil data. Clim Dyn 15:227–240

Vanhanen S (2010) Charred seeds in Mikkeli Orijärvi: A study of subsistence strategies of an Iron Age settlement in East Finland. Master's thesis, University of Helsinki. https://helda.helsinki.fi/handle/10138/26504

Vanhanen S (2014) Kasvimakrofossiilitutkimus Rovaniemi Koskenniska 2013 (Archaeobotanical report in the Archives of the National Board of Antiquities, Helsinki)

Vanhanen S, Pesonen P (2016) Wild plant gathering in Stone Age Finland. Quaternary Int 404:43–55

Vishnevski VI (1994) Novye materialy o zemledelii plemen d'yakovskoi kul'tury Verkhnego Povolzh'ya. Rossiiskaya Arkheologiya 3:123–126

Vuorela I, Lempiäinen T, Saarnisto M (2001) Land use pollen record from the Island of Valamo, Russian Karelia. Annales Botanici Fennici 38(2):139–165

Vuorela I, Saarnisto M, Lempiäinen T, Taavitsainen J-P (2001) Stone Age to recent land-use history at Pegrema, northern Lake Onega, Russian Karelia. Veg Hist Archaeobot 10(3):121–138

Warren G, Davis S, McClatchie M, Sands R (2014) The potential role of humans in structuring the wooded landscapes of Mesolithic Ireland: A review of data and discussion of approaches. Veg Hist Archaeobot 23:629–646

Wohlfarth B, Tarasov P, Bennike O, Lacourse T, Subetto D, Torssander P, Romanenko F (2006) Late Glacial and Holocene palaeo-environmental changes in the Rostov-Yaroslavl' area, west central Russia. J Paleolimnol 35(3):543–569

Wolters S (2016) Die pflanzlichen Reste der Mesolithstation Friesack. In: Benecke N, Gramsch B, Jahns S (eds) Subsistenz und Umwelt der Feuchtbodenstation Friesack 4 im Havelland: Ergebnisse der naturwissenschaftlichen Untersuchungen. Brandenburgisches Landesamt für Denkmalpflege und Archäologisches Landesmuseum, Wünsdorf, pp 189–203

Yeremeyev II, Dzyuba OF, Litisyna OV (2007) Drevnee rasselenie i zemledelie na vostochnom beregu ozera Il'men' po dannym arkheologii i paleogeografii. Zapiski Instituta istorii material'noi kul'tury RAN 2, 215–256

Zech W, Schad P, Hintermaier-Erhard G (2014) Böden der Welt: Ein Bildatlas. Springer Spektrum, Berlin, Heidelberg

Gardening at medieval farmsteads: Archaeobotanical indications of horticulture in Denmark and southern Sweden, AD 1000–1500

Kirstine Krath Helweg

Author address:
Department of Archaeology, Museum Thy, Thisted, Denmark
Jernbanegade 6, 7700 Thisted, Denmark

E-mail: (kkn@museumthy.dk)

Abstract

This paper investigates the indications of horticulture on farmsteads in southern Scandinavia (Denmark and southern Sweden) during the Middle Ages (AD 1000–1500). It is well known that, in theory, the common peasants would have maintained a garden as part of their farmsteads. However, archaeological evidence of this gardening activity has been lacking, due to poor preservation conditions, excavation practices and sampling strategies and the small number of available assemblages. This study interprets the archaeobotanical macrofossils (seed and fruits) available, taking into account taphonomic factors, and discusses the problem of distinguishing between horticulture and the collection of wild plants.

Introduction

During the Middle Ages, the majority of the population in Denmark and southern Sweden were occupied in agriculture. Gardens would have provided farmstead households with important supplements to their everyday diet, along with medical remedies. Thus, gardens were an important aspect of the individual farmer's nutrition and survival strategy (Hansson 1997). Despite the significant importance of farms and farmers during this period, they are underrepresented and underinvestigated as a research subject, both in archaeology and in archaeobotany. In contrast to the well-preserved and well-sampled urban assemblages, farmstead assemblages of this period are challenged by poor preservation conditions and insufficient sampling.

The research questions of this study are:

1. Is it possible to recognise gardening on farmsteads in Denmark and southern Sweden during the Middle Ages by using of a combination of macrofossils and written sources?
2. How does taphonomy challenge the current research and can context and preservation conditions of the existing assemblages provide information on the best sampling strategy for recognising the Danish and southern Swedish medieval farmstead garden?
3. Is it possible to distinguish horticulture from the collection of wild plants?

Medieval farmstead gardens

Gardens were connected to all levels of society during the Middle Ages. Gardens produced vegetables and herbs, but also provided pleasure to both farmsteads and seigneurial households (Taylor 2000). In order to better understand the function and structure of medieval farmstead gardens in particular, we need to understand the different kinds of gardens which were present in the medieval landscape.

In earlier literature, monks were hypothesised to have introduced gardening to Scandinavia (Karling 1931; Hjelmqvist 1991). However, this is today considered a partly outdated hypothesis. The garden historian Kjell Lundquist (1994) proposes that gardens in general were closely connected to the agricultural system and that they were introduced to southern Scandinavia during the Iron Age. Lundquist does, however, stress that even though the tradition of gardening may be older than the Middle Ages in southern Scandinavia, it is likely that many plant species were indeed introduced by the monks. Archaeobotanist Ann-Marie Hansson investigated the multiple aspects of diet and which plants were consumed during the Viking Age in southern Scandinavia. Hansson (1997) argues that these gardens may have been more advanced than previously thought and may have featured flowers for aesthetic reasons, as well as purely practical ones. Both linguist Inger Larsson and cultural geographer Clas Tollin question this. They claim that the gar-

Figure 1. Map showing location of the studied sites. 1 Bramdrup; 2 Havgård; 3 Lykkegård Midt; 4 Sejerslev; 5 Røjgård; 6 Tværvej; 7 Bondemosevej; 8 Rynkeby Mølle; 9 Bramstrup; 10 Hollenæs; 11 Tårnby Torv; 12 Herstedøster kirke; 13 Smedegård Nordøst; 14 Lindängelund; 15 Kiaby; 16 Örja; 17 Ängelholm; 18 Trottaberg; 19 Halmstad.

dens held a merely economical function and that it was not until the late medieval period that flowerbeds started as a practice in southern Scandinavia (Larsson 2010; Tollin 2005). In recent years, the idea of the monasteries as catalyst in the process of introduction of horticulture in southern Scandinavia has become antiquated (Heimdahl 2010). Based on archaeological investigations, Karin Lindeblad (2006) suggests that cities were distribution centres of knowledge on plant use and gardening during the Middle Ages. In contrast, Sabine Karg (2007) argues that the monasteries were the innovative centres of knowledge regarding horticulture, despite earlier articles where she and others (Sloth et al. 2012) described gardening in a Viking Age rural context. Quaternary geologist and archaeobotanist Jens Heimdahl dates the origin of gardens back to pre-Roman times, based on the presence of garden plants, such as dill (*Anethum graveolens*), celery (*Apium graveolens*), cabbage (*Brassica oleracea*) and turnip (*Brassica rapa*). He argues that the urban garden predates the monastery gardens in Sweden. Furthermore, Heimdahl (2010) has identified arable horizons and remains of fertiliser during excavations of medieval remains. It is, however, important to remember that some of the divergence in the theories is due to differences in source material and in definitions of the garden phenomenon at the time the research was undertaken.

Material and methods

The study comprises previously published datasets as well as unpublished datasets from the Moesgaard Museum, Nationalmuseet (National Museum of Denmark), and Arkeologerna, Statens historiska museer (The Archaeologists, National Historical Museums). The locations of the nineteen sites are given in Figure 1 and the publication and report sources are given in Table 1. Figure 2 shows time periods the sites date to. The archaeobotanical material discussed here consist of seeds, fruits, stems and flowers; cereals are not included in this paper. The archaeobotanical material was all collected and sam-

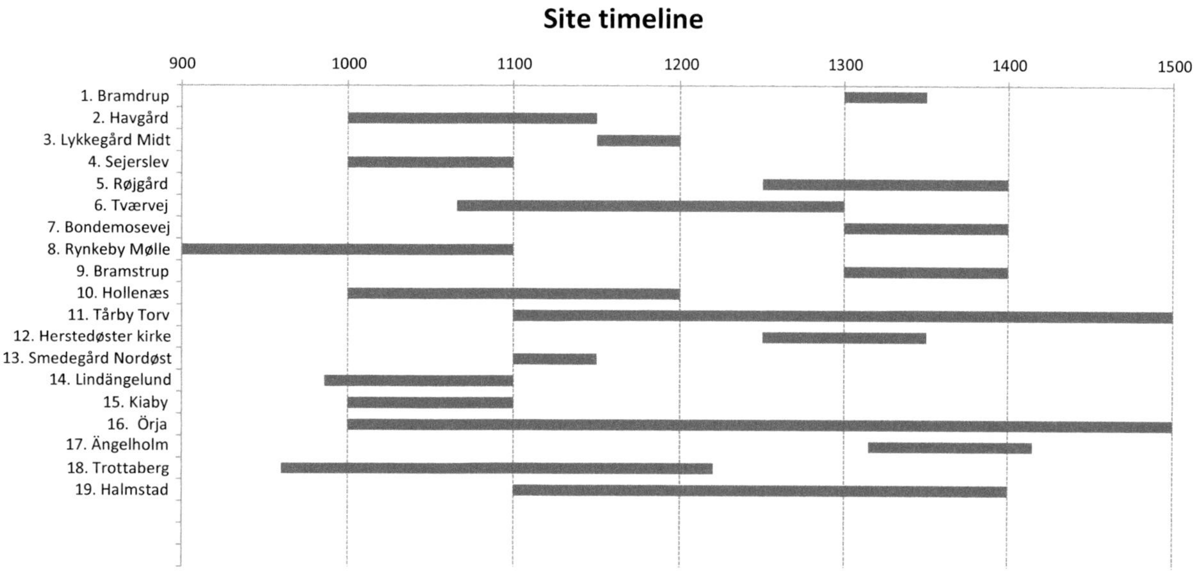

Figure 2. Timeline for the investigated sites.

pled by the local museum's archaeologists and analysed by different archaeobotanists and botanists. The selected archaeobotanical material originates from southern Scandinavia and includes the regions of Bornholm, Funen, Halland, Jutland, Lolland, Scania and Zealand. This area represents the Kingdom of Denmark in the period AD 1000–1500 (Roesdahl 1999). Blekinge is not included in the investigations due to the lack of access to archaeobotanical reports on rural settlements in this region. The northern German duchies Schleswig and Holstein are also not included, even though they had a strong connection to the Danish Kingdom during the Middle Ages.

This paper relies on eight historical written sources, dating from the 13th century AD to AD 1648 (Table 2) for comparison with the archaeobotanical record. Together, these sources contain more than 100 mentions of different plant species.

The archaeobotanical and written records do not provide enough data to make any statistically significant quantitative analysis. They do, however, have great qualitative possibilities (Viklund 1998, 30). This paper thus provides a qualitative picture of garden plant selection and garden practices at individual sites. A comparison of the archaeobotanical and written records illustrates which species of plants can or cannot be found through archaeobotanical investigations. The written and archaeobotanical records can help us understand plant usage and give us better insight into medieval society's understanding of plants and food. However, this does not come easy. There are a lot of pitfalls to be aware of when working with written sources, due to their varying level of accuracy. Larsson believes that they reflect different social contexts. This is important to keep in mind when working with these types of culturally biased sources (Larsson 2010, 54f).

One of the larger methodological questions is how to differentiate garden plants from cultivated arable crop plants and wild gathered plants. This subject is quite difficult since knowledge about plant use and cultivation methods in the medieval period is poorly documented. In this paper, the garden is defined by scale and placement. It is likely that the medieval peasant garden was placed within the boundaries of the parcel and close to the living quarters. This is also seen in contemporary sources on the nobility (Landsberg 2003). However, there are a lot of other factors which should be kept in mind, such as time and culture, when discussing the concept of a garden (Gleason 1994, 2). In this study, the plants which are referred to as garden plants are plants which must either be eaten fresh or be properly stored or preserved to give them a longer shelf life.

The plants present in the archaeobotanical and written sources are here divided into three artificial categories, namely, herbs, vegetables, and fruit and nuts. The categorisation is primarily based on Brøndegaard's ethnobotanical work *Folk og Flora* 1–4 (1979–1980a, b, c, d). This categorisation does not represent the diversity in plant use. Poisonous plants, in this case henbane (*Hyoscyamus niger*) and deadly nightshade (*Atropa belladonna*), are some of the few plants only used as medicinal herbs. But almost all the gastronomical herbs have multiple uses. Sage (*Salvia* sp.), which is native to Europe, can be used for both culinary and medicinal purposes. In *The Forme of Cury,* sage is used in the recipe '*Pigs in sage sauce*', but it is also known from the *Henrik Harpestreng* manuscripts as a medicinal plant that can help with liver conditions, stop bleeding, and help cleanse the blood (Brade 1999, 61).

The contexts investigated in this study are as follows: cultural layers; cultivation layer; ditch; faeces; fire-

Table 1. Overview of dating, contexts, and region for the studied sites. The numbering corresponds to Figure 1. The definition of individual contexts and stratigraphic phases, information on scientific dating of the phases, and the location are taken directly from the cited reports or from supplementary information provided by the museums. H/C-Typo = House and Ceramic Typology; Dendro = Dendrochronology.

No.	Site	Date	Dating Method	Context	Region	Reference
1	Bramdrup	1300–1350	^{14}C	posthole	Jutland	Nielsen 2017
2	Havgård	1000–1050	H/C-Typo	pit, posthole	Jutland	Harild 1999
3	Lykkegård Midt	1150–1200	Dendro	posthole	Jutland	Jensen 2009a
4	Sejerslev	late Viking Age/ early Middle Ages	^{14}C	burnt pit, floor	Jutland	Jensen 2009b
5	Røjgård	1250–1400	^{14}C	well	Jutland	Andreasen 2012
6	Tværvej	1066–1300	^{14}C	well	Funen	Thomsen 2008
7	Bondemosevej	1300–1400	H/C-Typo	well	Funen	Henriksen unpub.
8	Rynkeby Mølle	late Viking Age/ early Middle Ages	H/C-Typo	well	Funen	Henriksen and Harild 2008
9	Bramstrup	1321–1440	^{14}C	well	Funen	Thastrup 2015
10	Hollenæs	1100–1300	H/C-Typo	well, pit	Lolland	Alsleben unpub.
11	Tårnby Torv	1050/1150–1500	^{14}C	well, fill, pit floor, burnt layer, cultural layer	Zealand	Robinson and Harild 2005
12	Herstedøster kirke	1250–1350	H/C-Typo	well	Zealand	Karg 2000
13	Smedegård Nordøst	1100–1150	H/C-Typo	cultural layer, posthole	Bornholm	Henriksen and Jessen 2016
14	Lindängelund	1000–1100	^{14}C	posthole, pit	Scania	Lind 2014
15	Kiaby	late Viking Age/ early Middle Ages	H/C-Typo	pit	Scania	Regnell 2014
16	Örja	1050–1500	^{14}C	well, cultural layer, pit, deposit, oven, ditch, burnt layer, floor	Scania	Lagerås 2013
17	Ängelholm	1295–1425	^{14}C	well	Scania	Regnell 2013
18	Trottaberg	960–1220	^{14}C	posthole	Halland	Ranheden 2015
19	Halmstad	early Middle Ages–1400	^{14}C	cultivation layer, fill	Halland	Gustafsson 2016

place floor; oven/fireplace; pit; posthole; and wells (Figure 3). Some of the features in the studied reports have been merged into one single context type (e.g. 'layer' has been merged with 'cultural layer' and 'fill' has been merged with 'posthole'). Taphonomic processes have a great impact on the archaeobotanical remains, and these will differ by context.

The normenclature of Scientific plant names and common names follow Rostrup and Jørgesen 1979 and Hansen 1993.

Results

There is a clear connection between preservation condition, plant category and context type (Figure 3). Wells were by far the best context for the preservation of herbs. The category vegetables is almost equally distributed across several context types (pit, postholes and well). Herbs are the best represented garden plant category in the analysed material.

The best represented sampled context is pit (55 samples), followed by cultural layer (26 samples), well

Table 2. Overview of date, type of book, and geographical coverage at the time of their publication of selected written sources used in the analysis.

Name	Date	Type	Area	Reference
The Fromond List	c. 1525 AD	gardening list	England	Harvey 1989
The Feate of Gardening	c. 1400 AD	poem	England	Harvey 1985
Forme of Cury	c. 1390 AD	cookbook	England/France	Pegge 1780
Henrik Harpestengs urtebog & kogebog	12th century AD	herbarium/cookbook	Denmark	Harpestreng 1300, Brade 1999, Molbech 1844
Henrik Smid	c. 1546 AD	herbarium/medicine book	Denmark	Smid 1577, Brade 1999
Horticultura Danica	c. 1647–49 AD	gardening book	Denmark	Block 1647
Koge Bog	1616 AD	cookbook	Denmark/Germany	Sartorio 1616

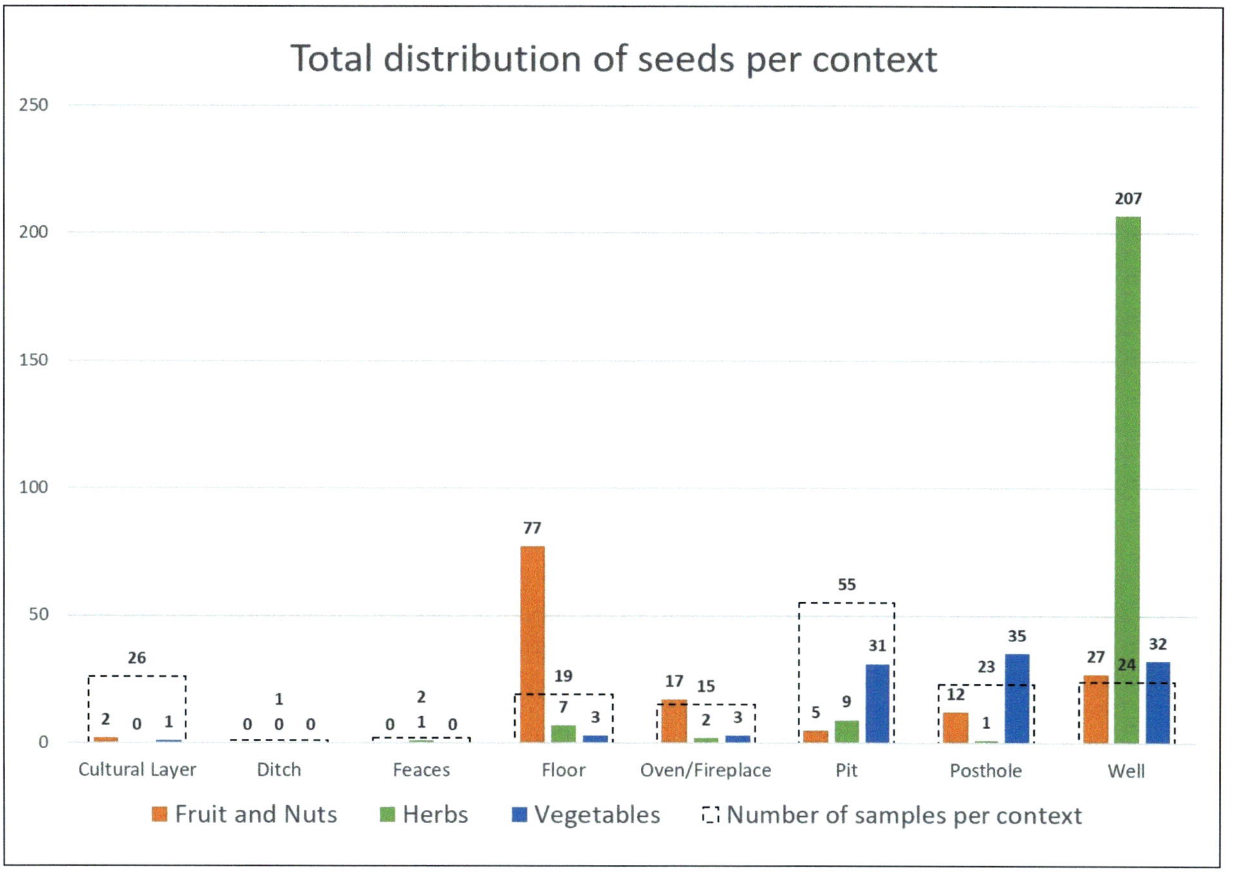

Figure 3. Number of samples per context and total number of seeds per context type, all 19 sites combined. Some contexts have been grouped together, due to similarity in preservation and context type. Number of samples = 165.

(24 samples), and posthole (23 samples) (Figure 3). The waterlogged material originates from wells, pits and floors. Wells have the greatest number of waterlogged materials, whereas pits and floors can contain material preserved either by carbonisation or by waterlogging.

Of the 478 plant macrofossils identified below the taxonomic level of genus, 454 were identified to the species level and 24 to a higher level. The written sources list 87 plant taxa as garden plants. Only 40 of these taxa were recognised among the macroremains. The archaeobotanical material includes 5 taxa that

Table 4 (right). Overview of investigated plant taxa and their presence in the written records listed in Table 2 and in the studied archaeobotanical assemblages, all 19 sites combined. Grey shading indicates that the taxon is present in one or more of the archaeological sites for that period. Black shading indicates that the taxon is mentioned in the written sources listed in Table 2.

Table 3. The distribution of taxa represented in the macrofossil assemblages and the written sources. Both categories and number of seeds is presented. A chronological view has been made across the sites. N = 454.

	Number of taxa	Number of seeds
Fruits & Nuts	23 ; 26 %	130 ; 29 %
Herbs	46 ; 53 %	224 ; 49 %
Vegetables	87 ; 21 %	100 ; 22 %
In Total	87 ; 100 %	454 ; 100 %

could have been included in the medieval cultivation, but that may also have been gathered. Table 3 provides an overview of the presence and number of macrofossils and the taxa in the written sources.

The table 4 shows the general distribution per taxon in the written records and in the archaeobotanical material over time. These data reveal chronological trends in plant introduction and presence, namely, that there is an increase in the introduction of plant species over the course of the Middle Ages in both the archaeobotanical assemblages and the written sources. The best represented period in both the archaeobotanical and written material is 1200–1299 AD. The second-best represented period is 1501–1700 AD.

From the archaeobotanical assemblage, it is deduced that the following plants could have been cultivated on farmsteads in southern Scandinavia: onion (*Allium cepa*), leek (*Allium porrum*), cabbage, pea (*Pisum sativa*), broad bean (*Vicia faba*), coriander (*Coriandrum sativum*), parsley (*Petroselinum crispum*), dill, black/white mustard (*Brassica nigra/ Sinapis alba*), hop (*Humulus lupulus*), hemp (*Cannabis sativa*) and cress (*Lepidum sativum*). All of these are thought to have been cultivated during the Middle Ages by the farmers and clergy (Karg 2007). Other taxa are more uncertain. Goutweed (*Aegopodium podagraria*), wild beet/sea beet (*Beta vulgaris* subsp. *maritima*), wild carrot (*Daucus carota*), fool's parsley (*Aethusa cynapium*), bog-myrtle (*Myrica gale*), henbane, white horehound (*Marrubium vulgare*), valerian (*Valeriana officinalis*), hazel (*Corylus avellana*) and other fruits and nuts could have been gathered from the surrounding landscape and could have come to be included in the household's food consumption in that way.

Discussion

The existing research on medieval rural garden plant selection and use relies almost singularly on the historical written sources from the medieval period. Although these sources can help archaeobotanists recognise signs of horticulture in the archaeobotanical material, they provide a very biased understanding of plant use and practices in the Middle Ages, whereby the upper class and the clergy are the best represented. The use of written sources as the main informant on farmstead horticulture during the Middle Ages can be problematic because the occupants of these sites were neither upper class nor members of the clergy.

Only a small number of plants appear in both datasets. The absence of a taxon in the archaeobotanical record can be a real absence or it can be an effect of faulty sampling strategies or, more likely, poor deposition, formation and preservation conditions. The historical sources can tell us which plants we might expect to find in the archaeobotanical screenings, and thus can guide us in selecting the best contexts to sample from. Table 4 illustrates that there is a correlation between the introduction of plant species in the archaeobotanical and written material. The oldest period is only represented by the archaeobotanical material, which is due to the lack of written sources from that period. The best represented period in both datasets is the period AD 1200–1299. At first it may seem like there were more plant species available in this period. However, this increased representation may reflect the introduction or preservation of medicinal and cookbooks among the upper class and the clergy. The archaeobotanical material is also well represented in the period AD 1200–1299. This appears to be linked to the number of sites investigated in this period.

Vegetables are typically harvested for their fresh and ripe leaves and roots, before they produce seeds and flower, with the exception of legumes, which are allowed to produce seeds before harvesting. The consequence of harvesting before the plant has set seed or produced flowers is that very few seeds are preserved in the archaeobotanical record (Karg 2010). This is why vegetables are one of the less prominent plant species preserved in the archaeobotanical assemblages (Figure 4).

The analysed material suggests that plants categorised as fruits and nuts are most likely to be found in carbonised condition in the context of floors. This might be due to the above-mentioned formation processes. There is a possibility of fruits and nuts being thrown into fires for waste management, e.g. farmers

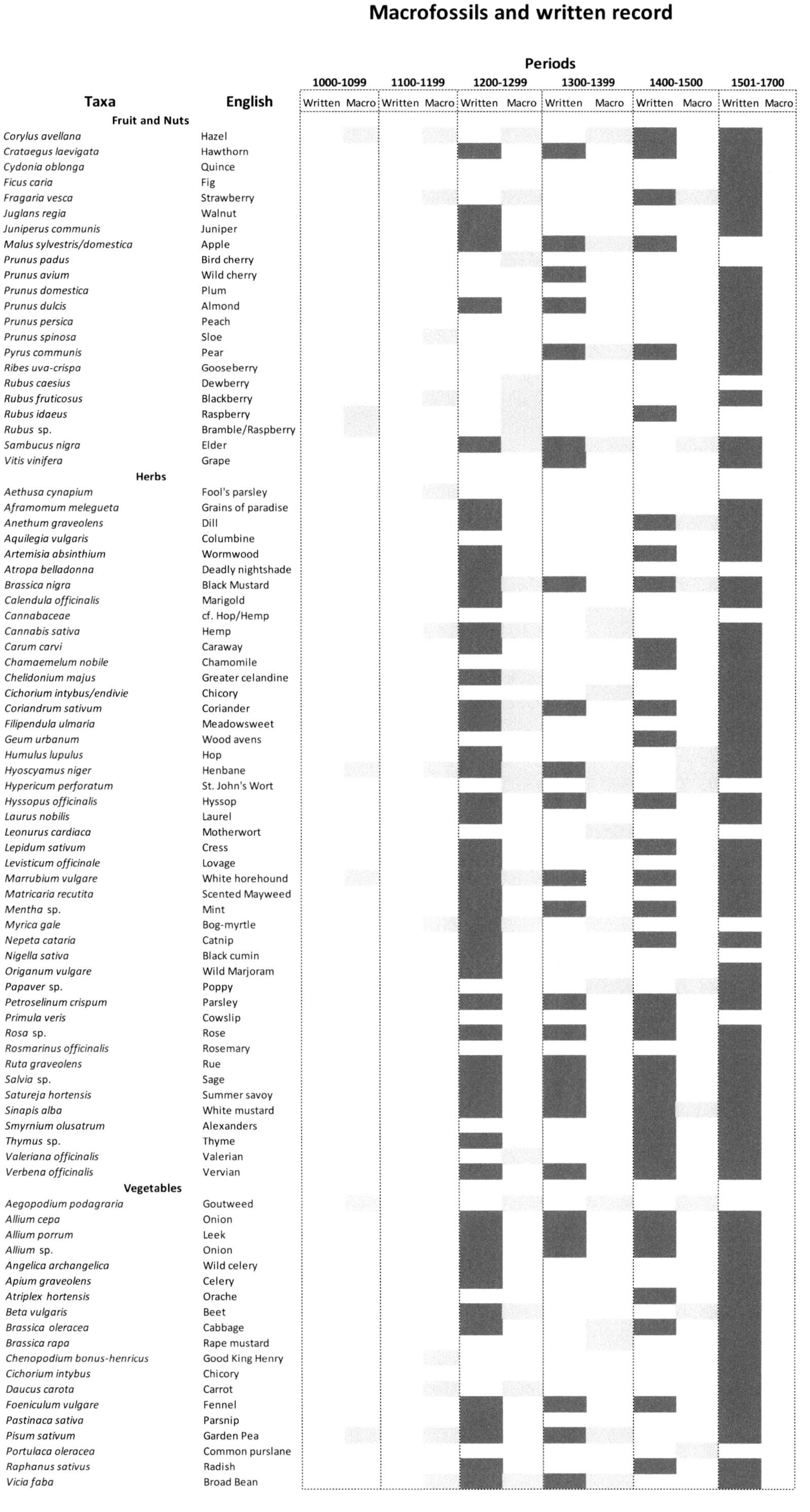
Macrofossils and written record
Periods
1000-1099
1100-1199
1200-1299
1300-1399
1400-1500
1501-1700
Written Macro
Taxa
English
Fruit and Nuts
Corylus avellana Hazel
Crataegus laevigata Hawthorn
Cydonia oblonga Quince
Ficus caria Fig
Fragaria vesca Strawberry
Juglans regia Walnut
Juniperus communis Juniper
Malus sylvestris/domestica Apple
Prunus padus Bird cherry
Prunus avium Wild cherry
Prunus domestica Plum
Prunus dulcis Almond
Prunus persica Peach
Prunus spinosa Sloe
Pyrus communis Pear
Ribes uva-crispa Gooseberry
Rubus caesius Dewberry
Rubus fruticosus Blackberry
Rubus idaeus Raspberry
Rubus sp. Bramble/Raspberry
Sambucus nigra Elder
Vitis vinifera Grape
Herbs
Aethusa cynapium Fool's parsley
Aframomum melegueta Grains of paradise
Anethum graveolens Dill
Aquilegia vulgaris Columbine
Artemisia absinthium Wormwood
Atropa belladonna Deadly nightshade
Brassica nigra Black Mustard
Calendula officinalis Marigold
Cannabaceae cf. Hop/Hemp
Cannabis sativa Hemp
Carum carvi Caraway
Chamaemelum nobile Chamomile
Chelidonium majus Greater celandine
Cichorium intybus/endivie Chicory
Coriandrum sativum Coriander
Filipendula ulmaria Meadowsweet
Geum urbanum Wood avens
Humulus lupulus Hop
Hyoscyamus niger Henbane
Hypericum perforatum St. John's Wort
Hyssopus officinalis Hyssop
Laurus nobilis Laurel
Leonurus cardiaca Motherwort
Lepidum sativum Cress
Levisticum officinale Lovage
Marrubium vulgare White horehound
Matricaria recutita Scented Mayweed
Mentha sp. Mint
Myrica gale Bog-myrtle
Nepeta cataria Catnip
Nigella sativa Black cumin
Origanum vulgare Wild Marjoram
Papaver sp. Poppy
Petroselinum crispum Parsley
Primula veris Cowslip
Rosa sp. Rose
Rosmarinus officinalis Rosemary
Ruta graveolens Rue
Salvia sp. Sage
Satureja hortensis Summer savoy
Sinapis alba White mustard
Smyrnium olusatrum Alexanders
Thymus sp. Thyme
Valeriana officinalis Valerian
Verbena officinalis Vervian
Vegetables
Aegopodium podagraria Goutweed
Allium cepa Onion
Allium porrum Leek
Allium sp. Onion
Angelica archangelica Wild celery
Apium graveolens Celery
Atriplex hortensis Orache
Beta vulgaris Beet
Brassica oleracea Cabbage
Brassica rapa Rape mustard
Chenopodium bonus-henricus Good King Henry
Cichorium intybus Chicory
Daucus carota Carrot
Foeniculum vulgare Fennel
Pastinaca sativa Parsnip
Pisum sativum Garden Pea
Portulaca oleracea Common purslane
Raphanus sativus Radish
Vicia faba Broad Bean

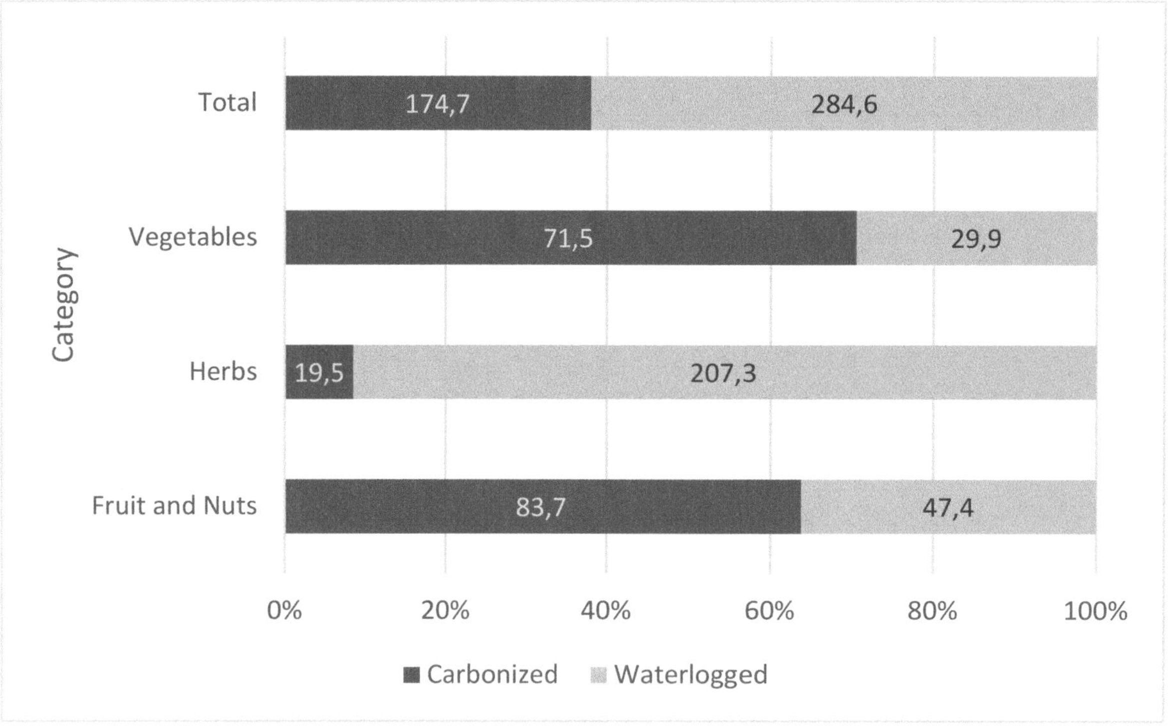

Figure 4. Percentage distribution of waterlogged and carbonised material, all 19 sites combined. Number of seeds = 459,3.

during the medieval period in southern Scandinavia could have managed their waste by burning it (Miller 1982, 135). The remains found in postholes can primarily be interpreted as remnants of house fires or food preparation. Whereas the Well, Posthole and Pit contexts can be interpreted as being waste deposits. Karg and Robinson (2002, 133) claim that fruits & nuts were the most likely to survive in waterlogged condition. However, the material makes it clear that there is no single best preservation condition when working with the category Fruits & Nuts. Nuts are most likely to be preserved as shell fragments. Hazel is preserved only in this way. Some fruits have numerous small seeds on a single berry, such as wild strawberry (*Fragaria vesca*), and others have large stones (Reitz and Shackley 2012, 209; Antolín et al. 2017). There are no archaeobotanical remains of plums (*Prunus domestica*) and other big stone fruits in the investigated records. It is nevertheless a known plant during the Middle Ages (Karg 2007). It can be preserved through both waterlogging and carbonisation. During plant processing, the fruit stone would have been separated from the fruit meat. This would undoubtedly have resulted in the stones being deposited, into, for example, a fire, waste pit, midden or abandoned well. The number of archaeological finds does not provide information about the number of fruits consumed. The fact that the number of finds does not match the number of consumed plants could be the outcome of the chosen sampling strategy. It could be that large stone fruits are difficult to find when working with limited sampling material (Antolín et al. 2017). In order to have the best conditions for finding them, one must sample at least 2–5 litres of soil (Karg and Robinson 2002, 137). This is, however, not always possible. It may be therefore that the samples do not contain many large plant macro remains such as fruit stones. Just because we do not find fruit stones in the macro remains does not mean they were not present in the past. We first recognise plums in post-medieval deposits. The written sources do not mention plum until the period AD 1501–1700, and this may reflect that they were not cultivated in the medieval farmstead garden.

Legumes are usually cultivated for their fleshy seeds (e.g. *Vicia faba*, *Pisum sativa)*, and they are the best represented kind of vegetable on the investigated sites. Broad beans were found at Bramdrup in large numbers in the posthole context, preserved by carbonisation. The material studied here shows that legumes are typically preserved in a carbonised state. The taphonomic process of the legumes is closely connected to the reason why they almost only appear as carbonised macrofossils. The early life stages of the plant, during which it is sown and cared for in the garden, seem to be of the least importance in terms of their archaeological preservation. The data imply that the circumstances affecting preservation happen after harvesting, during storage, food preparation, food consumption, and their incorporation into latrine waste. It is imaginable that the dried legumes were stored together in the same room/location as the cereals, which require the same storage conditions (cool and dark). When the plants are cooked, they are subject to cooking processes, e.g. roasting, chopping, and boiling. All these processes have different effects on the plant. For example, boiling legumes can result in the loss of their defined structure and shape, resulting in them becoming unrecognizable (Babot et al. 2014, 39). Furthermore, mechanical wear from chewing (both human and animal) also makes it hard to identify food remains from latrines. It is seldom that archaeobotanists find leftovers from the vascular, soft plant parts. Perhaps these were given to animals as fodder or used as compost, if they were not fit for human consumption. Hypothetically is it possible that people would drop vegetables into the fire when preparing food. It is, however, most likely that these types of garden plants would be preserved due to house fires. The site of Bramdrup was possibly a result of this. This argument is based on the large number of remains from cereals which were found in postholes from the interpreted building (Nielsen 2017).

Plants in the category of herbs are best preserved in waterlogged conditions. Vegetative parts of vegetables are also most likely to be found in waterlogged features. Leafy parts of the plants are often eaten fresh or as potherbs (boiled). This means that it is unlikely that these parts of the plants will become carbonised. If they are burned, it is very doubtful that they will be preserved in carbonised form and be identifiable afterwards (Reitz and Shackley 2012, 58). Black mustard (*Brassica nigra*) is the best represented herb in a waterlogged state. Mustard has poor prerequisites for preservation in carbonised condition, even though it is usually the seed which is preserved. Mustard seeds can pop (like popcorn) under high temperatures and lose their identifiable shape and proportions (Reitz and Shackley 2012, 293). Most small seeds will have a tendency to do the same under intense heat when roasted (ibid.). This can explain why black mustard is not found in charred state at the studied sites. It is hard to make a clear interpretation of the goutweed as a cooking vegetable. However, we know from written sources that goutweed was used as both a vegetable and a medicinal herb in the Middle Ages (Brøndegaard 1979–1980c, 281–282). There is no evidence of the soft tissue plant parts in the archaeobotanical material, even though it should be possible to find it in wells. The macrofossils may have ended up in the well due to the use of abandoned wells as waste pits. Bakels and Jacomet (2005) argue that archaeobotanists should not compare carbonised and waterlogged material as preservation agents because some seeds will preserve better under conditions of waterlogging than after charring. Both modes of preservation favour the compact, dry part of the plants (typically the seeds) (Bakels and Jacomet 2005, 543). This is the main explanation for why it is rare to find the soft tissue plant parts preserved in the archaeobotanical assemblage.

It can be difficult to puzzle out which 'wild' plants may have been present in the settlement. Geobotanist Rüdiger Wittig's (2004) study on invasive plants in urban settlements suggests that all the plants present in the settlements had specific purposes, including as insect repellents, dyes, medicines, cleaning agents, and raw material for for fibre production, and as food. This assumes that the plants present in settlements were not merely tolerated but also had value. These selected plants were excluded from weeding and were protected by the gardener (Wittig 2004, 325–330). A lot of the plants which today are considered to be weeds would have been thought of as useful in the medieval period. By definition, a weed is a plant which is considered undesirable. What might be a weed in one garden can be viewed as a herb, vegetable or fruit in another garden. Some plants might have been preserved and cultivated for their beauty alone. Today, plants like goutweed are considered weeds. The status of the individual plant thus differs from garden to garden. But there are also regional aspects to take into consideration, namely, what is fashionable in the region. Wittig (2004, 325) suggest in relation to urban settlements that those plants still present in today's flora are primarily plants which were introduced by humans at one point in history (i.e. they were not native to the ecosystem).

When we as archaeobotanists select which plants might have been present in medieval farmstead gardens, we have no possibility of knowing whether they were gathered or cultivated, or perhaps both. Plants such as elder tree (*Sambucus nigra*), hazel bush or plot with St. John's wort (*Hypericum perforatum*), which is utilised year after year, might not to be considered a part of the farmstead economy. Although, the medieval peasant population might have considered that exact tree, bush or plot a part of their garden. They would have returned and cared for the plant continuously, but would not have moved it to within the boundaries of the parcel. Some plants are easy to categorise as cultivated garden plants. Plants such as hemp and grape (*Vitis vinifera*) are both seen primarily in human contexts and are known for both their culinary use and their use as medicinal remedies. Other plants, like black mustard, hazel, elder and fool's parsley, are present both in the medieval garden/herb descriptions and as part of the natural landscape (Hansen 1993). This is also illustrated in the investigated archaeobotanical material.

An aspect which is yet to be discussed is that of imports and class-society in connection to the farmstead gardens. Peasants would occasionally visit monasteries and urban markets, where it would have been possible to buy dried herbs and seeds. These types of goods can travel over long distances before reaching their final destination. Sloth et al. (2012, 10) mention that strontium analysis has been conducted on archaeobotanical remains from Fyrkat in Northern Jutland (AD 980) and proved that herbs like coriander were locally grown. Furthermore is there the aspect of fashion. The aristocracy, clergy and burghers are an important in making plants fashionable among the peasants. Plants like rose (*Rosa* sp.) and grape are considered connected to the clergy and aristocracy. Historian Montanari stated that even though the written sources often are a reflection of the high levels of society, they still have some testimonial value in connection to the peasants (Montanari 2015, 115).

Conclusions

This study has shown that it is possible to recognise gardening in farmsteads in Denmark and southern Sweden during the Middle Ages, with help from historical written sources and macrofossils. It has also shown that written sources must not be taken as the definitive answer. It is important to remember that they were written by the clergy and upper class, for themselves. They could reflect a tendency in horticulture in the general population, but may also exclusively represent the upper class of medieval southern Scandinavia. However, the written sources can help archaeobotanists understand which plants may be markers for horticulture and which level of society they would be connected to. The integration of archaeobotanical data with written sources can help unlock knowledge of plant selection and plant use on medieval farmsteads in southern Scandinavia.

It is not impossible to recognise farmstead gardens, and there is evidence of their existence in the archaeobotanical material. Archaeological context and mode of preservation are important clues to recognising farmstead gardens and may guide us in selecting the best sampling strategy. Based on this study, it is recommended to target waterlogged contexts, such as wells and pits, for sampling because soft tissues, such as seeds and leafy remains, have the best chance at being preserved under these conditions. Fireplaces and ovens are also recommended for sampling. These contexts, with their charred remains, provided the highest number of garden plants in the study. The results in this paper derive from only a small percentage of all excavated rural medieval settlements in the investigated regions. Only a small fraction of the total number of medieval rural settlements have been excavated, and not all of these have been archaeobotanically sampled. Multiple plants which were present in the written records were not identified in the archaeobotanical record. This, however, does not mean that these plant taxa were not cultivated. Preservation mode and taphonomy have been shown to have a significant impact on their archaeological survival. It should be stressed that none of the datasets would have worked on their own, and that the best insight was gained by allowing them to supplement each other. The problems of preservation are a limiting factor to archaeobotanical data. This, however, is an aspect which is completely out of our hands as archaeologists and archaeobotanists. Vegetative parts of the vegetables have a low possibility of surviving in the archaeobotanical record, which is the main reason it is not possible to estimate the original yield or contents of the farmstead garden. Further investigation and new sampling methods are required.

It is suggested that the peasant's resources may have extended beyond the garden wall. Some plants which can be found in the wild or at the commons may not have been cultivated. However, the gardeners/peasants could have considered the wild resource as a part of their garden. Although this is hard to prove archaeologically and archaeobotanically, it is worth taking into consideration when working with the concept of gardening.

The information on archaeobotanical material in southern Scandinavia connected to medieval farmsteads is scarce. Each year, the local museums with responsibility for archaeology sample numerous contexts, which might help us recognise gardens during the Middle Ages. However, it is rare that the museums are specifically looking for evidence of gardens. Therefore the samples most likely do not originate from contexts most suited to the investigation of gardens (Alanko et al. 2015). To recognise horticulture in the archaeological assemblage, it is important to keep context, volume and strategy in mind when sampling, depending on the research questions at hand. Since the rural medieval settlement can be poorly preserved, it is not always possible to gain knowledge on specific plants types, i.e. herbs and vegetables. If archaeobotanists and archaeologists want to find the physical garden, other methods have to be used. Methods such as geophysics, archaeological excavation, micromorphology and cartographic/written historical sources may prove to be useful (Nielsen 2018, 106). In general, this study of medieval farmstead gardens showed that the research field is in need of new methods in addition to the inter-disciplinary method of comparing written sources and plant macroremains.

References

Alanko T, Oinonen M, Schulman L (2015) Plant remains from early modern garden of the manor of Kumpula, Hensinki, Finland: An alternative sampling method for macrofossil analysis. Veg Hist Archaeobot 24:571–585

Alsleben A (unpub) Hollenæs. Archaeobotanical Database, National Museum of Denmark.

Andreasen MH (2012) Analyse af makrofossiler fra brønd fra slutningen af 1200-tallet/begyndelsen af 1300-tallet (Rapport Afdelingen for Konservering og Naturvidenskab, Moesgaard Museum nr. 9). Moesgaard Museum, Højbjerg

Antolín F, Steiner BL, Jacomet S (2017) The bigger the better? On sample volume and the representativeness of archaeobotanical data in waterlogged deposits. J Archaeol Sci: Reports 12:323–333

Babot MP, Lund J, Olmos AV (2014) Taphonomy in the kitchen: Culinary practices and processing residues of native tuberous plants of the south-central Andes. Intersecciones en Antropología 15(1):35–53

Bakels C, Jacomet S (2005) Access to luxury foods in Central Europe during the Roman period: The archaeobotanical evidence. World Arch 34(3):542–557

Block HR (1647) Horticultura Danica (Ældste Danske Havebog, reprinted 1984). Copenhagen,

Wormianum, http://www.kb.dk/permalink/2006/manus/732/eng/. Accessed 7 April 2019

Brade AE (1999) Henrik Harpestrengs lægeurter og mineraler (Medicinsk-Historisk Småskrifter nr. 8). Københavns Universitets Medicinsk-Historiske Museum, Copenhagen

Brøndegaard VJ (1979–1980a) Folk og flora, bind 1. Rosenkilde og Bagger, Tønder

Brøndegaard VJ (1979–1980b) Folk og flora, bind 2. Rosenkilde og Bagger, Tønder

Brøndegaard VJ (1979–1980c) Folk og flora, bind 3. Rosenkilde og Bagger, Tønder

Brøndegaard VJ (1979–1980d) Folk og flora, bind 4. Rosenkilde og Bagger, Tønder

Gustafsson S (2016) Arkeobotanisk analys av förkolnade växter från Halmstad 45:1. In: Larsson F (ed) Staden der blev till sand. Kulturmiljö Halland, Halmstad, pp 127–129

Gleason K L (1994) To Bound and Cultivate: An Introduction to the Archaeology of Gardens and Fields. In: Miller N (ed), Gleason K L (ed) The Archaeology of Garden and Field. University of Pennsylvania Press, Philadelphia, pp 1-24

Hansen K (1993) Dansk Feltflora. Gyldendal, Viborg

Hansson, AM (1997) On plant food in the Scandinavian peninsula in early medieval times. Dissertation, University of Stockholm

Harild JA (1999) Arkæobotaniske undersøgelser af flotationsprøver fra bebyggelsen Havgård (M&M rapport nr. 23). Nationalmuseet, Copenhagen

Harpestreng H (1300) Harpestrengs kogebog. Det Danske Sprog og Litteratur Selskab. https://tekstnet.dk/harpestreng-nks70r/metadata. Accessed 13 April 2020

Harvey J (1985) The first English garden book: Mayster Jon Gardener's Treatise and its background. Garden History 13(2):83–101

Harvey J (1989) Garden plants of around 1525: The Fromond List. Garden History 17(1):122–134

Heimdahl J (2010) Barbariska trädgårdsmästare: Nya perspektiv på hortikulturen i Sverige framtill 1200-talets slut. Fornvännen 105(4):265–280

Henriksen PM, Harild JA (2008) Rynkeby Mølle: Makroanalyser fra brønd DMT (M&M rapport 15). Nationalmuseet, Copenhagen

Henriksen PS, Jessen C (2016) Pollen- og makrofossilundersøgelser af prøver fra to gruber ved Smågårde 3, TAK 1668 (M&M 49). Nationalmuseet, Copenhagen

Henriksen PS (unpub) Bondemosevej, OBM 398. Archaeobotanical Database, National Museum of Denmark.

Hjelmqvist H (1991) Några trädgårdsväxter från Lunds medeltid. Sven Botanisk Tidskr 85:225–248

Jensen PM (2009a) Arkæobotanisk diskussion af to middelalderlige gårdsanlæg (Rapport Afdelingen for Konservering og Naturvidenskab, Moesgaard Museum nr. 8). Moesgaard Museum, Højbjerg

Jensen PM (2009b) Arkæobotanisk analyse af korn- og frøprøver fra sen vinkingetid/tidlige middelalder (Rapport Konservering og Naturvidenskab, Moesgaard Museum nr. 10). Moesgaard Museum, Højbjerg

Karg S (2000) Medieval plant remains from a well near Herstedøster Kirke in Herstedøster (M&M Rapport 27). Nationalmuseet, Copenhagen

Karg S (2007) Long term dietary traditions: Archaeobotanical records from Denmark dated to the Middle Ages and early modern times. In: Karg S (ed) Medieval food traditions in northern Europe (PNM Studies in Archaeology & History). Nationalmuseet, Copenhagen, pp 137–160

Karg S (2010) Food from gardens in northern Europe: Archaeobotanical and written records dated to the medieval period and early modern times. In: Bakels C (ed) Van planten en slakken/Of plants and snails: Festschrift for Wim Kuijper. Sidestone Press, Leiden, pp 115–125

Karg S, Robinson D (2002) Secondary food plants from medieval sites in Denmark: Fruits, nuts, vegetables, herbs and spices. In: Viklund K (ed) Nordic Archaeobotany—NAG 2000 in Umeå. Archaeology and Environment 15:133–142

Kristiansen MS (2011) Strategrafisk analyse, faser og datering. In: Kristiansen MS (ed) Tårnby- Gård og Landsby gennem 1000 år. Jysk Arkæologisk Selskabs Skrifter, Aarhus, pp 88–144

Lagerås P (2013) Den agrara ekonomin och växrutnyttjandet i Örja: Analyser av växtmakrofossil och pollen. In: Sabo KM (ed) Örja 9:1. Riksantikvarieämbetet, Lund, pp 207–216

Landsberg S (2003) The medieval garden, 2nd edn. British Museum Press, London

Larsson I (2010) Millefolium, rölika och näsegräs: Medeltidens svenska växtvärld i lärd tradition. Kungliga Skogs- och lantbruksakademien, Stockholm

Larsson M, Lagerås P (2014) Bilaga 5: Rapport avseende Paleoekologisk analys. In: Carlie A, Lagergren A (eds) Lindängelund 1, Bebyggelse och offerplats från yngre stenålder, bronsålder, järnålder samt vikingatid/tidig medeltid. Riksantikvarieämbetet, Lund, pp 167–183

Lindeblad K (2006) Den gröna staden. In: Larsson S (ed) Nya stadsarkeologiska horisonter. Riksantikvarieämbetet, Stockholm, pp 301–318

Lundquist K (1994) Bondens trädgård: En funktion av många nyttiga och några sköna växter. In: Larsson B (ed) Bondens självbild och natursyn. Nordiska museet, Stockholm, pp 146–148

Miller NF (1982) Economy and environment of Malyan, a third millennium B.C. urban center in southern Iran. Dissertation, University of MichiganS

Molbech C (1844) En dansk Kogebog fra det 14de Aarhundrede, meddeelt. Historisk Tidsskrift 1(5):537–546

Montanari M (2015) Medieval Tastes - Food, Cooking and the Table. Columbia University Press. New York

Nielsen KK (2017) Arkæobotnisk analyse af et hus fra middelalderlandsbyen ved Bramdrup. HAM 4950, Bramdrup (FHM 4296/763) (Rapport Afdelingen for Konservering og Naturvidenskab, Moesgaard Museum nr. 1). Moesgaard Museum, Højbjerg

Nielsen KK (2018) Dansk havearkæologi: Strategi og fremtid. Gefjon 3, pp 98–116

Pegge S (1780) The Forme of Cury. Project Gutenberg. http://www.gutenberg.org/cache/epub/8102/pg8102-images.html. Accessed 13 April 2020

Ranheden H (2015) Bilaga 3. In: Nordvall L (ed) Albinsro: Bebyggelsesutveckling från senneolitikum till idag. Kulturmiljö Halland, Halmstad, pp 139–145

Regnell M (2013) Bilaga. In: Lindgren CH (ed) Ängelholm 2:47 och 2:48. CMB Uppdragsarkeologi, Löddeköping

Regnell M (2014) Botaniska analyser av jordprover från Kiaby 90:5, Skåne. CMB Uppdragsarkeologi, Stockholm

Reitz E, Shackley M (2012) Environmental archaeology. Springer, New York

Robinson DE, Harild JA (2005) Agrarøkonomi og omgivende landskab. In: Kristiansen MS (ed) Tårby: Gård og landsby gennem 1000 år. Jysk Arkæologisk Selskab, Aarhus, pp 423–444

Roesdahl E (1999) Indledning. In: Roesdahl E (ed) Dagligliv i Danmarks Middelalder. Nota, Aarhus, pp 11–24

Rostrup E, Jørgensen CA (1979) Den danske flora. A. Hansen. København.

Sartorio S (1616) Indeholdendis et hundrede fornødene stycker, Som ere/ om Brygning/ Bagning/ Kogen/ Brendevijn oc Miød at berede/ saare nytteligt vdi Huszholding/ etc. Som tilforn icke paa vort Danske Sprock vdi Tryck er vdgaaen. Renæssance Sprog i Danmark. http://renæssancesprog.dk/tekstbase/Kogebog_1616/1/view?query_id=None. Accessed 13 April 2020

Sloth PR, Hansen UL, Karg S (2012) Viking Age garden plants from southern Scandinavia: Diversity, taphonomy and cultural aspects. Danish J Archaeol 1:6–12

Smid H (1577) Henrik Smiths Lægebog I–VI (reprinted 1976). Rosenkilde og Bagger, Copenhagen

Taylor C (2000) Medieval ornamental landscapes. Landscapes 1:38–55

Thastrup MB (2015) Arkæobotanisk analyse a kornprøve fra Middelaldern (Rapport Konservering og Naturvidenskab, Moesgaard Museum nr. 6). Moesgaard Museum, Højbjerg

Thomsen, MS (2008) OBM8377 Tværvej, Asperup sogn, Vends herred, tidl. Odense amt. Sted nr. 08.07.01. Sb.nr. 54. Odense Bys Museer, Odense

Tollin C (2005) Medeltida grönsaksodling. In: Tunób H (ed) Etnobiologi i Sverige 2. Wahlström & Widstrand, Stockholm, pp 306–309

Viklund K (1998) Cereals, weeds and crop processing in Iron Age Sweden: Methodological and interpretive aspects of archaeobotanical evidence. Dissertation, Umeå University

Wittig R (2004) The origin and development of the urban flora of Central Europe. Urban Ecosystems 7(4):323–329

Roman Iron Age and Migration period plant cultivation at Salo Isokylä, south-western Finland

Santeri Vanhanen

Author address:
Arkeologerna Lund, Statens historiska museer

E-mail: santeri.vanhanen@arkeologerna.com

Abstract

Iron Age plant cultivation in Finland is poorly known because archaeobotanical studies have been conducted on only a few sites. Marjatta Aalto studied and partly published the rich archaeobotanical material from Salo Isokylä (municipality of Salo), where three sites, Ketohaka 1 and 2 and Vanutehtaanmäki 1, were excavated between 1978 and 1982. This archaeobotanical material has not been previously published in its entirety, and the chronology has been based on conventional radiocarbon dates on charcoal. In this chapter, the plant material is revisited, presented in detail, and AMS radiocarbon dated. The aim is to get a better understanding of Iron Age plant cultivation in the area and to understand whether crop choices reflect contact networks manifested in artefact finds. AMS dates confirm that the plant remains originate from the Roman Iron Age and the Migration period, c. AD 100–500. As was the case at the majority of coeval sites in Finland, barley was the main crop. New identifications show for the first time that naked and hulled barley were cultivated in roughly equal amounts. Emmer/spelt and oat were minor crops. Rye and bread/club wheat appeared as minor crops during the Migration period. Arable weed flora from the sites indicate manured fields where spring-sown crops were cultivated and harvested low on the stem, with a sickle. Flax is present in Roman Iron Age contexts. A hemp seed was AMS dated to cal AD 258–425, which makes it the earliest hemp find in Finland. A comparison with other regions around the Baltic Sea shows that agricultural knowledge as well as hemp cultivation could have been brought to Isokylä via trade networks manifested in the archaeological record.

Introduction

Marjatta Aalto published on the rich and interesting archaeobotanical materials from three sites in Salo Isokylä (municipality of Salo): Ketohaka 1, Ketohaka 2, and Vanutehtaanmäki 1 (Aalto 1982, 1997). In these studies, she presented assemblages dated to the Roman Iron Age and the Migration period (for periodization see Table 1), consisting mainly of barley, but also emmer wheat, possible spelt wheat, bread wheat, rye, oat, flax, pea, various wild species, and even charred grapevine pips, which she took to indicate export from wine-growing regions in the form of raisins or wine lees (Aalto 1997). Unfortunately, these archaeobotanical assemblages were never fully published. A full plant list has only been available from Ketohaka 1 (Aalto 1982), and the papers do not provide full contextual information (Aalto 1982, 1997). However, this material, rich in charred archaeobotanical remains, is of great interest because only a few other assemblages dated to the Roman Iron Age and the Migration period are available in Finland and our knowledge of crop plants during these periods is very limited.

Table 1. Iron Age periodisation and date ranges used in the text.

Pre-Roman Iron Age	500–0 BC
Early Roman Iron Age	AD 0–200
Late Roman Iron Age	AD 200–400
Migration period	AD 400–550
Merovingian period	AD 550–800
Viking age	AD 800–1050

The archaeobotanical material from Isokylä was deposited at the Department of Archaeology at the University of Helsinki. Here, it was organised. Then funds were obtained to conduct AMS radiocarbon dating of the plant remains. During the organising of the collection and the choosing of material for AMS radiocarbon dating, it was realised that the identifications of the archaeobotanical remains should be checked because all taxa were not reported in the original publications (Aalto 1982, 1997), barley va-

Figure 1. Map showing Roman Iron Age and Migration period sites discussed in the text: Raasepori Bäljars 2 (Vanhanen and Koivisto 2015; Alenius et al. 2017), Vantaa Gubbacka (Vanhanen and Mikkanen 2013; Alenius et al. 2017), Paimio Spurila (Seppä-Heikka 1985; Aalto 1997), Tammela Lepokallio, Valkeakoski Rapola (Vikkula et al. 1994), Lempäälä Naarankalmanmäki (Raike and Seppälä 2005), Vammala Ristimäki (Luoto et al. 1983), Nokia Viik (Lempiäinen 1991), and Vöyri Pörnullbacken (Engelmark and Viklund 2002) (map: Santeri Vanhanen).

rieties (naked or hulled) were not defined, and there were remains, such as grapevine pips, that are noteworthy because of their rarity in Finland. There were also remains that were not identified, and their species identification was carried out, when possible. All of the remains were therefore re-examined.

The chronology of these sites was already relatively well understood because 47 conventional radiocarbon dates had been obtained from charcoal samples. However, charcoal can be subject to the 'old wood effect' and may thus not be contemporaneous with the other archaeobotanical remains. In the current study, a better chronological control for the plant remains was sought by obtaining nine new AMS radiocarbon dates from charred seeds and grains and a grapevine pip.

By providing new dates and full plant lists with contextual information for the three sites, the first aim of this chapter is to study what crops were cultivated at Isokylä during the Roman Iron Age and the Migration period, with the ultimate goal of better understanding poorly known Iron Age cultivation practices in Finland. At Isokylä, gold and other rich finds (Hirviluoto 1991) date to the Late Roman Iron Age and the Migration period, but habitation in the area continued into later periods (Schauman-Lönnqvist 1989). The rich finds have been interpreted as representing the presence of an elite who had contacts with Scandinavia and the Baltic countries (Raninen 2006). The second aim of the chapter is to discuss whether these contacts within the Baltic Sea region could be discerned through the archaeobotanical material.

Studied sites

Isokylä, in the community of Salo, is situated in south-western Finland (Fig. 1). The earliest habitation at Isokylä dates to the Stone Age, and occupation continues during the Bronze Age and the Pre-Roman Iron Age. Most of the area's numerous ancient monuments date to the Iron Age (Fig. 2). The sites of Ketohaka 1 and 2 and Vanutehtaanmäki 1 are situated on south-east-facing hills in the Salonjoki River valley. The river runs into a long bay of the Baltic Sea, which was closer to the site during the Iron Age than today, due to post-glacial land uplift. The river valley is characterised by a clayey valley floor that is surrounded by hills with exposed bedrock, till and sand (Matiskainen 1982).

A pollen core from a pond situated in the immediate vicinity of the Ketohaka 2 settlement shows evidence

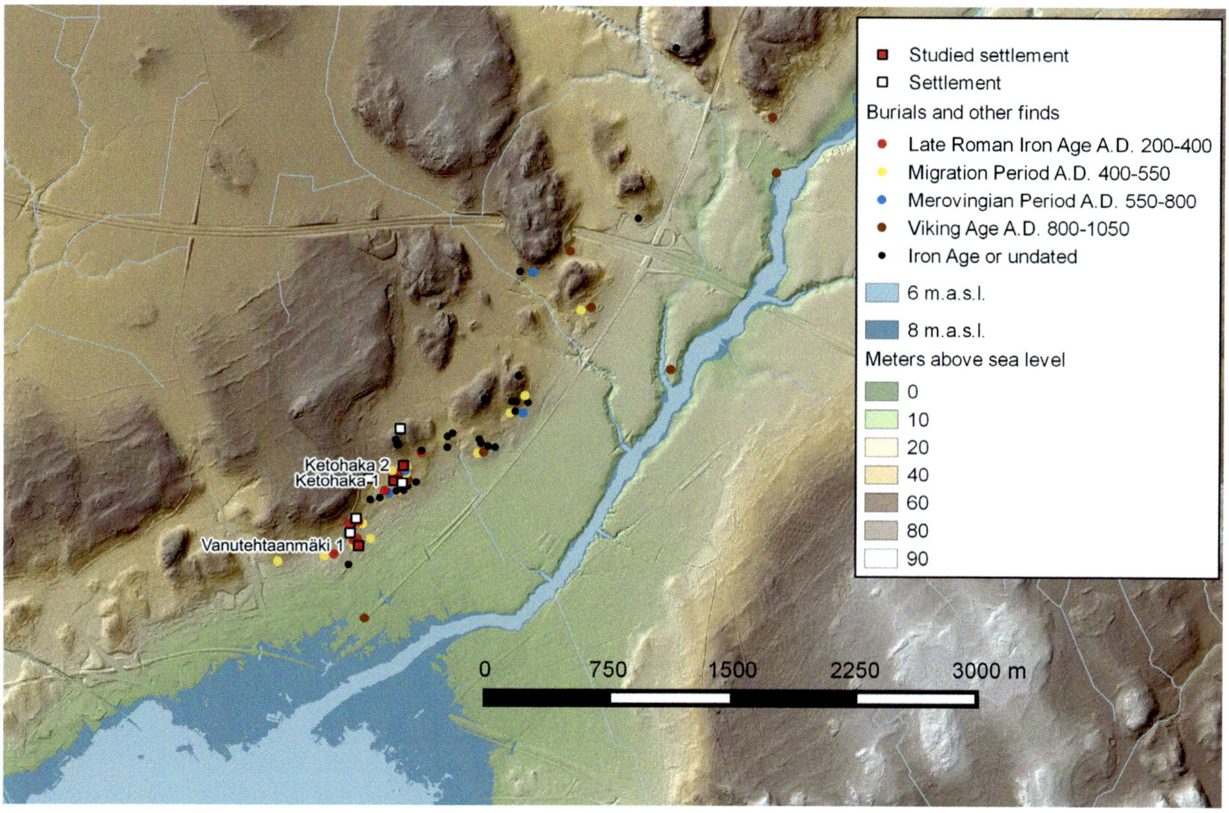

Figure 2. Map showing elevation around Isokylä and locations of the studied sites and other remains dating to the Iron Age (Schauman-Lönnqvist et al. 1986; Schauman-Lönnqvist 1989). Elevations of 8 m and 6 m above current sea level correspond to circa AD 100 and AD 500, respectively (Vuorela et al. 2009) (elevation model data: National Land Survey of Finland Topographic Database 07/2017; map: Santeri Vanhanen).

for plant cultivation and cattle grazing during the 1st millennium AD, as well as signs of clearance during the Bronze Age and the Pre-Roman Iron Age (Tolonen 1985). Although limited, Iron Age osteological materials from Isokylä show the keeping of pigs, horses, and sheep, supplemented by the hunting of seal (Hirviluoto and Vormisto 1984; Schauman-Lönnqvist 1989; Bläuer and Kantanen 2013). Animal husbandry is also manifested in the material culture by a find of shears in a Migration period burial at Ketomäki (no. 43) (Schauman-Lönnqvist 1989) and a horse bit at the Ketohaka 1 settlement (Hirviluoto 1991).

Settlement excavations at Isokylä formed part of a project called *The Iron Age society of Finland*, directed by Professor Carl Fredrik Meinander, of the Department of Archaeology at the University of Helsinki (Schauman-Lönnqvist et al. 1986). This project focused on settlements, which had at the time received much less attention than graves in Finnish archaeology (a bias that continues to persist today). The study aimed to research the social structure, organisation, and economic basis of the Iron Age communities. The archaeobotanical material discussed in this article was obtained during excavations in the period 1978–1982 from settlements and graves in Ketohaka 1 and 2 and Vanutehtaanmäki 1. According to these studies, the settlement sites Ketohaka 1 (also known as Katajamäki) and Ketohaka 2 date mainly to the Roman Iron Age (Uino 1986). Vanutehtaanmäki 1 and a burial cairn at Ketohaka 2 date to the Migration period (Schauman-Lönnqvist et al. 1986; Uino 1986). Archaeological material from the settlements at Ketohaka 1 and Ketohaka 2 has been published in detail by Pirjo Uino (1986), and a cairn studied at Ketohaka 2 has been published by Marianne Schauman-Lönnqvist (1986). For these sites, the botanical samples could be connected to the contexts presented in the publications. Archaeological material from Vanutehtaanmäki 1 has not been published in detail. There are, however, excavation reports from the site and notes in Aalto's material that enabled me to connect the botanical material with excavated structures.

Materials

This article is based on archaeobotanical material initially studied by Aalto, which includes several other sites, partly published and partly unpublished. A small amount of archaeobotanical material was also available from the settlement sites of Vanutehtaanmäki 2 and 3, but this is not included in this article. The material from Ketohaka 1 consists of 23 samples from discoloured soil, 24 samples from postholes, and 21 samples from hearths (Aalto

1997), of which 41 contain charred archaeobotanical material (Supplementary table 1). Material from Ketohaka 2 consists of 29 samples from cultural layers (discoloured soil), 22 samples from postholes, 3 samples from hearths, and 3 samples from a burial cairn (Aalto 1997), of which 40 contain charred archaeobotanical material (Supplementary table 2). At Vanutehtaanmäki 1, samples from more than 40 postholes and 1 soot-pit were studied (Aalto 1997), and here charred archaeobotanical material was found in 43 samples (Supplementary table 3).

Methods

Soil samples were floated and screened by Aalto and her field assistants. The samples were floated by using a salt water method modified from Núñez and Vuorela (1976). Using this method, the samples were poured into a bucket filled with saturated salt water, and floating particles were caught using a sieve with a mesh size of 0.7–1.0 mm (Aalto 1982). The salt water was reused after being sieved with a mesh of 0.3–0.5 mm. This method was used for samples from Ketohaka 1 and also appears to have been used for those from Ketohaka 2 and Vanutehtaanmäki 1. The soil samples were then screened, after which the plant material was determined with a stereomicroscope by Aalto. She also took some photographs and made line drawings of the plant material (Aalto 1982, 1997).
For this chapter, the archaeobotanical material was organised, the identifications were checked, and new AMS radiocarbon dates were obtained. All identifications were checked using literature (Cappers et al. 2006) and seed reference material stored at the Finnish Museum of Natural History. The taxonomic nomenclature in this chapter follows Hämet-Ahti et al. 1998. Some samples, containing only charred unidentified material or uncharred material, were excluded from this chapter. Nine remains were chosen for AMS radiocarbon dating. The aim was to pick seeds or grains from closed contexts, such as postholes or hearths, containing large amounts of charred botanical material. An uncharred grapevine seed was dated to find out whether it is a recent intrusion because this species has not been found in other prehistoric contexts in Finland. All dates presented were calibrated using OxCal 4.3. (Bronk Ramsey 2009) with the IntCal 13 atmospheric curve (Reimer et al. 2013). All calibrated ages are presented as the two standard deviations (2σ) range (Table 2).

Results

Results from archaeobotanical analysis and AMS radiocarbon dating from Ketohaka 1, Ketohaka 2, and Vanutehtaanmäki 1 are presented below and in Tables 2–3. Raw data is presented in Supplementary tables 1–3, which are available online only, see Acknowledgements. Excavation plans and selected contexts rich in archaeobotanical remains are presented in Figures 3–5, and images of selected plant remains are presented in Figures 6 and 7.

AMS radiocarbon dates

The new AMS dates from Ketohaka 1 all gave results that fall within the Roman Iron Age (Table 2). AMS dates from charred wooden vessel 15 and hearth 95 are in accordance with the earlier, conventional radiocarbon dates. Material from Ketohaka 2 resulted in Roman Iron Age dates. AMS dates from ditch 150, posthole 165, and hearth 169–170 are in accordance with the conventional dates. Here the resolution of the new AMS dates ascertains that these remains date to the Roman Iron Age. Fewer dates were obtained for Vanutehtaanmäki 1. Both a single conventional date and the two new AMS dates point towards a Late Roman Iron Age or Migration period use of the site. An uncharred grapevine pip (*Vitis vinifera*) (Fig. 7) from a posthole in Vanutehtaanmäki 1 was dated to cal AD 1682–1935, thus confirming that it is a recent intrusion.

Archaeobotanical remains

Assemblages from the sites contain large amounts of charred plant remains: 477 were recovered at Ketohaka 1, 556 at Ketohaka 2, and 1260 at Vanutehtaanmäki 1. Cultivated plants make up the bulk of the plant remains at each site, whereas arable weeds and wild gathered plants were found in smaller numbers. Wetland/meadow plants were present at the sites of Ketohaka 1 and 2, but absent at Vanutehtaanmäki 1. In the following section, cultivated plants are discussed in more detail.

Cultivated plants

Species-level identifications of cultivated plants are summarised in Table 3. These show that barley (*Hordeum vulgare*) is the main crop in all assemblages. Both hulled and naked barley are present in the majority of assemblages, and there are no clear changes in the representation of the different barley varieties through time. Oat (*Avena* sp.) is present in both older and younger contexts, but no floret bases were found, which precludes establishing whether oat was cultivated or an arable weed. Emmer/spelt (*Triticum dicoccum/spelta*) wheat is present in both Roman Iron Age and Migration period contexts in small amounts. Rye (*Secale cereale*) is present in Late Roman Iron Age/Migration period assemblages and is most numerous in Migration period pit 142.

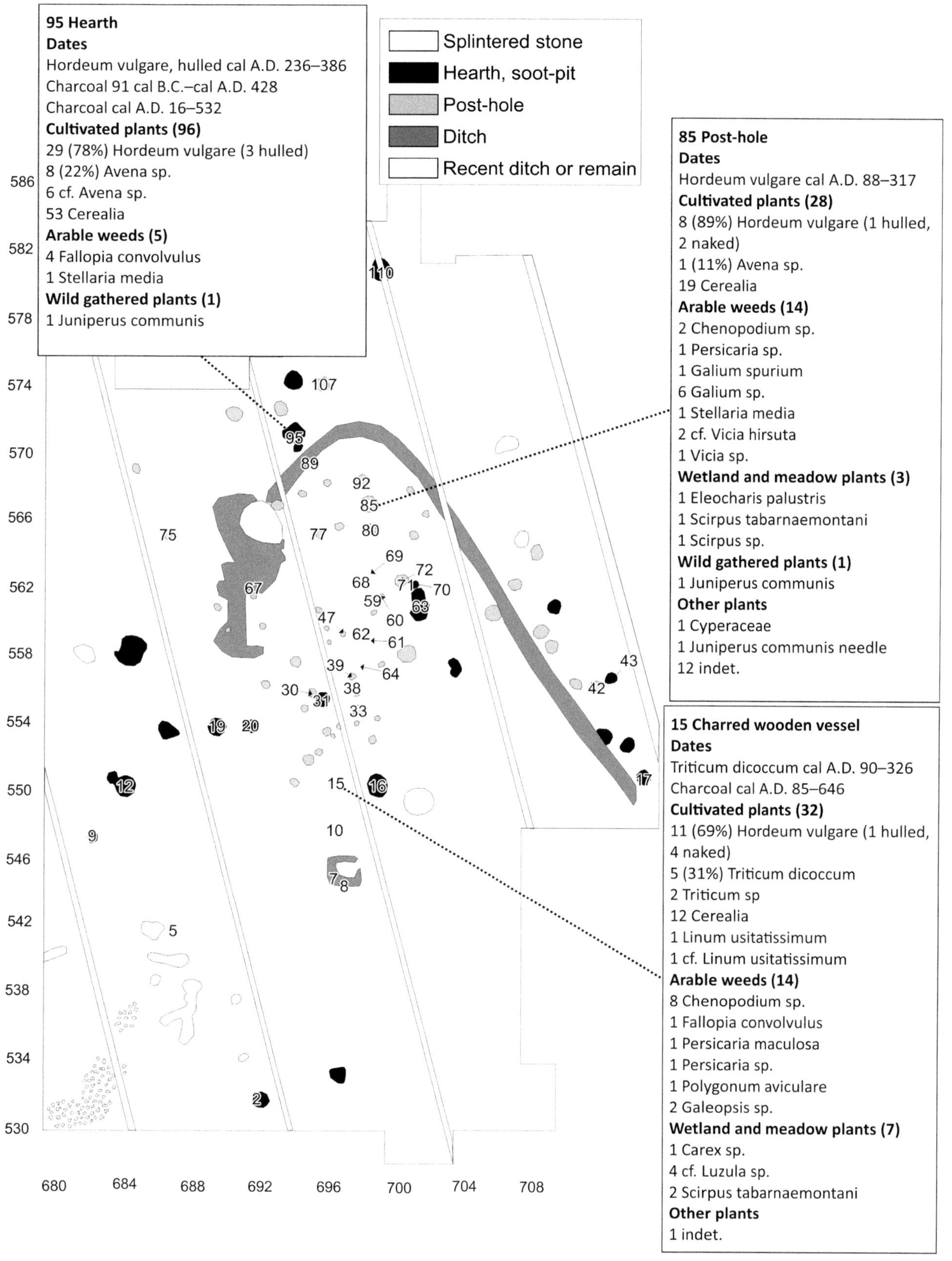

Figure 3. Excavation plan of Ketohaka 1 showing grid in metres. Context numbers refer to Supplementary table 1. Insets: Archaeobotanical results from contexts with the most numerous and best-dated plant remains. Percentages of cereals are based on the total number of identified cereals (map: Santeri Vanhanen, after Uino 1986).

Also, a single bread/club wheat (*Triticum aestivum* s.l.) grain was identified from the Late Roman Iron Age/Migration period site of Vanutehtaanmäki 1. Flax (*Linum usitatissimum*) occurs only in an assemblage dated to the Roman Iron Age, and one hemp (*Cannabis sativa*) is dated to the Late Roman Iron

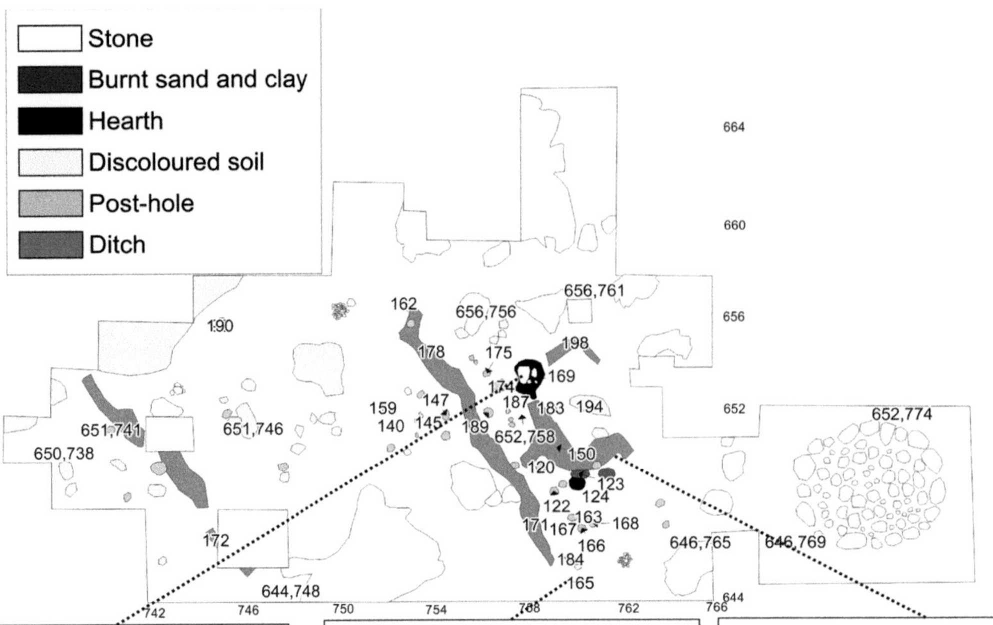

169-170 Hearth
Dates
Hordeum vulgare, hulled cal A.D. 128–323
Charcoal 386 cal B.C.–cal A.D. 221
Cultivated plants (44)
18 (100%) Hordeum vulgare (11 hulled)
26 Cerealia
Arable weeds (7)
7 Fallopia convolvulus
Wild gathered plants (4)
3 Corylus avellana
1 Juniperus communis
Other plants
2 stalk
3 indet.

165 Post-hole
Dates
Cannabis sativa cal A.D. 258–425
Charcoal 91 cal B.C.–cal A.D. 533
Cultivated plants (25)
6 (100%) Hordeum vulgare (4 hulled)
19 Cerealia
1 Cannabis sativa
Arable weeds (9)
1 Chenopodium sp.
5 Persicaria sp.
3 Stellaria media
Wild gathered plants (3)
3 Corylus avellana
Other plants
4 stalk
5 indet.

150 Ditch
Dates
Hordeum vulgare hulled cal A.D. 135–332
Charcoal cal A.D. 138–640
Cultivated plants (142)
58 (100%) Hordeum vulgare (25 hulled, 3 naked)
84 Cerealia
Arable weeds (13)
1 Fallopia convolvulus
6 Persicaria sp.
6 Galium sp.
Wild gathered plants (1)
1 Rubus saxatilis
Other plants
5 stalk
4 indet.

Figure 4. Excavation plan of Ketohaka 2 showing grid in metres. Context numbers refer to Supplementary table 2. The cairn in the eastern part of the excavation area is dated to the Migration period (Schauman-Lönnqvist 1989). Insets: Archaeobotanical results from contexts with the most numerous and best-dated plant remains. Percentages of cereals are based on total number of identified cereals (map: Santeri Vanhanen, after Uino 1986).

Age and four seeds are found in a context dated to the Migration period.

Roman Iron Age and Migration period agriculture at Isokylä

Based on the current study, the subsistence at Isokylä must have been based on crop cultivation and animal husbandry, supplemented by fibre and oil production and the gathering of wild plants. Domestic animals could have grazed on or been fed with plants growing on the wetlands situated in the estuary south of Isokylä (Fig. 2). Finds of wetland plants indicate that fodder was collected from these nearby grasslands (Supplementary tables 1 and 3). Grey club-rush (*Scirpus tabarnaemontani*) grows mainly in brackish water, so it was probably gathered for fodder from the estuary (Fig. 2), where common spike-rush (*Eleocharis palustris*), sedges (*Carex* sp.), and bur-reeds (*Sparganium* sp.) also could have grown and been collected.

What kind of cultivation methods were employed at Isokylä? The composition of the arable weed flora did not change significantly during the Late Roman and Migration periods. The same species of arable weeds—goosefoot (*Chenopodium* sp.), black bindweed (*Fallopia convolvulus*), knotweed (*Persicaria* sp.), false cleavers (*Galium spurium*), knotgrass (*Polygonum aviculare*), and common chickweed (*Stellaria media*)—occurred during the entire study period. Based on the nitrophilous arable weeds, such as goosefoot, black bindweed, common chickweed, and false cleavers, the arable fields at Isokylä appear

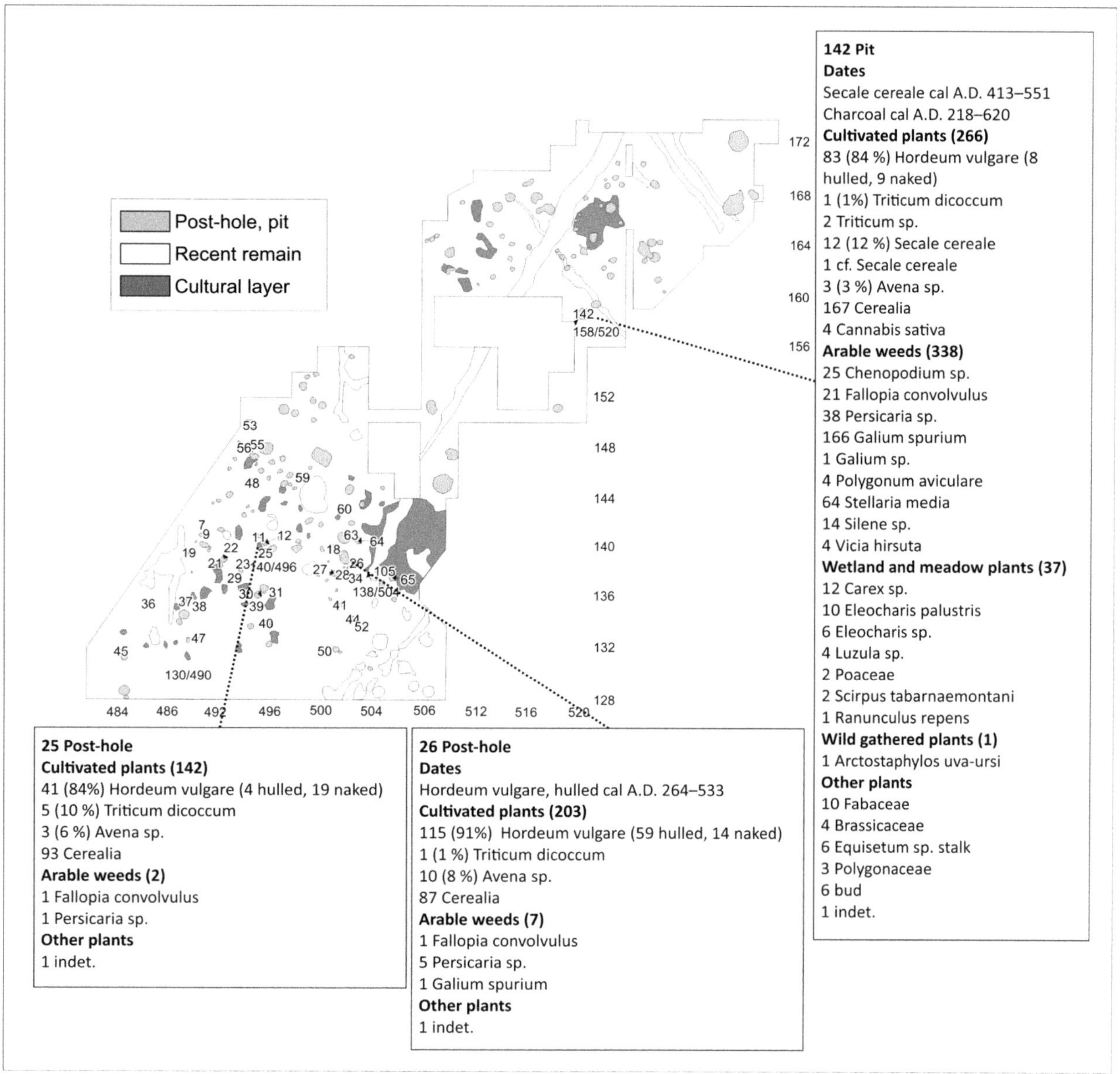

Figure 5. Excavation plan of Vanutehtaanmäki 1 showing grid in metres. Context numbers refer to Supplementary table 3. Insets: Archaeobotanical results from contexts with the most numerous and, except for 25, best-dated plant remains (25 has not been dated). Percentages of cereals are based on the total number of identified cereals (map: Santeri Vanhanen, after Linturi, unpublished manuscript).

to have been manured. There are no clear indications of winter annual weeds, and it thus seems that only summer annual crops were cultivated.

The development towards intentional rye cultivation has been interpreted as beginning when whole stems of cereals started to be cut near the ground, with sickles (Behre 1992). In Finland, sickles appeared for the first time during the Pre-Roman Iron Age, and they are common in Early Roman Iron Age cemeteries (Asplund 2008). Harvesting cereals near the ground would entail the incorporation of low-growing weeds in the archaeobotanical assemblages (Behre 1992). In this regard, pit 142 at Vanutehtaanmäki 1 is interesting because it contains a large amount of cultivated cereals and arable weeds (Fig. 5). Arable weeds are both medium-high-growing (41–80 cm) and low-growing (40 cm or less) species (Kreuz and Schäfer 2011). We have rye and what can be classified as low-growing arable weeds at Isokylä during the Migration period. This shows that cereals were possibly cut low on the stem with a sickle, but this issue should be studied in more detail with a larger archaeobotanical dataset.

Cubic stones suitable for grinding cereals have been found in a Migration period burial at Ketomäki (no. 43) and at the Ketohaka 1 settlement (Uino 1986; Schauman-Lönnqvist 1989). These stones are modest in size, their side lengths measuring 8.8 cm and 5.6 cm, and they represent the only type in use during the Roman period and Migration period in Finland. The small size of these grinding stones suggests that cereal grinding was conducted daily, as opposed to more periodically, involving grinding large amounts of grain on a single occasion.

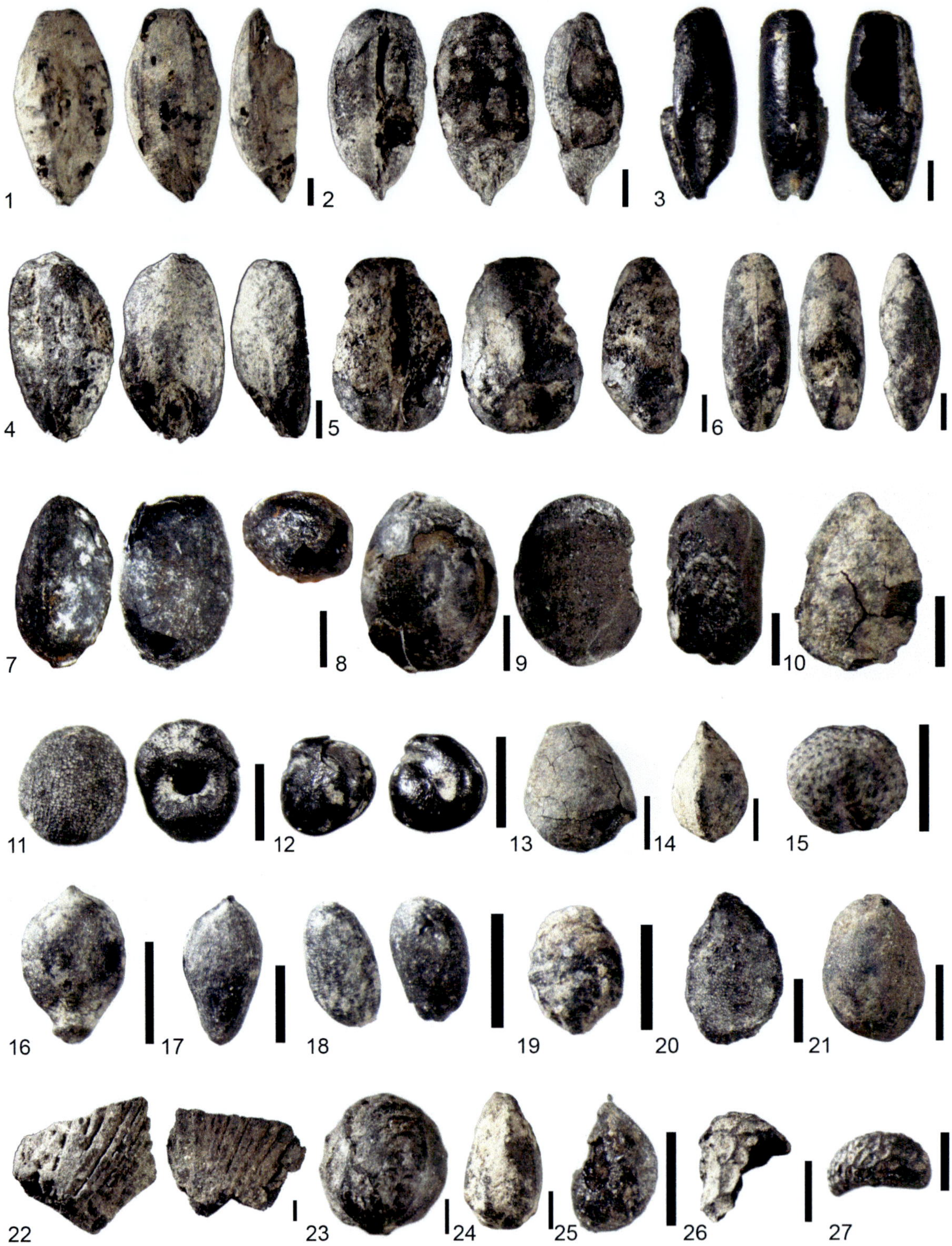

Figure 6. Charred plant remains from studied sites. Cereals photographed from ventral, dorsal, and lateral plans. Scale bars represent 1 mm. 1 hulled barley (*Hordeum vulgare*, hulled); 2 naked barley (*Hordeum vulgare*, naked); 3 rye (*Secale cereale*); 4 emmer/spelt wheat (*Triticum dicoccum/spelta*); 5 bread/club wheat (*Triticum aestivum* s.l.); 6 oat (*Avena* sp.); 7 hemp (*Cannabis sativa*) three different views; 8 hemp with preserved seed coat; 9 Fabaceae; 10 flax (*Linum usitatissimum*); 11 false cleavers (*Galium spurium*); 12 fat hen (*Chenopodium album*); 13 large flowered/common hemp nettle (*Galeopsis speciosa/tetrahit*); 14 black bindweed (*Fallopia convolvulus*); 15 common chickweed (*Stellaria media*); 16 common spike-rush (*Eleocharis palustris*); 17 grey club-rush (*Scirpus tabarnaemontani*); 18 wood-rush (*Luzula* sp.); 19 mint (*Mentha* sp.); 20 creeping buttercup (*Ranunculus repens*); 21 winter-cress (*Barbarea vulgaris*); 22 hazel (*Corylus avellana*); 23 bird cherry (*Prunus padus*); 24 juniper (*Juniperus communis*); 25 bog myrtle (*Myrica gale*); 26 stone bramble (*Rubus saxatilis*); 27 raspberry (*Rubus idaeus*) (photos: Santeri Vanhanen).

Table 2. Dates from contexts with archaeobotanical finds and AMS-dated cultivated plants from Finland from the Roman Iron Age and the Migration period. Dates from Ketohaka 1 and 2 published by Carpelan and Jugner (1982) and Uino (1986). Beta = Beta Analytic, Hel, Hela = Helsinki, Poz = Poznan, UA = Uppsala.

Site (Context no.)	Context/site type	Lab no.	BP	Calibrated date (2 sigma)	Dated material	Source
Ketohaka 1 (2)	Hearth	Hel-1187	1800±90	AD 24–417	Charcoal	Uino 1986
Ketohaka 1 (7)	Ceramics Pit	Hel-1255	2270±110	751–49 BC	Charcoal	Uino 1986
Ketohaka 1 (9)	Hearth	Hel-1185	1770±110	AD 21–536	Charcoal	Uino 1986
Ketohaka 1 (10)	Ceramics Pit	Hel-1618	2360±130	801–165 BC	Charcoal	Uino 1986
Ketohaka 1 (12)	Hearth	Hel-1257	1970±110	352 BC–AD 325	Charcoal	Uino 1986
Ketohaka 1 (12)	Hearth	Hel-1265	1720±110	AD 70–558	Charcoal	Uino 1986
Ketohaka 1 (15)	Charred wooden vessel	Poz-87879	1815±35	AD 90–326	Triticum dicoccum/spelta	current paper
Ketohaka 1 (15)	Charred wooden vessel	Hel-1613	1650±140	AD 85–646	Charcoal	Uino 1986
Ketohaka 1 (19)	Hearth	Hel-1262	1910±110	175 BC–AD 380	Charcoal	Uino 1986
Ketohaka 1 (20)	Hearth	Hel-1263	2000±110	357 BC–AD 243	Charcoal	Uino 1986
Ketohaka 1 (48)	Post-hole	Hel-1183	1940±110	338 BC–AD 340	Charcoal	Uino 1986
Ketohaka 1 (67)	Discoloured soil	Hel-1577	2670±100	1110–537 BC	Charcoal	Uino 1986
Ketohaka 1 (71)	Soot Pit	Hel-1578	1970±90	201 BC–AD 244	Charcoal	Uino 1986
Ketohaka 1 (85)	Post-hole	Poz-87880	1825±30	AD 88–317	Hordeum vulgare	current paper
Ketohaka 1 (92)	Post-hole	Hel-1614	1640±200	92 BC– AD 773	Charcoal	Uino 1986
Ketohaka 1 (95)	Hearth	Poz-87877	1740±30	AD 236–386	Hordeum vulgare hulled	current paper
Ketohaka 1 (95)	Hearth	Hel-1182	1830±110	91 BC–AD 428	Charcoal	Uino 1986
Ketohaka 1 (95)	Hearth	Hel-1189	1780±100	AD 16–532	Charcoal	Uino 1986
Ketohaka 1 (110)	Hearth	Hel-1261	1970±100	347 BC–AD 317	Charcoal	Uino 1986
Ketohaka 2 (11)	Cairn	Hel-1566	1490±80	AD 402–666	Charcoal	Uino 1986
Ketohaka 2 (120)	Post-hole	Hel-1531	1840±120	157 BC–AD 503	Charcoal	Uino 1986
Ketohaka 2 (150)	Ditch	Poz-87881	1785±30	AD 135–332	Hordeum vulgare hulled	current paper
Ketohaka 2 (150)	Ditch	Hel-1466	1630±120	AD 138–640	Charcoal	Uino 1986
Ketohaka 2 (159)	Post-hole	Hel-1516	1460±120	AD 262–862	Charcoal	Uino 1986
Ketohaka 2 (165)	Post-hole	Poz-87888	1675±30	AD 258–425	Cannabis sativa	current paper
Ketohaka 2 (165)	Post-hole	Hel-1573	1820±120	91 BC–AD 533	Charcoal	Uino 1986
Ketohaka 2 (168)	Post-hole	Hel-1615	1820±130	102 BC–AD 535	Charcoal	Uino 1986
Ketohaka 2 (169-170)	Hearth	Poz-87882	1810±30	AD 128–323	Hordeum vulgare hulled	current paper
Ketohaka 2 (169-170)	Hearth	Hel-1565	2050±130	386 BC–AD 221	Charcoal	Uino 1986
Ketohaka 2 (172 A)	Pit	Hel-1612	1800±100	36 BC–AD 427	Charcoal	Uino 1986
Ketohaka 2 (184)	Post-hole	Hel-1572	2080±90	361 BC–AD 76	Charcoal	Uino 1986
Ketohaka 2 (189)	Post-hole?	Hel-1611	1630±140	AD 87–656	Charcoal	Uino 1986
Ketohaka 2 (190)	Pit	Hel-1574	1900±120	181 BC–AD 392	Charcoal	Uino 1986
Ketohaka 2 (650/738)	Discoloured soil	Hel-1528	1850±120	160 BC–AD 424	Charcoal	Uino 1986
Ketohaka 2 (652/774)	Sooty soil, profile, Cairn	Hel-1571	1880±90	87 BC–AD 380	Charcoal	Uino 1986
Vanutehtaanmäki 1 (26)	Post-hole	Poz-87883	1650±30	AD 264–533	Hordeum vulgare hulled	current paper
Vanutehtaanmäki 1 (142)	Pit	Poz-87885	1575±30	AD 413–551	Secale cereale	current paper
Vanutehtaanmäki 1 (142)	Pit	Hel-1310	1630±90	AD 218–620	Charcoal	Linturi MS
Vanutehtaanmäki 1 (22/48/63?)	Post-hole	Poz-87884	100±30	AD 1682–1935	Vitis vinifera	current paper
Bäljars 2	Settlement	Poz-66756	1820±30	AD 90–321	Triticum dicoccum/spelta	Alenius et al. 2017
Gubbacka	Settlement	Hela-1996	1515±40	AD 427–622	Secale cereale	Alenius et al. 2017
Lepokallio	Grave	Ua-53044	1556±29	AD 422–566	Hordeum vulgare	excavation report Nyman 2016
Naarankalmanmäki	Grave	Hela-285	1725±70	AD 128–530	Hordeum vulgare	Raike & Seppälä 2005
Naarankalmanmäki	Grave	Hela-286	1685±70	AD 145–542	Hordeum vulgare	Raike & Seppälä 2005
Naarankalmanmäki	Settlement	Hela-283	1975±175	397 BC–AD 403	Secale cereale	Raike & Seppälä 2005
Naarankalmanmäki	Settlement	Hela-287	1525±70	AD 401–648	Secale cereale	Raike & Seppälä 2005
Pörnullbacken	Settlement/grave	Ua-14032	1855±80	36 BC–AD 377	Hordeum vulgare	Engelmark & Viklund 2002
Pörnullbacken	Settlement/grave	Ua-14030	1750±60	AD 131–406	Hordeum vulgare	Engelmark & Viklund 2002
Pörnullbacken	Settlement/grave	Ua-14035	1730±80	AD 87–534	Hordeum vulgare	Engelmark & Viklund 2002
Pörnullbacken	Settlement/grave	Ua-14029	1690±55	AD 222–535	Hordeum vulgare	Engelmark & Viklund 2002
Pörnullbacken	Settlement/grave	Beta-84095	1680±60	AD 233–536	Hordeum vulgare	Engelmark & Viklund 1997
Pörnullbacken	Settlement/grave	Ua-14037	1640±55	AD 256–546	Hordeum vulgare	Engelmark & Viklund 2002
Pörnullbacken	Settlement/grave	Ua-14033	1610±65	AD 259–595	Hordeum vulgare	Engelmark & Viklund 2002
Pörnullbacken	Settlement/grave	Beta-110633	1600±30	AD 399–539	Hordeum vulgare	Engelmark & Viklund 2002
Pörnullbacken	Settlement/grave	Ua-16454	1560±60	AD 385–623	Hordeum vulgare	Engelmark & Viklund 2002
Pörnullbacken	Settlement/grave	Ua-14031	1545±70	AD 386–646	Secale cereale	Engelmark & Viklund 2002
Rapola	Settlement	Ua-4149	1975±80	177 BC–AD 224	Triticum compactum	Vikkula et. al. 1994

Table 3. Archaeobotanical remains of cultivated plants from Ketohaka 1 (Keto1), Ketohaka 2 (Keto2), and Vanutehtaanmäki 1 (Vanu1). Selected contexts with more numerous archaeobotanical remains and/or AMS-dated remains are shown separate from other contexts for each site. Archaeobotanical remains and dates of the selected contexts are presented in Figures 3–5.

Site	Keto1	Keto2	Keto1	Keto1	Keto2	Keto2	Keto1	Vanu1	Vanu1	Keto2	Vanu1	Vanu1
Context number (Uino 1986)	Other	Other	85	15	150	169-170	95	Other	25	165	26	142
Context type			PH	V	D	H	H			PH		
Date cal AD 2 sigma	RIA	RIA	88–317	90–326	135–332	128–323	236–386	LRA–MP	LRA–MP	258–425	264–533	413–551
Amount of contexts	39	36						40				
Cereals	52	59	9	16	58	18	43	62	49	6	126	100
barley (*Hordeum vulgare*)	35	30	5	6	30	7	26	39	18	2	42	66
hulled barley (*Hordeum vulgare*, hulled)	8	16	1	1	25	11	3	10	4	4	59	8
naked barley, cf. (*Hordeum vulgare*, naked, cf.)	6	8	2	4	3			4	19		14	9
barley (*Hordeum vulgare*) total	49	54	8	11	58	18	29	53	41	6	115	83
rye, cf. (*Secale cereale*, cf.)								2				13
emmer/spelt (*Triticum dicoccum/spelta*)	3	1		5				2	5		1	1
bread/club wheat (*Triticum aestivum* s.l.)								1				
oat, cf. (*Avena* sp., cf.)		4	1				14	4	3		10	3
Oil/fibre plants				2						1		4
flax (*Linum usitatissimum*)				1								
cf. flax (cf. *Linum usitatissimum*)				1								
hemp (*Cannabis sativa*)										1		4

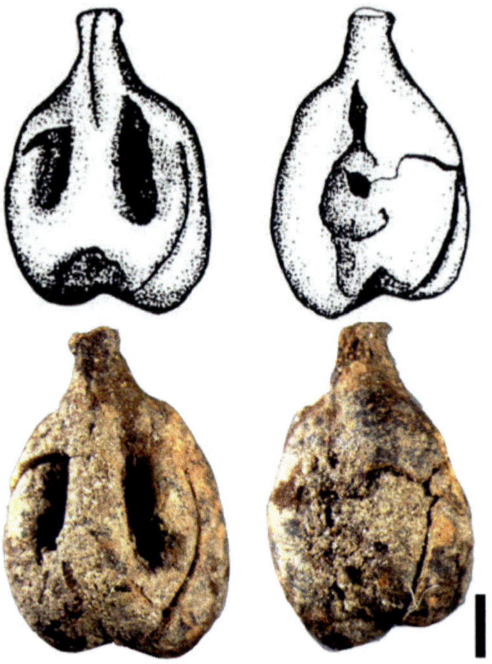

Figure 7. Grapevine (*Vitis vinifera*) pip from a posthole at Vanutehtaanmäki. Drawings from the original excavation report (top) and photos of the same pip, identified as such based on breakage patterns (bottom). The pip comes from posthole 22, 48, or 63 at Vanutehtaanmäki 1, of which 22 and 63 contained brick fragments and 63 had a hole possibly caused by a hay pole. The pip was AMS radiocarbon dated to cal AD 1682–1935 and is thus a recent intrusion. Scale bar represents 1 mm (drawings: Aalto (1997); photos: Santeri Vanhanen).

A simple spindle whorl made of bone has been discovered at Isokylä in a Migration/Merovingian period grave at Palomäki (no. 50) (Schauman-Lönnqvist 1989). Spindles could have been used for spinning thread from either animal or plant (flax and hemp) fibres. Hemp or hop (Cannabaceae) pollen has been recovered from a pollen core studied from a pond in the immediate vicinity of Ketohaka 2 (Tolonen 1985). It is probable that such a small pond was used for hemp retting during the Iron Age (Larsson and Lagerås 2015). Archaeological finds and archaeobotanical data thus show that both retting and spinning were conducted at Isokylä, although hemp and flax were most probably also used for producing oil.

Cultivation was supplemented by the gathering of wild plants. Of these, hazel must have been collected for its edible nuts. However, the majority of the discovered wild gathered plants have aromatic qualities: juniper, bog myrtle, bird cherry, raspberry, bearberry and stone bramble. According to historical sources, juniper, bog myrtle, mint species, and raspberry were used to flavour beer (Behre 1998), which could also have been done at Isokylä.

In conclusion, the combined archaeobotanical and archaeological material shows that cultivation was conducted on spring-sown, manured fields and concentrated on hulled and naked barley, supplemented with minor crops. Nearby wetlands provided fodder for domestic animals, which produced manure. Cereal growing was supplemented by the cultivation and processing of flax and hemp and the gathering of wild plants.

Roman and Migration period cultivation in Finland

Iron Age cultivation is poorly understood in Finland. To get an overview of Roman and Migration period cultivation, a list of archaeobotanical assemblages dated to these periods and cereal grains with a calibrated median date falling into the Roman Iron Age or the Migration period (AD 0–550) was compiled. The archaeobotanical assemblages are discussed in the following text, and the dates are presented in Table 2.

Barley is the dominant crop at nearly all sites of this period and the most common radiocarbon-dated plant (for locations of sites, see Fig. 1). Hulled barley was found in a Migration period mound at Ristimäki (Luoto et al. 1983) and naked barley in a Late Roman period and Migration period mound at Viik (Lempiäinen 1991). With these few identifications, it has not previously been possible to quantify the amount of hulled versus naked barley. The current study from Isokylä shows that both were cultivated in roughly equal amounts.

Based on materials from southernmost Sweden, Engelmark (1992) suggested that growing hulled barley on manured fields became the dominant form of agriculture during the period c. 1000–500 BC and that this form continued throughout most of the Iron Age. The model has been applied in Sweden and Finland (Welinder 1998; Engelmark and Viklund 2002; Holmblad 2010), but it has been questioned because the connection between manuring and hulled barley is not unambiguous and because hulled barley does not become dominant simultaneously in all regions (Lagerås and Regnell 1999). Currently, archaeobotanical data are limited for Finland, but at least at Isokylä we have both naked and hulled barley, showing that naked barley was a cultivated species during the Migration period. At Orijärvi, in eastern Finland, naked barley occurs in small amounts as recently as the Late Iron Age, cal AD 600–1200 (Vanhanen 2012). Thus far, no stable isotope analyses have been conducted for cereal grains found in Finland, but archaeobotanical finds of nitrophilous weeds indicate that fields were manured. Crop finds show no dominance of hulled barley, and the phasing out of naked barley in Finland was a slow process, lasting until the Late Iron Age.

Rye grains have been found at a small number of sites dating to the Roman Iron Age and the Migration period. What is possibly the earliest find of rye grains in Finland was discovered in a Pre-Roman period grave at Spurila (in Paimio commune) (Aalto 1997), but this material has not been AMS dated or properly published. At Naarankalmanmäki, a pit contained 21 rye grains, one of which was AMS-dated and calibrated to 397 BC–AD 403, showing that larger finds of rye were present at the latest during the Roman period (Raike and Seppälä 2005, Table 2). Additional Roman Iron Age rye has been found in a grave at Spurila (Aalto 1997). Migration period rye grains have been found in southern Finland and in southern Ostrobothnia, Finland, at the sites of Naarankalmanmäki, Spurila (Aalto 1997), Ristimäki (Luoto et al. 1983), Gubbacka, and Pörnullbacken (Table 2). Only during the Viking age and later did rye become a dominant crop at some sites in southern Finland (Lempiäinen-Avci et al. 2017).

Wheat occurs in small amounts at various sites in southern Finland during the Roman Iron Age and the Migration period. Emmer/spelt wheat is present at some sites in southern Finland. Emmer (*Triticum dicoccum*) is the dominant crop only at a Late Roman Iron Age settlement at Spurila (Aalto 1997), and emmer/spelt has been AMS-dated to the Roman period at Bäljars 2 (Alenius et al. 2017). Aalto (1982) reported a single grain of possible spelt wheat (*Triticum spelta*) at Ketohaka 1. She did not consider this a certain find, and no spelt grains were found in the current study. Spikelet remains of emmer have only been reported at Late Roman Iron Age Spurila (Seppä-Heikka 1985). Currently all the finds of emmer/spelt-type wheat in Finland should be considered to represent emmer or emmer/spelt wheat, and further finds will show whether spelt was also grown during this period. We do not have finds of emmer or spelt later than the Migration period, and their cultivation appears to have ceased during the Merovingian period, as suggested by Onnela et al. (1996).

Naked wheats dated to the Roman Iron Age and the Migration period have been found in small amounts at a few sites: club wheat (*T. compactum*) at Rapola (Table 2), cf. bread wheat (*T.* cf. *aestivum*) at Migration period Ristimäki (Luoto et al. 1983), and bread/club wheat (*T. aestivum* s.l.) at a Late Roman Iron Age settlement at Spurila (Aalto 1997).

In addition to the sites at Isokylä presented in the current article, oat has been found in small amounts at the sites of Pörnullbacken (Engelmark and Viklund 2002) and Spurila, both dated to the Roman Iron Age (Seppä-Heikka 1985). None of these sites report finds of floret bases, which would enable the secure identification of the oat species in question. Oat is already present during the Late Bronze Age and Pre-Roman Iron Age, but none of these finds are securely identified as cultivated oat (Holmblad 2010; Vanhanen and Koivisto 2015).

An impression of pea (*Pisum* sp.) and a silicified lentil (*Lens esculenta*) were discovered in the Migration period mound of Ristimäki (Luoto et al. 1983). These plants have yet to be found at other sites. Flax has been found at a Late Roman Iron Age settlement at Spurila (Aalto 1997), and this flax and that from Isokylä make up the oldest flax remains in Finland.

The crop assemblage at Isokylä shows similarities with the Baltic Sea region, where barley was the dominant crop and rye, emmer, spelt, oat and naked wheat were minor cereal crops (e.g. Grabowski 2011; Grikpėdis & Motuzaite Matuzeviciute, this volume). The limited data from Finland, however, make it difficult to compare Finnish material with the rest of the Baltic Sea region.

The AMS-dated hemp seed from Isokylä is the earliest hemp macrofossil find in Finland. The second oldest macrofossil remains of hemp in Finland date to the Late Iron Age (Lempiäinen 1999; Vanhanen 2012). In Sweden, hemp was introduced during the Early Roman Iron Age (Larsson and Lagerås 2015) and in Lithuania the earliest hemp macrofossil date is cal AD 430–620 (Stančikaite et al. 2008; Grikpėdis and Motuzaite Matuzeviciute in this volume). Thus, based on macrofossil plant remains, cultivated hemp was introduced into the Baltic Sea region during the Roman Iron Age, which has been labelled as an introductory phase of hemp retting in Scandinavia (Larsson and Lagerås 2015). Archaeological finds from Isokylä show that it was incorporated in a network with the Baltic Sea region (Raninen 2006), and it is thus possible that knowledge regarding hemp cultivation and use was brought to the area via this network.

Conclusions

The new analysis of archaeobotanical plant remains from Isokylä provided a detailed view of this rich material, dated to the Roman Iron Age and the Migration period. The Iron Age settlement was located near an estuary suitable for fodder collection and the grazing of animals, which in turn provided manure for the arable fields. Finds of arable weeds show that manure was spread on spring-sown fields to grow mainly hulled and naked barley, supplemented with emmer/spelt, oat and, later, during the Merovingian period, rye and bread/club wheat. Fibres were produced and spun not only from animal hairs, but also from hemp and flax, which also would have provided oils. The Roman Iron Age hemp find is the earliest in Finland, and an earlier pollen analysis shows that it was probably retted in a pond near the settlement. Archaeobotanical studies around the Baltic Sea show that the crop package found at Isokylä is similar to that elsewhere in the region. Networks discerned in the archaeological material could also have transported agricultural knowledge and the novelty of hemp cultivation to Isokylä.

Acknowledgements

I would like to thank the *Varsinais-Suomi Regional Fund* of the Finnish Cultural Foundation for funding the radiocarbon dates. I would like to thank Ilkka Pylkkö and Kristo Siig for organising Marjatta Aalto's material and Tuuli Timonen for providing access to this material. I would like to thank Terttu Lempiäinen for helping identify the hemp seeds. I am also grateful to Pirjo Uino for information on the samples and their contexts and for providing valuable comments on the article. I would like to thank Per Lagerås, Christian Carpelan and Marko Marila for giving valuable comments on the article. I thank the Doctoral Programme in History and Cultural Heritage at the University of Helsinki for funding my PhD and the writing of this article.

Electronic supplementary material for the paper is accessible at the data exchange platform of the Johanna Mestorf Academy at Kiel University (https://www.jma.uni-kiel.de/en/research-projects/data-exchange-platform).

References

Aalto M (1982) Archaeobotanical studies at Katajamäki, Isokylä, Salo, south-west Finland. In: Hackens T (ed) Pact 7. Council of Europe, pp 137–147

Aalto M (1997) The cultivated plants of Finnish Iron Age. In: Slavyane i finno-ugry. Dmitriy Bulanin, St. Petersburg, pp 47–59

Alenius T, Haggrén G, Koivisto S et al. (2017) Landscape dynamics in southern Finland during the Iron Age and the Early Modern Era: Pollen-based landscape reconstruction (LRA), macrofossil and historical data from western Uusimaa. J Archaeol Sci Reports 12:12–24. doi: 10.1016/j.jasrep.2016.11.041

Asplund H (2008) Kymittæ Sites, centrality and long-term settlement change in the Kemiönsaari region in SW Finland. University of Turku

Behre K-E (1992) The history of rye cultivation in Europe. Veg Hist Archaeobot 1:141–156

Behre K-E (1998) The history of beer additives in Europe: A review. Veg Hist Archaeobot 8:35–48. doi: 10.1007/BF02042841

Bläuer A, Kantanen J (2013) Transition from hunting to animal husbandry in southern, western and Eastern Finland: New dated osteological evidence. J Archaeol Sci 40:1646–1666. doi: 10.1016/j.jas.2012.10.033

Bronk Ramsey C (2009) Bayesian analysis of radiocarbon dates. Radiocarbon 51:337–360

Cappers RTJ, Bekker RM, Jans JEA (2006) Digital seed atlas of the Netherlands. In: Groningen Archaeol. Stud. 4. www.seed-atlas.nl

Carpelan C, Jugner H (1982) Radiocarbon and thermoluminescence dates from Iron Age dwelling places in Isokylä, Salo, south-west Finland. Pact 7: 149–155

Engelmark R (1992) A review of the farming economy in south Scania based on botanical evidence. In: Larsson L (ed) The archaeology of the culture landscape: Field work and research in a south Swedish rural region. Acta Archaeologica Lundensia Ser 4° 19:369–375

Engelmark R, Viklund K (2002) Korn i rågens rike! Pörnullbackens jordbruk och kulturlandskap. In: Viklund K, Gullberg K (eds) Från romartid till vikingatid. Scriptum, Vasa, pp 13–24

Grabowski R (2011) Changes in cereal cultivation during the Iron Age in southern Sweden: A compilation and interpretation of the archaeobotanical material. Veg Hist Archaeobot 20:479–494. doi: 10.1007/s00334-011-0283-5

Hirviluoto A-L (1991) Salon esihistoria. Salon kaupunki, Salo

Hirviluoto A-L, Vormisto T (1984) En massgrav från yngre romersk järnålder i Isokylä, Salo. Finskt Museum 1983:21–49

Holmblad P (2010) Coastal communities on the move: House and polity interaction in southern Ostrobothnia 1500 BC–AD 1. University of Umeå

Kreuz A, Schäfer E (2011) Weed finds as indicators for the cultivation regime of the early Neolithic Bandkeramik culture? Veg Hist Archaeobot 20:333–348. doi: 10.1007/s00334-011-0294-2

Lagerås P, Regnell M (1999) Agrar förändring under sydsvensk bronsålder: En diskussion om skenbara samband och olösta gåtor. In: Olausson M (ed) Spiralens öga: Tjugo artiklar kring aktuell bronsåldersforskning. Riksantikvarieämbetet, pp 263–276

Larsson M, Lagerås P (2015) New evidence on the introduction, cultivation and processing of hemp (*Cannabis sativa* L.) in southern Sweden. Environ Archaeol 20:111–119. doi: 10.1179/1749631414Y.0000000029

Lempiäinen-Avci M, Haggrén G, Rosendahl U et al. (2017) Archaeobotanical analysis of radiocarbon-dated plant remains with special attention to *Secale cereale* (rye) cultivation at the medieval village of Mankby in Espoo (Finland). Veg Hist Archaeobot 26:435–446. doi: 10.1007/s00334-017-0604-4

Lempiäinen T (1991) Nokian Viikin rautakautisen kumpukalmiston makrofossiilit. Karhunhammas 13:38–45

Lempiäinen T (1999) On the history of hemp (*Cannabis sativa* L.) in Finland and the archaeobotanical evidence. In: Selsing L, Lillehammer G (eds) Museumlandskap: Artikkelsamling til Kerstin Griffin på 60-årsdagen. Arkeologisk Museum, University of Stavanger, Stavanger, pp 71–78

Luoto J, Pärssinen M, Seppä-Heikka M (1983) Grain impressions in ceramics from Ristimäki, Vammala, Finland. Suom Mus 1981:5–33

Matiskainen H (1982) Anthropographic interpretation of the Isokylä area, Salo, south-west Finland. Pact 7: 129–136

Núñez MG, Vuorela I (1976) A field method for the retrieval of plant remains from archaeological sites. Memo Soc pro Fauna Flora Fenn 52:19–22

Onnela J, Lempiäinen T, Luoto J (1996) Viking age cereal cultivation in SW Finland: A study of charred grain from Pahamäki in Pahka, Lieto. Ann Bot Fenn 33:237–255

Raike E, Seppälä S-L (2005) Naarankalmanmäki: An Iron Age Complex in Lempäälä, southern Finland. Fennoscandia Archaeol 22:43–78

Raninen S (2006) Il duce Salon seudulla? Rikkaiden hautausten probleema, läntiset kontaktit ja nuorempi roomalainen rautakausi. In: Immonen V, Haimila M (eds) Mustaa valkoisella: Ystäväkirja arkeologian lehtori Kristiina Korkeakoski-Väisäselle. Turun yliopisto, arkeologia, Turku, pp 199–222

Reimer P, Bard E, Bayliss A et al. (2013) Intcal13 and marine13 radiocarbon age calibration curves 0–50,000 years cal bp. Radiocarbon 55:1869–1887. doi: 10.2458/rc.v51i4.3569

Schauman-Lönnqvist M (1989) The development of Iron Age settlement in the Isokylä Area in Salo. In: Edgren T (ed) Iron Age studies in Salo. Suomen muinaismuistoyhdistyksen aikakauskirja 89:2. Suomen muinaismuistoyhdistys, Helsinki, pp 1–99

Schauman-Lönnqvist M, Hirviluoto A-L, Linturi E, Uino P (1986) The research history of the Isokylä area in Salo. In: Linturi E (ed) Iron Age studies in Salo. Suomen muinaismuistoyhdistyksen aikakauskirja 89:1. Suomen muinaismuistoyhdistys, Helsinki, pp 12–24

Seppä-Heikka M (1985) Grains and seeds from younger Roman Iron Age excavations in Spurila. Iskos 5:460–461

Stančikaite M, Kisieliene D, Mažeika J, Blaževičius P (2008) Environmental conditions and human interference during the 6th and 13th–15th centuries A.D. at Vilnius Lower Castle, east Lithuania. Veg Hist Archaeobot 17:S239–S250. doi: 10.1007/s00334-008-0181-7

Tolonen M (1985) Palaeoecological reconstruction of vegetation in a prehistoric settlement area, Salo, SW Finland. Ann Bot Fenn 22(2):101–116

Uino P (1986) An Iron Age community at Ketohaka in Salo and other remains of Metal period buildings in Finland. In: Linturi E (ed) Iron Age studies in Salo. Suomen muinaismuistoyhdistyksen aikakauskirja 89:1. Suomen muinaismuistoyhdistys, Helsinki, pp 25–201

Vanhanen S (2012) Archaeobotanical study of a Late Iron Age agricultural complex at Orijärvi, Eastern Finland. Fennoscandia Archaeol 29:55–72

Vanhanen S, Koivisto S (2015) Pre-Roman Iron Age settlement continuity and cereal cultivation in coastal Finland as shown by multiproxy evidence at Bäljars 2 site in SW Finland. J Archaeol Sci Reports 1:38–52. doi: 10.1016/j.jasrep.2014.10.003

Vanhanen S, Mikkanen T (2013) New radiocarbon dates for a medieval oven with botanical remains from the hamlet of Gubbacka, Vantaa, southern Finland. SKAS 2013:16–25

Vikkula A, Seppälä S-L, Lempiäinen T (1994) The ancient field of Rapola. Fennoscandia Archaeol 11:41–59

Vuorela A, Penttinen T, Lahdenperä A (2009) Review of Bothnian Sea shore-level displacement data and use of a GIS tool to estimate isostatic uplift. Posiva Working report 2009-17.

Welinder S (1998) Neoliticum–Bronsålder 3900–500 f.Kr. In: Welinder S, Pedersen EA, Widgren M (eds) Det Svenska jordbrukets historia, Bd. I: Jordbrukets första femtusen år 4000 f.Kr.–1000 e.Kr. Natur och Kultur/LTs Förlag, Stockholm, pp 11–236

New cereal grain finds from southern Satakunta, Finland, dated from the Late Bronze Age to the Middle Ages

Terttu Lempiäinen[1], Maija Helamaa[3], Heli Lehto[2,3], Ulla Moilanen[1,3], Markku Oinonen[2], Sakari Salomaa[1,3], Kari Uotila[1,2,3]

Author addresses:
[1] University of Turku, [2] University of Helsinki, [3] Muuritutkimus Oy

E-mail: terlem@utu.fi

Abstract

New cereal grains were discovered from three archaeological excavations conducted during the years 2012–2015 in southern Satakunta, south-western Finland. The oldest of these, identified as *Hordeum vulgare*, were found at the site of Eura Luistari, between the stones of a cooking pit. With a Late Bronze Age ^{14}C date they are the oldest cereal grain finds from southern Satakunta to date. In total, nine cereal grains were dated from Luistari. These resulted in Late Bronze-Age, Pre-Roman Iron Age, Late Iron Age, and modern dates. Also, ard marks indicating ancient fields were noted at this site and at the site of Eura Kauttua. Cereal grains of *Hordeum vulgare*, *Secale cereale* and *Triticum aestivum* cf. *compactum* were found at the sites of Kauttua and Rauma Ellinniitty. Single grains of wheat and barley from Ellinniitty were ^{14}C dated to the Pre-Roman Iron Age.

Study areas and site descriptions

The three studied sites: Eura Kauttua Old village, Eura Luistari Luistarintie and Rauma Ellinniitty are all are situated in the southern part of Satakunta, a province of western Finland (Fig. 1). The known history of the village of Eura Kauttua extends back to the Middle Ages, at which time it was one of the largest villages in the municipality of Eura. Archaeologically, the history of the village stretches back in time to at least 900–1100 AD. Currently, parts of the settlement remains are under a 30–50 cm thick layer of field soil. Foundations of buildings as well as cultural layers from the village have survived for a millennium. From the area of Eura Luistari (municipality of Eura), cooking pits and other signs of a settlement from the Early Metal Age are known. In the same area are cairns that date to the Bronze Age, as well as the Luistari cemetery, which was in use from the 5th century AD. There was still a cemetery in the same area during the early Middle Ages. The cemetery area of Luistari and the village of Kauttua seem to form a single, larger entity. The Pre-Roman Iron Age site (500–0 BC) of Rauma Ellinniitty contains hundreds of low stone mounds. Two of them have been identified as burial cairns covered with red sandstone slabs. One of them contained two iron bracelets. There are also settlement structures at the Ellinniitty knoll (municipality of Rauma), but here no ard marks indicating ancient fields were found (Uotila and Helamaa 2015; Uotila and Lehto 2016).

The sites were studied archaeobotanically, concentrating on the cultivation history of southern Satakunta. A few radiocarbon-dated macrofossil cereal grain finds from the Eura Luistari cemetery area have already been published (Lehtosalo-Hilander 1999, 2000). A charred grain of barley (*Hordeum vulgare*) was dated to cal AD 780–562. Another barley grain was dated to the Viking age, cal AD 699–841, whereas a rye grain (*Secale cereale*) resulted in a much later dating, to the 18th century AD (Lehtosalo-Hilander 1999). The Viking age cereals were accompanied by weed seeds, namely, fat-hen (*Chenopodium album*), black-bindweed (*Fallopia convolvulus*), corn spurrey (*Spergula arvensis*), and even seeds of flax (*Linum usitatissimum*) and *Camelina linicola*. Lehtosalo-Hilander (1999) dated charcoal from some graves near the present study site within the Luistari cemetery area, and the oldest date was 1308–1078 cal BC. In this chapter, new ^{14}C dates on cereal grains found from Eura Luistari and Rauma Ellinniity will be presented.

Studied samples and archaeobotanical methods

Soil samples were collected by the field archaeologists from every studied site. The archaeobotanical studies comprised an analysis of soil samples, which

Figure 1. Map showing location of the municipalities of Rauma and Eura discussed in the text (map: Santeri Vanhanen).

Figure 2. Aerial view of Eura Kauttua Old village site (left) (reproduced with permission from Antti Kylänpää 1977) and view of the archaeological excavation with ard marks (right) (reproduced with permission from Eeva Raike 2012).

varied in volume from 0.4 dl to 2 litres. A total of 107 samples from various contexts were collected, and 102 of these are included in the current chapter. Detailed results are described in archaeobotanical and archaeological reports (Lempiäinen 2012, 2013, 2014a, b, 2015; Uotila 2013; Uotila and Helamaa 2015; Uotila and Lehto 2016):

At Eura Kauttua (Fig. 2), a total of 36 samples were collected at a depth of 10–60 cm from the settlement layers, consisting of with dark and sooty to sandy soil with clay and charcoal, and a total of 8 samples were collected from the ancient plough layers/ard marks of an ancient field, approximately 34.5 m above sea level.

At Eura Luistari Luistarintie (Fig. 3), a total of 45 soil samples were retrieved near and in connection to the large Luistari cemetery area (Lehtosalo-Hilander 1982, 2000). Samples were collected from cooking pits (10 samples, at a depth of c. 37 cm), graves (20 samples), the ancient plough layers/ard marks of an ancient field (4 samples), the settlement area (5 samples), and postholes within the settlement area (6 samples). Nine cereal grains from Luistarintie were ^{14}C dated.

Figure 3. View of Eura Luistari Luistarintie showing the area excavated in 2013 with a fireplace or a cooking pit on the lower left corner (photo: Terttu Lempiäinen).

Figure 4. Views of Rauma Ellinniitty showing a posthole area before (left) and after excavation, with dirty, sooty, sandy soil with small stones, pieces of plants and charcoal (right) (photos: Kari Uotila).

At Rauma Ellinniitty (Fig. 4), a total of 21 samples were collected from small postholes filled with dark, sooty and sandy soils, small stones, pieces of plants, wood and charcoal. Two cereal grains, one each of wheat and barley, were ^{14}C dated.
This chapter concentrates on those samples from which macrofossil cereal grains were identified. Soil samples were floated in tap water, after which plant remains were picked from the plant mass with the help of a microscope (Olympus SZX 9; at magnification of 40–100×). After drying, the charred cereal grains were stored in small bottles and housed in the macrofossil collection of the Botanical Museum, University of Turku, Finland. Plant names are according to Hämet-Ahti et al. (1998) and identifications are based on Cappers et al. (2006).

Radiocarbon dates

Nine samples collected from different layers and contexts in the Eura Luistari settlement area and the Luistarintie settlement/cemetery area were radiocarbon dated. These comprised eight charred cereal grains, of barley, bread wheat (*Triticum aestivum*) and rye, and one seed of the field weed fat-hen. Two cereal grains, one each of wheat and barley, were ^{14}C dated from Rauma Ellinniitty. All samples were analysed at the Laboratory of Chronology of the Finnish Museum of Natural History. The results are presented in Table 1, sorted by the sample laboratory number.

Results of the archaeobotanical analysis and ^{14}C dating

Tables 2 and 3 present the results of the macrofossil analysis at the three studied sites, Eura Kauttua Old village, Eura Luistari Luistarintie and Rauma Ellinniitty, during the years 2012–2015.

Eura Kauttua

Only one charred cereal grain of *H. vulgare* was found in the 2012 samples, from a soil sample taken from an ard mark (Table 2). A total of eight charred cereal grains were found in the 2013 samples, from soil samples taken from cultural layers or settlement soils, namely, five *H. vulgare*, two *S. cereale* and one *T. aestivum*. The dating of the remains is based on archaeological observations, because the *H. vulgare* grain from the ard mark was too small for ^{14}C dating. Archaeologically, the history of the village stretches back to at least 900–1000 AD. The cereals were probably from early medieval layers, and at that period *Hordeum vulgare, Secale cereale* and *Triticum aestivum* were being cultivated in southern Finland and in the Eura Luistari area (Lehtosalo-Hilander 1999, 2000; Alenius et al. 2017; Grönlund and Simola 2011; Pukkila 2011). The find of rye brome (*Bromus secalinus*) from this site is interesting because it is a field weed associated with rye cultivation (Lempiäinen 2010). Other weeds accompanying the cereal grains, typical of cultivated fields and settlement sites, are *Chenopodium album*, common orache (*Atriplex patula*), common fumitory (*Fumaria officinalis*), bush vetch (*Vicia sepium*) and *Alchemilla* sp.

Eura Luistari Luistarintie

The cereal grains found in the different archaeological contexts are presented in Figure 5. Altogether, 19 macrofossil finds of charred cereal grains were found. The cereal species were *H. vulgare*, *S. cereale* and *T.* cf. *aestivum* (Fig. 6). One grain of *H. vulgare*, from the soil sample from the cooking pit, was ^{14}C dated to 894–786 cal BCE (Table 1), and barley grains were dated to 755–400 cal BCE, 471–195 cal BCE and cal AD 891–1026. *T. aestivum* was dated to cal AD 1461–1643 and *S. cereale* to cal AD 1680–1939. Five

Table 1. Radiocarbon dates on cereal grains and seeds of weeds from Eura Luistari and Rauma Ellinniitty.

Lab. ID	Site	Sample	Material dated	d13C (‰)	Radiocarbon date (BP)	±	Calibrated date (2σ range)
Hela-3328	Eura Luistari	cooking pit	barley grains (*Hordeum vulgare*)	-26,8	2641	31	900-780 cal BC
Hela-3402	Eura Luistarintie	Mn 10 or 52	barley grain (*Hordeum vulgare*)	-25	1065	40	cal AD 890-1030
Hela-3408	Eura Luistarintie	Mn 7	barley grain (*Hordeum vulgare*)	-25,7	2419	49	760-390 cal BC
Hela-3409	Eura Luistarintie	Mn 8	barley grain (*Hordeum vulgare*)	-26,6	>MODERN (101,35 pMC)		Modern
Hela-3410	Eura Luistarintie	Mn 17	wheat grain (*Triticum aestivum*)	-26,2	338	41	cal AD 1460-1650
Hela-3411	Eura Luistarintie	Mn 17	barley grain (*Hordeum vulgare*)	-26	>MODERN (100,34 pMC)		Modern
Hela-3413	Eura Luistarintie	Mn 19	barley grain (*Hordeum vulgare*)	-25,9	>MODERN (100,82 pMC)		Modern
Hela-3414	Eura Luistarintie	Mn 23	barley grain (*Hordeum vulgare*)	-24,4	2281	54	480-190 cal BC
Hela-3415	Eura Luistarintie	Mn 27	rye grain (*Secale cereale*)	-22,7	84 (98,96 pMC)	45	cal AD 1680-1940
Hela-3980	Rauma Ellinniitty	Mn 6	wheat grains (*Triticum aestivum*)		2341	37	540-250 cal BC
Hela-3981	Rauma Ellinniitty	Mn 7	barley grains (*Hordeum vulgare*)		2285	44	410-200 cal BC

Table 2. The number and contents of soil samples taken from Eura Kauttua Old village and Eura Luistari Luistarintie. * charred.

Studied site	Soil samples	Sample size and contents	Number of macrofossil remains	Number of cereal grains	Cereal species	Weeds
Kauttua Old village 2012	29	0.1 -1.5 l clayey sandy settlement soil	39	1	***Hordeum vulgare*/barley***	***Chenopodium album**, 38** ***Atriplex patula**, 1** ***Fumaria officinalis**, 1** ***Alchemilla* sp.*, 1**
Kauttua Old village 2013	7	0.1-1.5 l clayey sandy settlement soil	17	8	***Hordeum vulgare*/barley*** ***Secale cereale*/rye*** ***Triticum aestivum*/bread wheat***	***Bromus secalinus**, 1** ***Chenopodium album**, 8** ***Vicia sepium**, 1** ***Picea abies*/needles* ,**
Eura Luistarintie 2013	45	0.1-1.2 l sooty sandy soil with charcoal	159	19	***Hordeum vulgare*/barley*** ***Secale cereale*/rye*** ***Triticum* cf. *aestivum*/bread wheat***	***Bromus secalinus**, 1** ***Chenopodium album**, 138** ***Scleranthus annuus**, 1** ***Picea abies*/needles*, +**

of the nine dated cereals and the single dated fat-hen Chenopodium album seed were found to be of modern date (not included in Table 1).

The ^{14}C dates indicate that the soil was very mixed, containing also modern material. The oldest cereal date, on *H. vulgare*, is 894–786 cal BCE, which makes it the oldest cereal grain find thus far from southern Satakunta. The nearest location subjected to pollen analyses is the Eura Hyväsuo bog, about 1 km south-east of the centre of Eura. The first and earliest signs of cereal cultivation in the pollen core date from the 4th century BCE (Vuorela 2003; Grönlund and Simola 2011), and the more or less permanent field cultivation dates from as late as the beginning of the Middle Ages. However, according to pollen assemblages from the same time as the earliest cereal grain finds (i.e. of the 9th century BCE), spruce (*Picea abies*) and some deciduous trees (*Tilia cordata, Ulmus glabra* and *Corylus avellana*) were drastically diminished in relation to other pollen types. Two of the dated cereal grains of *H. vulgare* come from the Late Bronze Age and Early Iron Age, respectively. The grains of *T. aestivum* and *S. cereale* were both dated to the end of the Middle Ages to Modern times (Table 1). The weeds with the cereal finds were *Chenopodium album*, annual knawel (*Scleranthus annuus*), *Bromus secalinus* and *Fallopia convolvulus*.

Rauma Ellinniitty

Archaeologically, the site was dated to the Pre-Roman Iron Age. Total of 155 macrofossil remains were found, but only 14 of these were charred (Table 3), including 12 charred cereal grains, all from six sooty and sandy soil samples taken from a small pit of dark soil full of burnt bones. The cereals comprised 11 grains of *H. vulgare* and 1 grain of *T. aestivum* cf. *compactum*, all in very poor condition (Fig. 7). The ^{14}C date of one *H. vulgare* grain was 409–206 cal BCE, and that of one *T. aestivum* cf. *compactum* grain was 540–259 cal BCE. The *Triticum* grain is one of the oldest wheat finds in Finland. There are some finds from the Roman period and the Late Iron Age (e.g. Alenius et al. 2017; Matiskainen 1984; Vikkula et al. 1994). Other *Triticum* finds date archaeologically to the Early Iron Age, for instance those of Salo Katajamäki (Aalto 1982), Salo Ketohaka (Aalto 1997), Vanutehtaanmäki (Aalto 1997) and Paimio Spurila (Seppä-Heikka 1985). The dating of the

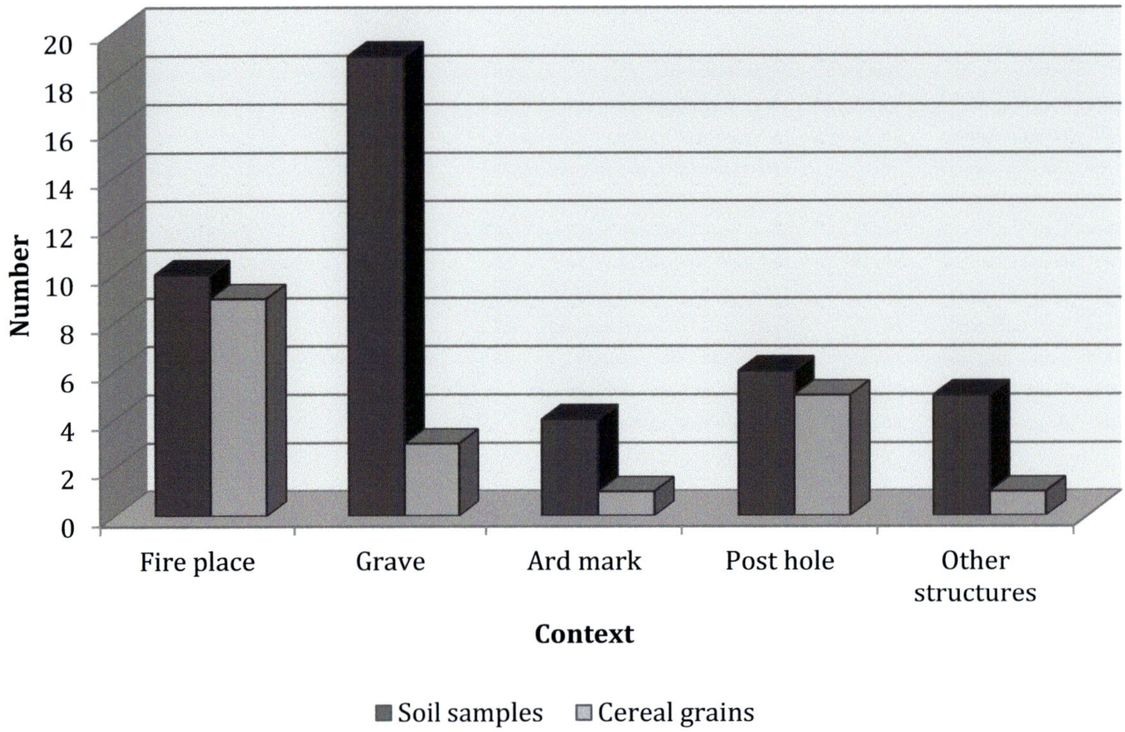

Figure 5. The distribution of soil samples and charred cereal grain finds among different contexts in the Eura Luistari Luistarintie 2013 material.

Figure 6. Charred cereal grains of barley (Hordeum vulgare) from an oven at Eura Luistari Luistarintie (photos: Mikael Kukkonen).

Table 3. Macro-remains identified from samples taken from five different excavation areas and from and from 21 soil samples at Rauma Ellinniitty in 2015. * charred; +++ plenty of charcoal in soil samples.

Layer	Area A2, cairns with sooty sandy soil	Area B1 with sooty sandy soil	Area F1 with sooty sandy soil	Area G1 with sooty sandy soil	Area D1 /M 416 with sooty sandy soil	Sum
Number of samples	12	1	4	2	2	21
Plant species						
Cereals						
Hordeum vulgare	11*					11
Triticum aestivum cf. *compactum*	1*					1
Weeds						
Fallopia convolvulus	1*					1
Trees and shrubs						
Rubus idaeus	1*					1
Other remains						
Charcoal	+++	+++	+++	+++	+++	
Sum	14	-	-	-	-	14

Figure 7. Charred cereal grains of barley (Hordeum vulgare) (left) and club wheat (Triticum aestivum cf. compactum) (right) from a posthole at Rauma Ellinniitty (photos: Mikael Kukkonen).

Triticum at Ketohaka was confirmed by radiocarbon dating (Vanhanen, this volume). *Fallopia convolvulus* represented settlement weeds. The cereal grain finds relate to some kind of cultivation at the site or near the area during the Pre-Roman Iron Age, although barley and wheat were already being cultivated in southern Satakunta some centuries earlier (see above).

Discussion

According to written documents, the history of the village of Kauttua extends back to the Middle Ages, at which time it was one of the largest villages in the municipality of Eura. Archaeologically, the history of the village stretches back to at least 900–1100 AD. The macrofossil cereal grain found at Kauttua consist of barley, rye and wheat. They were dated archaeologically to the 10th–12th century AD. All three cereals were cultivated in southern Finland at that time, but *Hordeum vulgare* was the most commonly cultivated (Salo 1981; Lempiäinen and Häkkinen 1995; Häkkinen and Lempiäinen 1996; Vuorela 2003; Grönlund and Simola 2011).

Cooking pits and other signs of a settlement from the Early Metal Age are known from the area of Eura Luistari (Lehtosalo-Hilander 1982, 2000). In the same area, there are cairns that date to the Bronze Age and a cemetery that was in use from the 5th century AD. There was a cemetery in the same area during the Middle Ages (Lehtosalo-Hilander 1982, 2000; Pukkila 2011; Uotila 2011). The cemetery area of Luistari and the village of Kauttua seem to form one larger entity. The oldest cereal grains, identified as *H. vulgare*, come from Eura Luistari Luistarintie, between the stones of a cooking pit, and were directly ^{14}C dated to 894–786 cal BCE and 755–400 cal BCE. They are the oldest cereal grain finds from southern Satakunta to date. In total, nine dates were obtained on cereal grains from Luistari; one *H. vulgare* was dated to the Early Iron Age and another to the Late Iron Age. The grains of *T. compactum*–type, and *S. cereale,* were found to be much younger, dating to the late Middle Ages and to Modern times, respectively.

At the Pre-Roman Iron Age site of Rauma Ellinniitty, there are hundreds of low stone mounds. Two of them have been identified as cairns covered with red sandstone slabs. In total, 12 cereal grains come from six sooty and sandy soil samples from the area, which was interpreted as a settlement site. The cereals consisted of 11 *H. vulgare* grains and 1 *T. aestivum* cf. *compactum* grain. The barley grain was ^{14}C dated to 409–206 cal BCE. The wheat grain was dated to 540–259 cal BCE. It is one of the oldest wheat finds in Finland. The cereal finds relate to some kind of cultivation at the site or near the area during the Pre-Roman Iron Age. Both barley and wheat cultivation were carried out in southern Satakunta (cf. Alenius et al. 2017; Salo 1981; Lehtosalo-Hilander 2000; Vuorela 2003; Grönlund and Simola 2011). There are very few macrofossil finds of cereals from the studied area. The nearest cereal finds that date to the same age come from Laihia in southern Ostrobothnia—far away from the present studied sites. At Laihia, the dates for *H. vulgare* range between 1000–830 cal BCE and 380–190 cal BCE, and the date for *Avena* sp. is 770–400 cal BCE (Holmblad 2010). Engelmark and Viklund (2002) published grain finds of *H. vulgare* from Vöyri in southern Ostrobothnia dated to the Early Roman Iron Age and some younger finds dating from the Late Roman Age to the Viking age. Previously published dates from Eura Luistari cemetery on barley grains show that they date to the Pre-Roman Iron Age and the Viking age, whereas a rye grain dates to the 18th century (Lehtosalo-Hilander 1999, 2000, pp. 265–266).

Acknowledgements

The authors would like to thank the editors and reviewers, especially Santeri Vanhanen, for useful comments on the manuscript. Marja Vieno revised the English text.

References

Aalto M (1982) Archaeobotanical studies at Katajamäki, Isokylä, Salo, south-west Finland. In: Hackens T (ed) PACT 7/1. Council of Europe, Strasbourg, pp 137–147

Aalto M (1997) The cultivated plants of finnish Iron Age. In: Кирпичников АН, Рябинин ЕА, Сакс АИ (eds) Славяне и финно-угры: археология, история, культура. С.-Петербург, pp 47–61

Alenius T, Haggrén G, Koivisto S, Vanhanen S, Sugita S (2017) Landscape dynamics in southern Finland during the Iron Age and the Early Modern Era: Pollen-based landscape reconstruction (LRA), macrofossil and historical data from western Uusimaa. J Archaeol Sci Reports 12:12–24

Engelmark R, Viklund K (2002) Korn i rågens rike! Pörnullbackens jordbruk och kulturlandskap. In: Viklund K, Gullberg K (eds) Från romartid till vikingatid. Scriptum, Vasa, pp 13–24

Grönlund E, Simola H (2011) Suoturvetta ja järvenpohjaa: Euran Hyväsuon ja Köyliönjärven paleoekologiset tutkimukset. In: Uotila K (ed) Avauksia Ala-Satakunnan esihistoriaan. Eura Print, Eura, pp 81–110

Cappers RTJ, Bekker RM, Jans JEA (2006) Digitale Zadenatlas van Nederland. Barkhuis Publishing & Groningen University Library, Groningen

Häkkinen K, Lempiäinen T (1996) Die älteste finnische Getreide und ihre Namen. Finno-Ugrische Forschung 53(1–3):115–182

Hämet-Ahti L, Suominen J, Ulvinen T, Uotila P (1998) Retkeilykasvio. Luonnontieteellinen keskusmuseo, Helsinki

Holmblad P (2010) Coastal communities on the move: House and polity interaction in southern Ostrobothnia 1500 BC–AD 1. Dissertation, University of Umeå

Lehtosalo-Hilander P-L (1982) Luistari III. Vammalan Kirjapaino, Vammala

Lehtosalo-Hilander P-L (1999) Dates. In: Huurre M (ed) Dig it all: Papers dedicated to Ari Siiriäinen. Archaeological Society of Finland, Helsinki, pp 39–43

Lehtosalo-Hilander PL (2000) Kalastajista kauppanaisiin: Euran esihistoria. Euran kunta, Vammala

Lempiäinen T (2010) Muinaisten peltojen rikkaruohot makrofossiiliaineistoissa. In: Hirvilammi, J (ed) Varhainen viljely Suomessa. Suomen Maatalousmuseo Sarka, Loimaa, pp 3–19

Lempiäinen T (2012) Eura Kauttua, vanhan kylän paikka (Arkeologinen kaivaus, makrofossiilitutkimusraportti). Unpublished archaeobotanical report, Biodiversiteettiyksikkö, Turun yliopisto

Lempiäinen T (2013) Eura Luistari Luistarintie (Arkeologinen kaivaus 2013, makrofossiilitutkimusraportti). Unpublished archaeobotanical report, Biodiversiteettiyksikkö, Turun yliopisto

Lempiäinen T (2014a) Euran Kauttuan vanhan kylän paikan kaivaukset 2013. Unpublished archaeobotanical report, Biodiversiteettiyksikkö, Turun yliopisto

Lempiäinen T (2014b) Rauma Ellinniitty: Arkeologinen kaivaus 2014. Unpublished archaeobotanical report, Turun yliopisto

Lempiäinen T (2015) Rauma Ellinniitty: Arkeologinen kaivaus 2015. Unpublished archaeobotanical report, Biodiversiteettiyksikkö, Turun yliopisto

Lempiäinen T, Häkkinen K (1995) Koska Suomessa on alettu viljellä maata? Hiidenkivi 6:11–14

Matiskainen H (1984) Getreidekörner aus der spätzeitlichen Siedlungsksmmer Domargård 1 in Karjaa, Südfinnland. Fennoscandia Archaeologica I:43–50

Pukkila J (2011) Euran viikinkiaika. In: Uotila K (ed) Avauksia Ala-Satakunnan esihistoriaan. Eura Print, Eura, pp 111–132

Salo U (1981) Satakunnan historia, osa 1, 2: Satakunnan pronssikausi. Satakunnan maakuntaliitto, Rauma.

Seppä-Heikka M (1985) Grains and seeds from younger Roman Iron Age excavations in Spurila. Iskos 5:460–461

Uotila K (2011) Avauksia Ala-Satakunnan esihistoriaan. Eura Print, Eura

Uotila K (2013) Eura Luistarintie, arkeologinen kaivaus 2013. Unpublished excavation report, Muuritutkimus ry.

Uotila K, Helamaa M (2015) Rauma, Ellinniitty 2014: Varhaisrautakautinen röykkiökaivaus. Unpublished excavation report, Muuritutkimus ky.

Uotila K, Lehto H (2016) Rauma, Ellinniitty 2015: Unpublished excavation report, Muuritutkimus ky.

Vikkula A, Seppälä S-L, Lempiäinen T (1994) The ancient field of Rapola. Fennoscandia Archaeologica 11:42–57, Archaeological Society o f Finland.

Vuorela I (2003) Satakunnan kasvillisuuden historia ja siitepölyanalyysit. In: Sarka 2001–2002, Satakunnan Museon vuosikirja. Satakunnan Museo, pp 68–76. Pori

From barley to buckwheat: Plants cultivated in the Eastern Baltic region until the 13th–14th century AD

Mindaugas Grikpėdis, Giedre Motuzaite Matuzeviciute

Author addresses:
Department of Archaeology, Vilnius university, Vilnius, Lithuania

E-mail: mindaugas@archeologas.lt

Abstract

For a long time, the Eastern Baltic region was *terra incognita* within the international literature on cultivated plants. With the intensification of archaeobotanical studies and direct radiocarbon dating of cultivated plant species, the data on the introduction of various crop species in this region have started to accumulate, although they are still scattered in reports and papers in local languages. Therefore, it seems timely to present a synthesis of all existing macrobotanical data. In addition, we present a few AMS radiocarbon dates of directly dated remains of crops, as the primary data for this article. Our review of currently available archaeobotanical material shows that there is no firm evidence to suggest plant cultivation in the Eastern Baltic region during the Neolithic period. The earliest radiocarbon-dated macroremains of *Hordeum vulgare* come from the middle of the Bronze Age. During the 1st millennium BCE, a much broader spectrum of domestic plants starts to be cultivated, including *H. vulgare*, *Triticum spelta*, *T. dicoccum*, *Panicum miliaceum*, *Pisum sativum*, *Vicia faba* and *Camelina sativa*. *Secale cereale* and naked wheats *(T. aestivum/durum, T. compactum)* are found only from the Roman period onwards. During the Roman and Migration periods, a few sites already have rye and naked wheats as the dominant crop species. However, rye does not start to dominate the broader economy of the Eastern Baltic population until the end of the 1st millennium AD. *Linum usitatissimum* and *Cannabis sativa* are poorly represented during the 1st millennium AD; however, they may have been cultivated since the Roman period. *Avena sativa* was present during the Roman period, unfortunately, the current dataset is too poor to justify speculating on its importance at that time. Finally, crops such as *Fagopyrum esculentum, Lens culinaris* and few others appear from c. the 13th–14th century AD, completing the array of cultivated plant species until the Columbian crop arrival.

Introduction

The first appearance and development of agriculture in the Eastern Baltic, although much debated, are still only very superficially understood. To date, little is known about the earliest farmers in the region or how agricultural systems developed over time. Macrobotanical investigations have been carried out sporadically since the first half of the 20th century; however, most of the data are scattered in various reports and papers in local languages and are therefore not accessible to a wider audience. With the intensification of archaeobotanical studies and direct radiocarbon dating of cultivated plant species, it seems timely to present a synthesis of macrobotanical data from the Eastern Baltic.

In recent years, some publications have drawn attention to archaeobotanical material from medieval Eastern Baltic sites (Banerjea et al. 2017; Sillasoo and Hiie 2007; Stančikaitė et al. 2008), but the crop species cultivated and consumed during the earlier periods remain poorly investigated. Therefore, in this chapter, we focus on domesticated plant species introduced in the region before the 13th–14th century AD. We aim to determine the timing of the earliest appearance of different crop species, mainly cereals, pulses and oil plants, showing their individual history of origin and distribution in the Eastern Baltic region.

Material and methods

This chapter was prepared and submitted by November, 2017. We present all data from publications and unpublished excavation reports from the Eastern Baltic with a publication date of 2015 or earlier known to us. We also include some data published in 2016 and 2017. The Eastern Baltic region encompasses Lithuania, Latvia, Estonia and north-eastern Poland. The northernmost part of the Eastern Baltic region – Finland is not investigated in this paper, however we occasionally refer to it for a broader

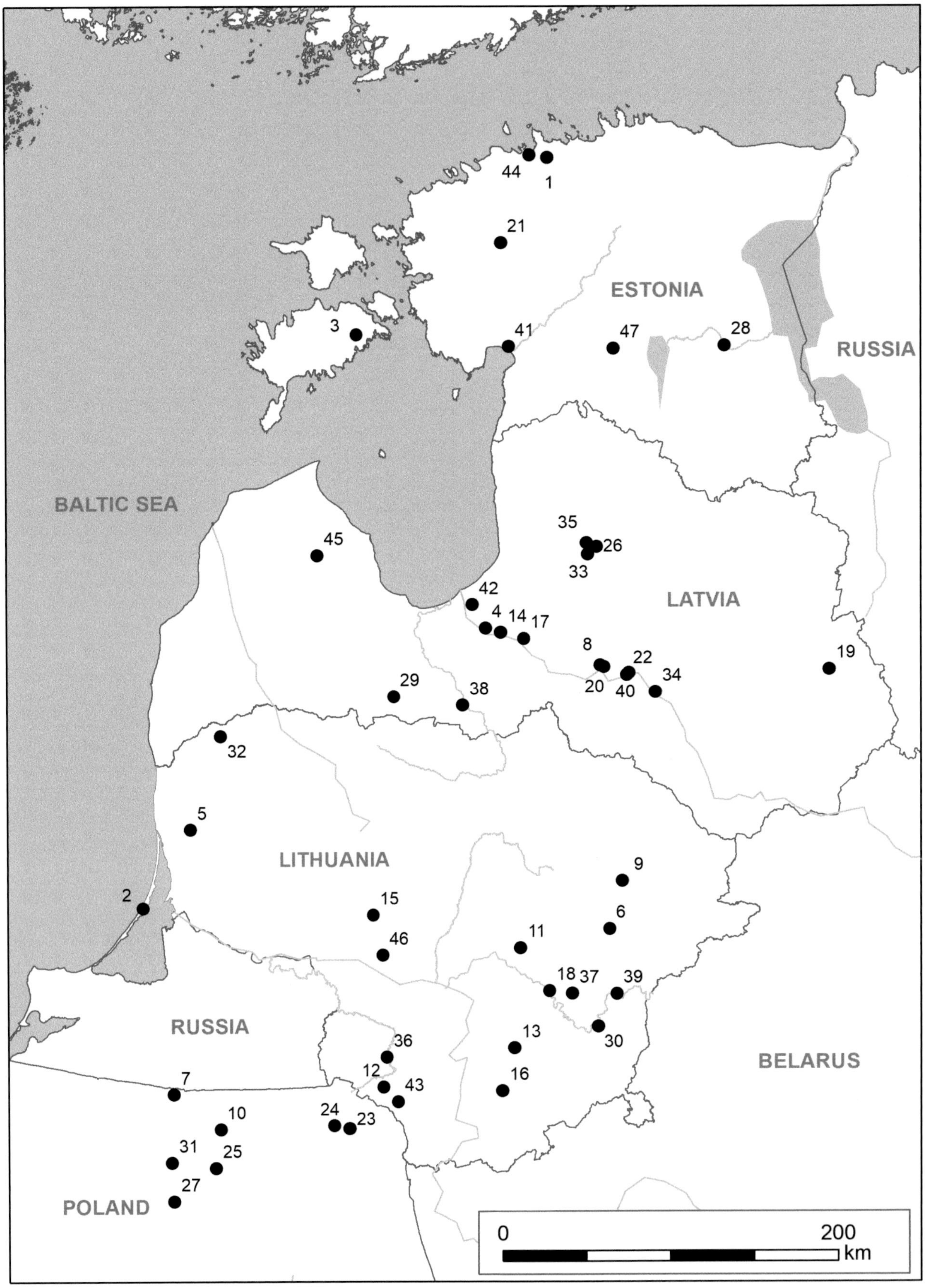

Figure 1. Map showing sites in the Eastern Baltic region with macrobotanical remains discussed in this chapter. Information about the sites is given in Table 1.

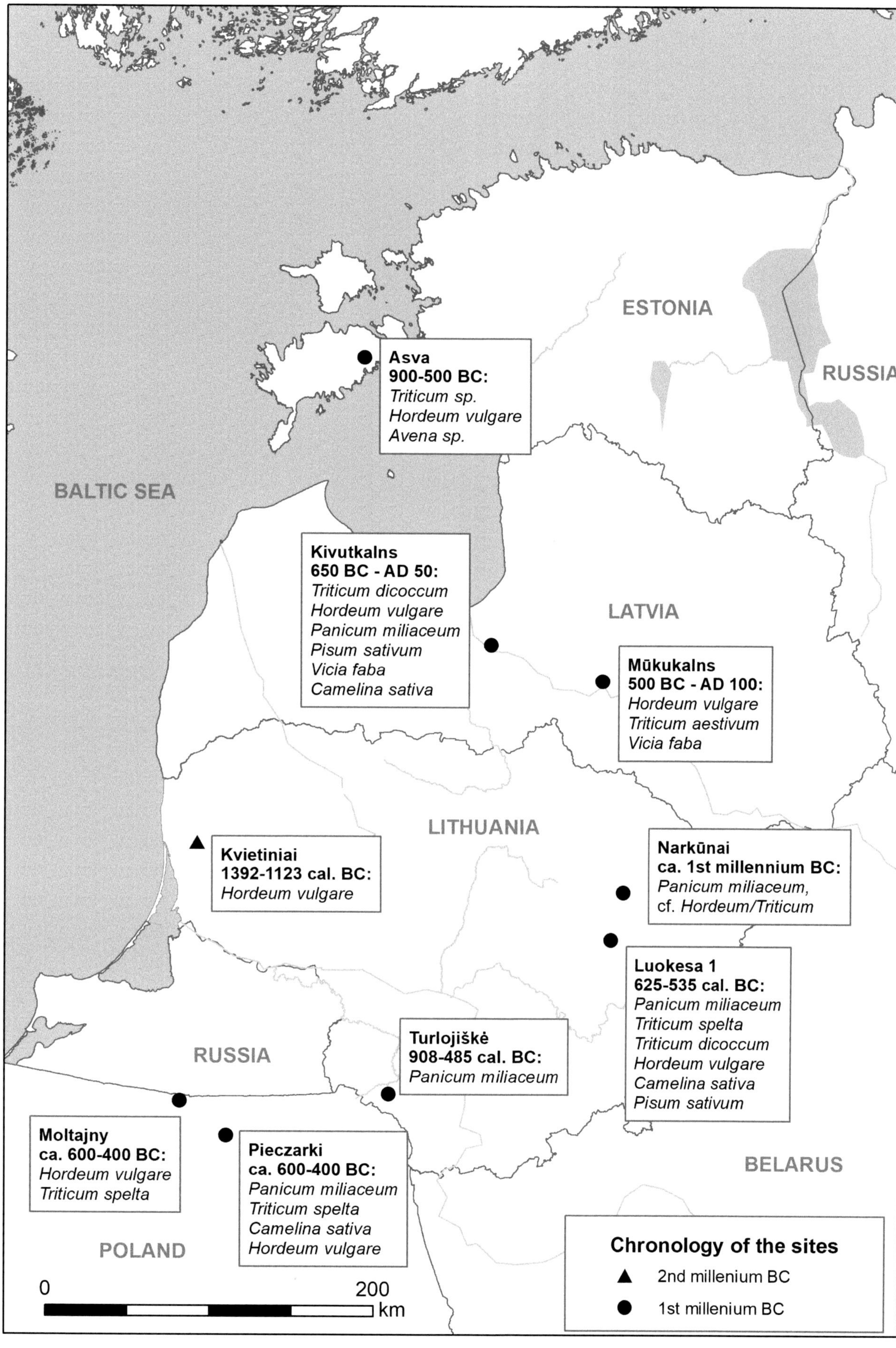

Figure 2. Map showing sites in the Eastern Baltic region pre-dating the Roman period with reliably dated and identified crop taxa.

Table 1. Eastern Baltic region sites with macroremains dating up to the 13th–14th century AD. References and comments for the sites are provided in Table 2. Also see section 'Material and methods' for more information. The quantity of macroremains is provided only if it was given in the original publication, otherwise the taxon is simply marked as present '+'. In cases where presence is marked '*', additional information is provided in the Table 2 or in the text. Grey cell background indicates that the macroremain of that species was subjected to radiocarbon dating; sp. means identified only to species level; cf. means species identified with uncertainty; M means remains of this taxon were the most numerous in the samples. If in earlier publications *Avena* is mentioned only as 'oat' in the local language, and not as *Avena sativa* (common oat), then it is here added to the column *Avena* sp.

No. on map	Site and chronology	*Avena* sp.	*A. fatua*	*A. sativa*	*Fagopyrum esculentum*	*Hordeum vulgare*	*Panicum miliaceum*	*Secale cereale*	*Setaria* sp.	*S. italica*	*Triticum* sp.	*Tr.* cf. *monococcum*	*Tr. dicoccum/spelta*	*Tr. spelta*	*Tr. dicoccum*	*Tr. aestivum/durum*	*Tr. compactum*	*Pisum sativum*	*Vicia faba*	*Vicia sativa*	*Lens culinaris*	*Brassica rapa/campestris*	*Camelina sativa*	*Cannabis sativa*	*Linum usitatissimum*	*Papaver somniferum*	Other
	3 millennium BC: Neolithic sites with macrobotanical remains																										
1	Iru, c. 3 mill. BC					*																					
2	Nida & Pilkopa, c. 3 mill. BC					*									*												
	2–1 millennium BC: Bronze Age and Early Iron Age sites with macrobotanical remains																										
3	Asva, 900–500 BC	1*				>10*					1*																
4	Ķivutkalns, 650 BC–AD 150		+			+	+				+				+			+	+				+				
5	Kvietiniai, 1392–1123 cal BC					1*																					
6	Luokesa I, 625–535 cal BC					155	6073					75	3040	589	3363			54					1925*				
7	Moltajny, c. 600–400 BC					1602								1542													
8	Mūkukalns, 500 BC–AD 100					+										+			+								
9	Narkūnai, c. 1 mill. BC					*	+				*																
10	Pieczarki, c. 600–400 BC					27	576							107								4*	139				+
11	Samantonys, c. 1000–500 cal BC					*																					
12	Turlojiškė, 908–485 cal BC						293																				
	Roman Period (1-4 centuries AD) and Middle Iron Age (5-8 centuries AD) sites with macrobotanical remains[aa]																										
13	Aukštadvaris, 1–2 cent. AD					+								+	+				+								
14	Daugmale, 5 cent. AD		+			M	+	+	+						+	+											+
15	Gabrieliškės, cal AD 85–250			+		+	+	M						+	+	+	cf.										
16	Kaukai, *V. faba* = cal AD 649–775					*												*	+							*	
17	Ķenteskalns, 5–9 cent. AD	+				+	+	+			+				+	+		+	+								
18	Kernavė, Aukuro kalnas, 5 cent. AD	+				+	+							+	+	+		+	+								
	Kernavė, 3–4 cent. AD					+																					
19	Kivtu, 2–4 cent. AD					+									+	+		+	+						+		
20	Koknese, 1–4 cent. AD	+				M	+	+							M			+	+								
	Koknese, 5–9 cent. AD		+			+		+							+				+			*					
21	Linnaaluste, cal AD 640–900					M																					
22	Lokstene, 5–9 cent. AD					+									+	+		+				*					
23	Osinki, 2–3 cent. AD														1												
	Osinki, 5–6 cent. AD					31		50							8	40	8										
	Osinki, 2–6 cent. AD					362								85	46	329	97	363	22								

24	Osowa, 3–5 cent. AD			32		1695		290			49			89	103	1414	76	227	12								
25	Paprotki Kolona 1, 1–3 cent. AD					+	+				+	*			*												
	P. Kolona 41, Roman period	1					1	1			+			1													
	P. Kolona 41, Migration period	1 cf.				8	1	11			+																
	P. Kolona 41, 3–6 cent. AD					1 cf.						1															
26	Sarumkalns, 2–4 cent. AD					M										+											
27	Stare Kielbonki, 2–5 cent. AD				+			+			+																
28	Tartu Fort, cal AD 670–875	2				155		65								7		4	2								
29	Tērvete, 5–6 cent. AD					+	+				+																
30	Vilnius Lower castle, cal AD 430–620																							1			
31	Wyszembork 1, 2–3 cent. AD					12	2	7														*5 cf.		24*			
	Wyszembork 1, 6–7 cent. AD					30	13	22														559* cf.					
	Wyszembork 4, 3–6 cent. AD					16		681										1				*18 cf.					
	Sites dating to 9–13/14 century AD discussed in this paper																										
32	Apuolė, cal AD 1224–1282			+		M		+						+		+	+	+	+	cf.							
33	Āraiši, cal AD 780–1050					+		+			+				+	+		+	+			*	+		+		
34	Asote, 9–13 cent. AD	+		+		+	+	+							+	+		+	+		+	*			+		+
13	Aukštadvaris, cal AD 1310–1419	*a				+		+		+				+	+	+	+	+	+	*b	+						
35	Cēsis castle, 14 cent. AD			+	+	+		+			+													+			+
14	Daugmale, 11–12 cent. AD					+		+			+				+			+				*a				*b	
16	Kaukai, c. 5–14* cent. AD					*												*	*							*	
18	Kernavė, 14 cent. AD	*			+	+		+								+											
36	Kumelionys, c. 10–11 cent. AD						+											+	+	+							
37	Maišiagala, 13–14 cent. AD	+			+	+	+	+			+			+		+		+	+	*	+			+		+	+
38	Mežotne, 11–13 cent. AD	+				+	+	+	+						+	+		+	+			*	+		+		
39	Nemenčinė, 1–2 and 13–14 cent. AD	*a				+	+	+						+		+		+	+	*b				+			
40	Oliņkalns, 11 cent. AD							+															+				
41	Pärnu, 14 cent. AD			+	+	+		+							cf.	+		+	cf.			*		+	+	+	+
42	Riga, 13 cent. AD	+				+	+	+		+	+					+		+				*			+	+	+
43	Rudamina, c. 10–13 cent. AD	*a				+	+								+	+		+	+	*b							
44	Tallinn, 13–14 cent. AD				+	+		+								+						*			+	+	+
45	Talsi, 11–13 cent. AD	+		+		+		+							+	+		+	+			*	+	+	+		
28	Tartu settlement, AD 1030–1061					+										+								+			
	Tartu, 14 cent. AD				+		+	+														*		+	+	+	+
29	Tervete, 11–13 cent. AD	+	+			+	+	+	+		+				+			+	+		+	*	+		+	sp.	+
46	Veliuona, c. 13–14 cent. AD			+	*	+	+	+							+	+		+									
47	Viljandi, 13–14 cent. AD					+		+								+						*			+		
30	Vilnius Lower castle, 14 cent. AD						4																		2		

[aa]At few more sites, such as the Nemenčinė and Velykuškės hillforts in Lithuania, archaeobotanical material dating to the Roman period has previously been reported (Kulikauskas 1955; Michelbertas 1986). However, its chronological attribution to this period is not firm, these are multiperiod sites with mixed contextual data, therefore we excluded it. Two other hillforts—Mūkukalns and Ķivutkalns—are dated from the end of the Bronze Age up to the first centuries AD. Recent chronological studies have shown that Ķivutkalns ceased to exist sometime during the 1st–2nd century AD (Oinonen et al. 2013). The same spectrum of crops was identified in earlier and later chronological periods at Ķivutkalns, while at Mūkukalns different phases were not separated (Rasiņš and Tauriņa 1983); therefore, data from these hillforts are included in the 1st millennium BCE section.

Table 2. References and comments for the data provided in Table 1.

Site no.	Sources for analysis; sources for chronology (preceded by 'Chr:'); notes
	3 millennium BC: Neolithic sites with macrobotanical remains
1	Kriiska 2009; Lang 1996. *discussed on page 162
2	Heydeck 1909; Chr: Piličiauskas 2016. *discussed on page 162
	2–1 millennium BC: Bronze Age and Early Iron Age sites with macrobotanical remains
3	Lang 2007: 63, 111; *Number of grain imprints on pottery
4	Rasinš and Taurina 1983; Chr: Vasks and Zarina 2014
5	Grikpėdis and Motuzaite Matuzeviciute 2017; **Hordeum vulgare* = 1 grain; cf. *Triticum/Hordeum* sp. = 7 grains
6	Pollmann 2014; **Camelina sativa* includes *Camelina* sp. remains
7	Polcyn 1995; Chr: Gackowski 2000
8	Rasinš and Taurina 1983
9	Podėnas et al. 2016; *imprints of *P. miliaceum* and *Triticum* or *Hordeum* grains
10	Polcyn 2000; Chr: Gackowski 2000; *identified as *Brassica campestris;* 'other' = seeds of *Brassica nigra*
11	Rimantienė 1960; Chr: Podėnas et al. 2016; *needs re-examining
12	Antanaitis-Jacobs et al. 2004; Chr: Antanaitis and Ogrinc 2000
	Roman Period (1-4 centuries AD) and Middle Iron Age (5-8 centuries AD) sites with macrobotanical remains[aa]
13	Daugudis 1962; Some grains were re-inspected by authors at the National museum of Lithuania
14	Rasinš and Taurina 1983; 'Other' = *Sesamum indicum* and wild seeds
15	Matlakowna 1926, 1929; Lideikytė-Šopauskienė 1935; Grikpėdis and Motuzaitė Matuzevičiūtė 2016
16	Kulikauskas 1982; *According to the typological chronology, site was inhabited during ca. 5th–14th century AD. Macrobotanical data is mentioned as originating from the later cultural horizon, dated to ca. 9th–14 century AD (Kulikauskas 1970). In the National museum of Lithuania, we managed to find only charred macroremains of *V. faba*, which was subjected to AMS dating (Table 3). Chronology of other crop species remains unknown.
17	Rasinš and Taurina 1983
18	Luchtanas 2015; Motuzaitė Matuzevičiūtė and Rusteikytė 2017
19	Rasinš and Taurina 1983
20	Rasinš and Taurina 1983; *identified as *Brassica campestris*
21	Konsa and Kivi 2012
22	Rasinš and Taurina 1983; *identified as *Brassica campestris*
23	Litynska-Zajac 1997
24	Litynska-Zajac 1997; Chr: Czeczuga and Kossacka 1973
25	Karczewski 2012; Karczewski et al. 2009; *identified as *Triticum* cf. *monococcum/dicoccum* and *Tr.* cf. *dicoccum*
26	Rasinš and Taurina 1983
27	Mellin-Wyczolkowska 2007
28	Tvauri and Vanhanen 2016
29	Rasinš and Taurina 1983
30	Stančikaitė et al. 2008
31	Litynska-Zajac 1997; **Cannabis sativa* and *Brassica campestris* not charred, while other material was charred
	Sites dating to 9–13/14 century AD discussed in this paper
32	Lideikytė-Šopauskienė 1935; Hillfort was inhabited from the first centuries AD; however, most artefacts are dated to 9th–13th century AD. No contextual information is available for macro-remains. Chr: Table 3
33	Rasinš and Taurina 1983; Chr: Stivrins et al. 2015; *identified as *Brassica campestris*
34	Rasinš and Taurina 1983; 'Other' = 2 grains of *Oryza sativa;* *identified as *Brassica campestris*
13	Daugudis 1962; Šimkūnaitė 1957; Rasinš 1958; Chr: Table 3, also Vengalis 2008; *a *Avena* specified as 'oat'; *b *Vicia sativa* specified as 'vetch' (see discussion on page 167)
35	Brown and Badura 2017; 'Other' = *Brassica nigra, Humulus lupulus, Ficus carica*
14	Rasinš and Taurina 1983; *a identified as *Brassica campestris*; *b identified as *Papaver* sp.
16	*see note above - site No. 16
18	Kulikauskas and Luchtanas 1980; Šimkūnaitė 1981; **Avena* specified as 'oat'
36	Kulikauskas 1982; Chr: Volkaitė-Kulikauskienė 1978, Baubonis and Zabiela 2005b; See discussion about *V. sativa* on page 167.
37	Volkaitė-Kulikauskienė 1974; Also reported macroremains of *Carum carvi, Humulus lupulus, Crambe maritima, Sinapis alba, Cerasus vulgaris.* *see discussion about *V. sativa* in on page 167
38	Rasinš and Taurina 1983; *identified as *Brassica campestris*
39	Kulikauskas 1958; Chr: Vengalis 2008; *a *Avena* specified as 'oat'; *b *Vicia sativa* specified as 'vetch'; Some of the macrobotanical material is coming from the early habitation stage (Roman period), however, it's not specified exactly which.
40	Rasinš and Taurina 1983
41	Sillasoo and Hiie 2007; *identified as *Brassica rapa*; 'Other' includes *Lactuca sativa, Pastinaca sativa, Ficus carica,* among others

Table 2 Cont.

42	Banerjea et al. 2017; *identified as *Brassica rapa*; 'Other' includes *Brassica nigra, B. oleracea, Foeniculum vulgare, Malus* sp., *Pastinaca sativa,* among others
43	Kulikauskas 1982; *a *Avena* specified as 'oat'; *b *Vicia sativa* mentioned as 'vetch'; See discussion about *V. sativa* on page 167.
44	Sillasoo and Hiie 2007; *identified as *Brassica rapa*; 'Other' includes *C. carvi, H. Lupulus, F. carica,* among others
45	Rasinš and Taurina 1983; *identified as *Brassica campestris*
28	Tvauri and Vanhanen 2016
	Kihno and Hiie 2008; Sillasoo and Hiie 2007; *identified as *Brassica rapa*; Also remains of *F. carica*
29	Rasinš and Taurina 1983; *identified as *Brassica campestris*; 'Other' includes cf. *Sesamum indicum*, among others
46	Lideikytė-Šopauskienė 1935; Matlakowna 1926; Swederski 1926; Chr: Baubonis and Zabiela 2005a. *two charred fruits of *F. esculentum* mentioned among weeds.
47	Sillasoo and Hiie 2007; *identified as *Brassica rapa*
30	Stančikaitė et al. 2008; More data from 15th–16th century AD

geographical comparison. *Hordeum vulgare* var. *vulgare* (hulled barley) and *H. vulgare* var. *nudum* (naked barley) are not separated in this review because most of the reports do not differentiate between these two varieties. All data are presented in Table 1 with additional information in Table 2. It includes all sites with macroremains of crops dated prior to the 8th century AD; later sites are only included if they yielded crops poorly represented or absent in earlier periods, including *Fagopyrum esculentum* (buckwheat), *Lens culinaris* (lentil) and *Linum usitatissimum* (flax), among others.

While preparing such a broad review and synthesis, one has to critically evaluate the available data. We excluded some material due to uncertain chronology or doubtful botanical identification. In order to clarify earlier reported species from Lithuanian archaeological sites, we re-examined some of the macroremains currently stored at the National Museum of Lithuania and at Vytautas the Great War Museum. Unfortunately, in some cases, the actual cereal remains have been lost and thus cannot be revaluated for species clarification or directly dated. Others lack precise information about their context. In addition, it is important to note that much of the archaeobotanical data come from multi-period sites with a wide chronological span based on typological dating of artefacts and not on radiocarbon dates. In order to obtain a more accurate chronology, we directly radiocarbon dated grains from three Lithuanian hillforts (Table 3). Finally, we note that some of the reports or publications mention only the crops that were present, without giving the exact numbers or percentages of recovered species. Therefore, it is impossible to conduct a quantitative analysis of the specific crop species for most of the sites.

Our collated data on the crops in the Eastern Baltic are divided into two groups. The first group (Fig. 1, Table 1) represents all the data we were able to collate. The second group represents only what we consider to be 'reliable data', i.e. securely dated and reliably identified crop species cultivated in the region before the Roman period (Fig. 2). Please see our arguments below why we think not all the data in Figure 1 and Table 1 fit the 'reliable data' criteria.

Earliest evidence of crops in the Eastern Baltic region

The earliest evidence of agriculture in the Eastern Baltic is made up of solitary or small numbers of Cerealia-type pollen grains in peat layers dated from c. the 4th millennium cal BCE (Daugnora and Girininkas 2004; Loze 1997; Stančikaitė et al. 2002). The limitations of using pollen data for identifying early agriculture in the region have already been discussed (Grikpėdis and Motuzaite Matuzeviciute 2017; Piličiauskas et al. 2017). We believe that currently the most reliable method for identifying early agriculture in the region is the presence of macroremains of cultivated crops backed up by radiocarbon dates.

Ostensibly the earliest macroremains of domesticated plants in the region—*Triticum dicoccum* (emmer), *Panicum miliaceum* (broomcorn millet), *Setaria italica* (foxtail millet) and *Cannabis sativa* (hemp) seeds dating to the 3rd millennium BCE—have been report-

Table 3. New dates for plant macroremains from three Lithuanian hillforts excavated in the 20th century AD. Calibrated using OxCal version 4.3 with IntCal13 atmospheric curve (Reimer et al. 2013). UBA = 14CHRONO Centre, Queen's University Belfast.

Site	Dated material	Lab code and number	^{14}C age BP	Calibrated age, 2σ range
Aukštadvaris	*Lens culinaris,* charred grain	UBA-30597	569 ± 24	cal AD 1310–1419
Kaukai	*Vicia faba,* charred grain	UBA-30598	1301 ± 39	cal AD 649–775
Apuolė	*Triticum compactum,* charred grain	UBA-30599	759 ± 23	cal AD 1224–1282

ed by Rimantienė (1992). However, recently it has been shown that all those finds were either mistakenly identified as domestic species or date to a much later period (Grikpėdis and Motuzaite Matuzeviciute 2017; Piličiauskas et al. 2017).

Sparse data on possibly early farming come from grain imprints on potsherds. Evidence of *Hordeum vulgare* (barley) and *T. dicoccum* dated to c. 3rd millennium BCE has been reported from three sites—Iru, Nida and Pilkopa (Table 1). In the case of Nida and Pilkopa, the potsherds with crop impressions are now lost and thus unavailable for re-evaluation. Moreover, it is very difficult to define the beginnings of crop cultivation from pot impressions alone without confirmation in the form of actual macroremains. The occasional grain imprint on pottery does not necessarily imply local cultivation, as either the pots or the grains may have been traded into the Eastern Baltic from farming communities. Currently there are no actual macroremains of cultivated plants in the Eastern Baltic reliably dated to earlier than the 2nd millennium BCE.

The earliest AMS-dated evidence of crops in the region comes from the Middle Bronze Age settlement of Kvietiniai, in western Lithuania (Fig. 2). A few caryopses of charred barley were recovered from a fireplace flotation sample (Grikpėdis and Motuzaite Matuzeviciute 2017). One barley grain was AMS-dated to 1392–1123 cal BCE (Table 1). Together with cereals, also a few seeds of wild plants and 35 fragments of *Corylus avellana* (hazel) nutshells were found. The absence of cereal chaff, ruderal plants and weeds, together with the presence of only a few cereal grains, could lead to speculation that the people at Kvietiniai were supplementing wild foods with some domesticated crops, but it remains unclear whether the crops were grown at the site or were acquired from elsewhere (Grikpėdis and Motuzaite Matuzeviciute 2017). Similarly, in Finland, the earliest macroremains of cultivated plants were identified as *H. vulgare* and radiocarbon dated to the 2nd millennium cal BCE (Lavento 1998; Vuorela and Lempiäinen 1988). The sparse data on macrobotanical crop remains during the 2nd millennium BCE do not allow us to speculate whether the earliest barley from the Eastern Baltic was cultivated locally. Moreover, the number of seeds is so small that cereals do not seem to have been an important staple food for the local population during this period.

First millennium BCE

The Late Bronze Age and the Early Iron Age in the region are associated with the appearance of hilltop settlement sites (Baubonis and Zabiela 2005a; Lang 2007). Moreover, the number of fossil field systems increases compared with earlier periods (Lang 2010). In addition, continuous Cerealia curves appear in the pollen data (Stančikaitė et al. 2004), leading archaeologists to identify the existence of plant cultivation across the region (Antanaitis-Jacobs and Girininkas 2002; Lang 2007). Therefore, it is not surprising that macroremains of cultivated plants are also numerous during this period (Fig. 2, Table 1).

Cereals

Barley

The most widespread crop during the 1st millennium BCE was barley, which was present in the majority of the archaeobotanically investigated sites (Table 1). However, it is difficult to assess the importance of barley at different sites because data about recovered grain quantities is known from only a few locations. For example, at the two sites with the most intensive archaeobotanical investigations from the 1st millennium BCE—the lakeshore settlements Luokesa 1 and Pieczarki—barley constitutes only a very small proportion of the identified crops (Polcyn 2000; Pollmann 2014). However, at Ķivutkalns and Mūkukalns, barley grains dominated some assemblages (Rasiņš and Tauriņa 1983). This could be explained by two completely different preservation conditions, namely, waterlogging (at Luokesa 1 and Pieczarki) and charring (at Ķivutkalns and Mūkukalns). The waterlogging preserves a much wider plant spectrum, which otherwise would have been lost to fire, in that way 'diluting' the barley percentages. In general, during the 1st millennium BCE, barley was one of the main crops of central and northern Europe. For example, in Poland during the Hallstatt and La Tène periods, barley was the most widespread cereal, identified in more than half of the sites (Wasylikowa et al. 1991). Similarly, in Finland, barley seems to have been the main crop during the 1st millennium BCE (Vanhanen and Koivisto 2015).

Hulled wheats

There is firm evidence for the cultivation of hulled wheats during the 1st millennium BCE. *Triticum spelta* (spelt) remains were found at three lakeshore dwellings dated to c. the middle of the 1st millennium BCE, and *T. dicoccum* was identified at two sites (Table 1). Spelt is represented mainly by waterlogged glume bases and spikelet forks or, indirectly, by impressions on pottery and daub (Polcyn 2000; Pollmann 2014). Spelt cultivation in Europe is thought to have started by the end of the Neolithic (Akeret 2005; Nesbitt and Samuel 1996). During the Bronze Age, it spread to northern Europe (Nesbitt and Samuel 1996). The thick and tough glumes of hulled wheats provide protection for the grains in the field and during storage (Nesbitt and Samuel 1996). These glumes may have been an important attribute for the early establishment of agriculture in the Eastern Baltic region.

Einkorn is represented only by some waterlogged

glume bases and spikelet forks, identified as *Triticum* cf. *monococcum* at Luokesa 1 (Pollmann 2014). It is important to note that throughout the prehistory of the region no clear cultivation of einkorn can be identified. Macroremains identified as *T.* cf. *monococcum* were present at only one other site, Paprotki Kolona 41 (Karczewski 2012), dated to Late Roman and Migration periods (Table 1). As evidenced by the data from Poland (Wasylikowa et al. 1991) and other regions of Europe (Zohary et al. 2013), the importance of einkorn decreased from the Neolithic onwards. Therefore, by the time agriculture was introduced in the Eastern Baltic, einkorn was already a rarely cultivated crop. Similarly, no einkorn remains have been found in Finland (Aalto 1997; Vanhanen and Koivisto 2015).

Free-threshing wheats

Remains of *Triticum aestivum* (bread wheat) are known from only one site, Mūkukalns, which is dated to a very broad period (Table 1). As no free-threshing wheats were present in the other 1st millennium BCE sites, we suggest that the grains of *T. aestivum* found at Mūkukalns probably come from the end of the 1st millennium BCE or even the beginning of the 1st millennium AD. Similarly, in Finland, the earliest grain of the free-threshing wheat *Triticum compactum* (club wheat), from Sääksmäki Rapola, was radiocarbon dated to c. 200 cal BC–cal AD 200 (Vikkula et al. 1994). Therefore, *T. aestivum* may have appeared in the region only by the end of the 1st millennium BCE or even later.

Broomcorn millet

P. miliaceum was found in five Eastern Baltic sites dated to the 1st millennium BCE, and in three cases (Turlojiškė, Luokesa 1 and Pieczarki) it was the most numerous crop (Table 1). The newest research on the arrival of broomcorn millet in Europe has shown that it most likely appeared during the mid-2nd millennium BCE (Motuzaite Matuzeviciute et al. 2013). In the 1st millennium BCE, its importance in Poland was growing—from 29 per cent of Hallstatt sites to 72 per cent of Roman period sites (Wasylikowa et al. 1991). Broomcorn millet was not cultivated in the northern part of the Eastern Baltic. In fact, the Daugava River in Latvia was the northern boundary for broomcorn millet cultivation, as no remains of this plant are found either in Estonia or in Finland until the medieval period. Moreover, during the medieval period, millet was not cultivated locally in Livonia but, rather, was imported (Sillasoo and Hiie 2007).

Pulses and oil plants

Pisum sativum (pea) and *Vicia faba* (broad bean) have been found at two sites, in Latvia and Lithuania (Fig. 2). No legumes are known from 1st millennium BCE Estonia (Lang 2007). However, not much macrobotanical data from this period in Estonia is published; therefore, legumes may be underrepresented. In Finland there are more data, however also no remains of pulses are known from the 1st millennium BCE (Aalto 1997; Vanhanen and Koivisto 2015).

Camelina sativa (gold-of-pleasure) is the only oilseed crop that was cultivated during the 1st millennium BCE (Fig. 2). The Luokesa 1 lake dwelling site contains abundant remains of *C. sativa,* preserved not only as seeds but also as pods and stems (Pollmann 2014). At Pieczarki, gold-of-pleasure constituted 13 per cent of the crop remains (Polcyn 2000), and at Ķivutkalns, it dominated two samples and was present in another 2 out of 7 analysed (Rasiņš and Tauriņa 1983). It is a valuable and undemanding plant, with a short vegetative period, and has been cultivated in north-eastern Europe mainly from the Late Bronze Age onwards. In Poland, it appeared during the Hallstatt period (Wasylikowa et al. 1991), and in Finland, only in the second half of the 1st millennium AD (Aalto 1997).

The Roman period and the Middle Iron Age (1st–8th century AD)

Barley was the most widespread cereal, being present at more than 80 per cent of the sites (Fig. 3). Broomcorn millet and emmer were found at approximately half of the sites (Fig. 3, Table 1). New crops, such as *T. aestivum, T. compactum, Secale cereale* (rye), *Avena* sp. (oat)*, Linum usitatissimum* and *Cannabis sativa* (Table 1), joined the crop assemblage. In addition, one unusual plant for 1st millennium AD northern Europe is mentioned from Daugmale hillfort. Seeds of *Sesamum indicum* (sesame) were reported from the layers dated to the 5th century AD (Rasiņš and Tauriņa 1983). There are no other finds of this exotic plant from the Eastern Baltic during the Iron Age; it is present only later, in medieval contexts (Rasiņš and Tauriņa 1983). Daugmale hillfort was inhabited until medieval times; therefore, this seed may be a contamination from younger layers or may have been misidentified.

Cereals

Wheat

A variety of wheat species—*T. spelta*, *T. dicoccum, T. aestivum*, *T. compactum* and possibly *T. monococcum*—were identified for the Roman and Middle Iron Ages in the Eastern Baltic (Fig. 3, Table 1). Emmer was the most frequent among the hulled wheats. In Latvia during the Roman period and the Middle Iron Age, emmer is mentioned from five sites, but no spelt remains were identified (Rasiņš and Tauriņa 1983). In Lithuania and north-eastern Poland, both emmer and

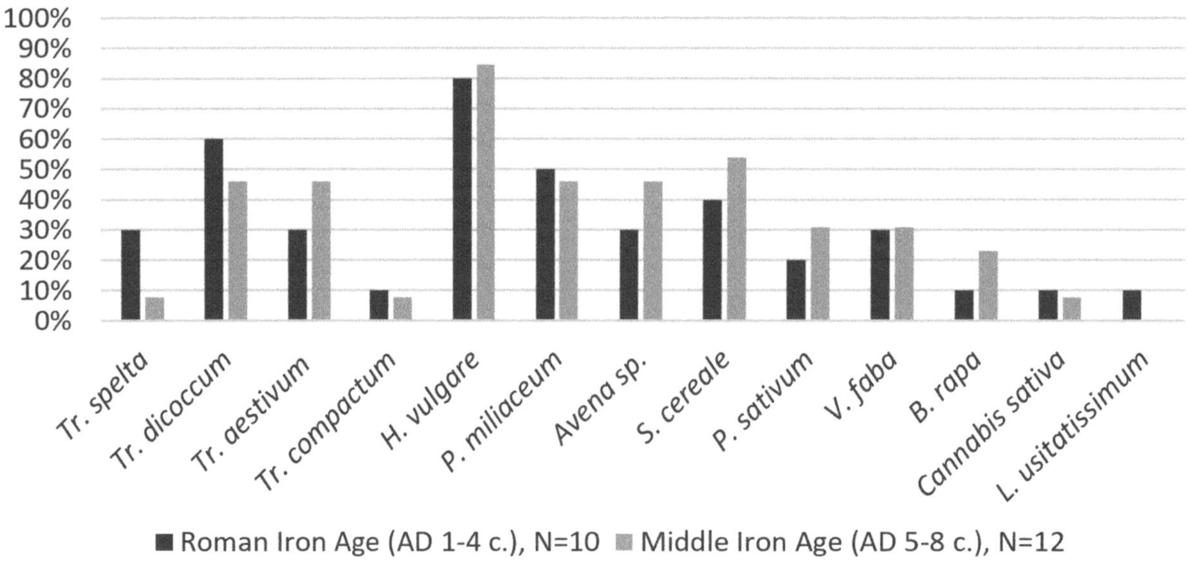

Figure 3. Presence of cultural plant species in Eastern Baltic region sites dated to the 1st–8th century AD. Percentage refers to proportion of the total number of sites with one or more identifications to that taxon. N refers to total number of sites per period. Sites with macrobotanical material dated to transitional periods, e.g. 3rd–5th century AD, have been excluded. Kaukai has also been excluded due to chronological uncertainty.

spelt are mentioned (Table 1). However, it is very difficult to differentiate between spelt and emmer on the basis of only charred caryopses. And most of the time chaff is not present. Therefore, spelt may be underrepresented in the distribution data (Fig. 3). Another hulled wheat, einkorn, is represented only by one grain, identified as *T.* cf. *monococcum,* at Paprotki Kolona 41 (Table 1). Therefore, its cultivation seems doubtful.

In the Roman period and Middle Iron Age, free-threshing wheats are mostly represented by *T. aestivum* species, while *T. compactum* is recorded in small numbers, from just a few sites (Table 1). In most parts of Europe, free-threshing wheats became widespread and dominated over hulled wheats during the 1st millennium AD (van der Veen 2007). The importance of free-threshing wheats in the Eastern Baltic increased throughout the 1st millennium AD—it was present in 30 per cent of sites during the Roman period and in 46 per cent of sites during the Middle Iron Age (Fig. 3). Moreover, in some Roman and Migration period sites, such as Osowa and Osinki, *T. aestivum* was already among the most numerous crops (Table 1). It has been proposed that free-threshing wheats may have been preferred only when the need for higher yields outweighed the greater investment required in time and fertiliser (van der Veen 2007).

Rye

The appearance of *S. cereale* in the Eastern Baltic, at least in its southern parts, can be traced back to the Roman period (Grikpėdis and Motuzaitė Matuzevičiūtė 2016). Charred rye grains were found at 40 per cent of archaeobotanically investigated sites dated to the 1st–4th century AD and at 54 per cent of sites dated to the 5th–8th century (Fig. 3). Rye spread across Europe as a weed among other crops, and only by the end of the 1st millennium BCE did it attain cultivated crop status (Behre 1992). In most Eastern Baltic sites dated to the 1st–8th century, rye constitute only a small portion of crops; however, in two sites it was the most numerous crop, namely, Gabrieliškės (Grikpėdis and Motuzaitė Matuzevičiūtė 2016) and Wyszembork 4 (Litynska-Zajac 1997). Moreover, the beginning of rye cultivation in this period is supported by the rising *S. cereale* pollen curve (Grikpėdis and Motuzaitė Matuzevičiūtė 2016). The gradual northern spread of rye can be observed in Poland. During the Late Pre-Roman Iron Age, it was common in southern Poland, and during the Late Roman Iron Age it became important in north-eastern Poland (Litynska-Zajac 1997). In Finland, the earliest grains of rye come from Pre-Roman or Roman period sites; however, it remained of minor importance until the Viking age and later times (Vanhanen in press).

Oat

Avena sativa (common oat) underwent a similar process of development as rye. In central Europe, it became an important crop only in the Roman period; during earlier times it may have been only a weed among other crops (Zohary et al. 2013). It is difficult and sometimes impossible to differentiate between the cultivated species *A. sativa* and wild species, such as *Avena fatua* (common wild oat), if only charred grains and no parts of the florets or glume bases are present (Jacomet 2006). While preparing the dataset for this chapter, we found out that *A. sativa* is mentioned from only two Roman period sites (Table 1). At Gabrieliškės, not only charred grains (Lideikytė-Šopauskienė 1935; Matlakówna 1926) but also remains of chaff were present, allowing an identification of *A. sativa* (Matlakówna 1929). At the site of Osowa, 32 grains of *A. sativa* have been reported (Table 1). At six more sites, oat remains were identified as *Avena* sp. or *A. fatua* (Table 1).

To conclude, *Avena* sp. was a common element of the Eastern Baltic crop fields from the Roman period onwards, and possibly even earlier; however, the current dataset is too poor to speculate on its importance. By the 11th–12th century, common oat seems to appear much more often in the macrobotanical assemblages of the region (Table 1). Therefore, Rasiņš and Tauriņa (1983) have suggested that common oat cultivation in Latvia started around that time. In Estonia, the earliest *A. sativa* grains are reported from the 11th–12th century AD; however, the beginning of its cultivation was proposed to have started sometime in the second half of the 1st millennium AD, when the sediments of Lake Maardu show the uninterrupted presence of *Avena* sp. pollen (Tvauri and Vanhanen 2016). In Poland, the cultivation of common oat spread from c. the second half of the 2nd century AD (Litynska-Zajac 1997). In Finland oat is considered to be a weed throughout the Iron Age (Aalto 1997). Oat grains are present from the Late Bronze Age onwards; however, no floret bases come from the Roman period or earlier deposits (Vanhanen in press).

Oil and fibre plants

Linum usitatissimum seeds have been found in the Roman period settlement of Kivtu, in Latvia (Table 1). Unfortunately, they were not radiocarbon dated. Moreover, remains of linen cloth are reported from 6th century AD deposits in Estonia (Kriiska et al. 2005). However, it was not specified what methods were used for the identification of the fibre as linen. During the Roman period, flax was present in Poland (Litynska-Zajac 1997) and Finland (Aalto 1997; Vanhanen in press). Therefore, it is possible that flax was already in use during the Roman period in the Eastern Baltic; however, further investigation is needed to prove it.

Cannabis sativa seeds are reported from two sites (Table 1). At Wyszembork, uncharred hemp seeds were identified among charred cereals dated to the 2nd–3rd century AD (Litynska-Zajac 1997). At Vilnius Lower Castle, one hemp seed was recovered from the bottom of a cultural layer, but no other crop remains were found in this horizon (Stančikaitė et al. 2008). The sediment was radiocarbon dated to cal AD 430–620 (Stančikaitė et al. 2008). Macroremains of hemp were found only in a few Polish Roman period sites (Litynska-Zajac 1997). In Finland, the earliest radiocarbon-dated hemp seeds come from the late Roman and Migration periods (Vanhanen in press). To conclude, it is very likely that hemp was used in the 1st millennium AD Eastern Baltic. However, currently we do not have reliably dated evidence to prove it.

Tubers

Brassica rapa (white turnip) may have been an important food plant in the Eastern Baltic. It was found at several medieval sites (Table 1). However, remains of the closely related plant *Brassica campestris* are mentioned already from the mid-1st millennium BCE (Polcyn 2000). At Wyszembork 1 and 4, macroremains identified as *B.* cf. *campestris* were present in Roman and Migration period contexts (Table 1). However, most of the *B.* cf. *campestris* seeds were found in uncharred condition among charred cereal remains (Litynska-Zajac 1997). Moreover, some seeds identified as *B. campestris* were present in two Middle Iron Age hillforts in Latvia (Table 1). Originally, Carl Linnaeus used the term *B. campestris* for the wild weed and the term *B. rapa* for cultivated turnip-type plant varieties (Reiner et al. 1995). Later it was realised that both varieties have to be classified as the same species, introducing confusion in the nomenclature (Reiner et al. 1995). Reiner et al. (1995) noticed that it is always problematic to distinguish the seeds of wild-type and turnip-type *B. rapa*. Therefore, it is possible that turnips in the Eastern Baltic could have been consumed already during the 1st millennium AD. However, a more thorough investigation is needed.

Late Iron Age (9th–12th century AD) and medieval period

During the Late Iron Age, important economic and societal transformations changed the cultural landscape of the Eastern Baltic and led to the appearance of states in the 13th century AD. In total, macrobotanical remains have been reported from c. 50 Eastern Baltic sites dated to the 9th–14th century AD. However, in this section, we will present only a generalised, short overview of the crops cultivated during this period, concentrating more on the plants that were absent or poorly represented in earlier periods, i.e. *Fagopyrum esculentum, L. culinaris, Vicia sativa* (common vetch), *Camelina sativa, Cannabis sativa, L. usitatissimum* and *Papaver somniferum* (opium poppy) (Table 1).

Cereals

Cereals are represented by *S. cereale, H. vulgare, A. sativa, F. esculentum, T. aestivum, T. compactum, T. dicoccum, T. spelta, P. miliaceum* and possibly *S. italica.* Barley remained among the most numerous

Figure 4. AMS-dated macroremains of cultivated plants from Lithuanian hillforts presented in this chapter. Information about the dates is given in Table 3. Top row: *Vicia faba* (broad bean) from Kaukai; middle row: *Triticum compactum* (club wheat) from Apuolė; bottom row: *Lens culinaris* (lentil) from Aukštadvaris. Scale bar represents 1 mm.

crops. Moreover, rye became one of the most important crops in the region (Table 1). A similar situation is seen all over northern Europe, with rye cultivation increasing in the 8th–10th century and even more during the 11th century (Behre 1992). The third most numerous cereal, after barley and rye, seems to have been bread wheat. At the same time, the cultivation of hulled wheats decreased dramatically (Table 1).

Local cultivation of buckwheat started sometime during the 13th–14th century. Macroremains (Table 1) and pollen (Bliujienė et al. 2013; Stančikaitė et al. 2008) of buckwheat became widespread at the same time. Written sources mentions that by the end of the 14th century buckwheat was being grown in the fields around Riga (Sillasoo and Hiie 2007). From the 13th–14th century, buckwheat remains are present all over Europe (de Klerk et al. 2015). Thousands of buckwheat grains were found at the *Civitas Rutenica* quarter in Vilnius, dated to the end of the 14th to first half of the 15th century (Motuzaitė Matuzevičiūtė et al. 2017). It was inhabited by the earliest Orthodox Christians, who had arrived in Vilnius from Slavic cities.

Setaria italica has been reported from the 13th–14th century layers at the Aukštadvaris hillfort and at the site of Riga Old Town (Table 1). At a few other sites, seeds identified as *Setaria* sp. were found (Table 1).

In Riga, only a few remains of *S. italica* were present, and these were interpreted as 'cultivated or weed' (Banerjea et al. 2017). While the quantity of *S. italica* recovered from the Aukštadvaris hillfort is unknown (future material re-examination should answer this question), according to Latałowa et al. (2007), small amounts of this species should be interpreted as weed plants. Therefore, we can conclude that the current dataset does not allow us to speak of cultivation of *S. italica* in the Eastern Baltic in medieval times.

Pulses

In addition to *P. sativum* and *V. faba*, two new pulses, *Vicia sativa* and *Lens culinaris*, are mentioned (Table 1). *L. culinaris* was identified at four sites—two hillforts in Lithuania dated to 13th–14th century (Table 1) and two hillforts in Latvia dated to the 12th century and approximately to the 11th–13th century, respectively (Rasiņš and Tauriņa 1983). The above-mentioned chronology is based on typological dating, and not on radiocarbon dates. Therefore, we AMS-dated one grain of *L. culinaris* from the Aukštadvaris hillfort, which resulted in a date of cal AD 1310–1419 (Fig. 4, Table 3).

V. sativa is reported from six Lithuanian hillforts. In five cases, it was identified by the same person, and it was recorded either as '*V. sativa*' or as 'common vetch' (Table 1). In the sixth case, it was identified as '*Vicia sativa* L. (?)' (Lideikytė-Šopauskienė 1935). No remains of this plant are known from the other Eastern Baltic countries. Single seeds of *V. sativa* are reported from 14th century northern Poland; however, it was suggested that these seeds may have come from plants growing as weeds among other crops (Latałowa et al. 2007). While preparing this chapter, we re-examined some charred pulses from the mid-20th century excavations at Nemenčinė hillfort (currently stored at the National museum of Lithuania) previously identified as *V. sativa*. However, the only remains we were able to identify to the species level were of *V. faba* and *P. sativum*. Some other remains of pulses could only be identified as *Vicia*-type. Therefore, we believe that all macroremains from Lithuanian sites reported as *V. sativa* or 'common vetch' may have been mistakenly identified. A thorough examination of macrobotanical material stored at the National museum of Lithuania from excavations at the Aukštadvaris, Maišiagala and Nemenčinė hillforts may definitively confirm or reject these *V. sativa* identifications.

Oil and fibre plants and other cultivated plants

Oil and fibre plants are represented by *Camelina sativa, Cannabis sativa, L. usitatissimum* and *P. somniferum* (Table 1). The first three were already present in earlier periods; however, seeds of opium poppy definitely appear only from the 13th–14th century, as evidenced by data from Eastern Baltic towns and from the Maišiagala hillfort in Lithuania (Table 1). Seeds identified as *Papaver* sp. were found in Latvian hillforts dated to c. 11th–13th century based only on artefact typology, not on radiocarbon dating. While in Lithuania seeds of *P. somniferum* were reported from the Kaukai hillfort, dated to c. 5th–14th century (Table 1), unfortunately, only *V. faba* grains have survived in the National museum of Lithuania, which were AMS dated to cal AD 649–775 (Fig. 4, Table 3). Moreover, no contextual information about the *P. somniferum* seeds from Kaukai is known. Therefore, currently the appearance of *P. somniferum* can be identified only from the 13th–14th century.

From approximately the 13th–14th century, macroremains of horticultural products, such as *Brassica oleracea* (cabbage), *Foeniculum vulgare* (fennel), *Pastinaca sativa* (parsnip) and *Lactuca sativa* (lettuce), are present in Eastern Baltic archaeological contexts (Table 1). Vegetable gardens are mentioned in the written sources starting in the 14th century (Sillasoo and Hiie 2007). At the same time, exotic plants, such as *Ficus carica* (common fig), *Vitis vinifera* (common grape vine) and *Oryza sativa* (rice) (Table 1), were brought by merchants.

Discussion and conclusions

The development of plant cultivation in the Eastern Baltic is in part similar to and related to economic processes that occurred in other regions of northern Europe, but with local differences. Although the data are not equally distributed among the different chronological periods, general tendencies can be outlined. First, agriculture and crop cultivation in the region appeared much later than, for example, in southern Scandinavia or central Poland. Currently, the earliest secure evidence of cereal macroremains comes from a few grains of barley dated to approximately the 14th–12th century cal BCE. Established agriculture can be identified only from the 1st millennium BCE, when a broad spectrum of cultivated plants appears—*H. vulgare, T. dicoccum, T. spelta, P. sativum, V. faba, P. miliaceum* and *Camelina sativa*. Over the course of the 1st millennium AD, the importance of free-threshing wheats (*T. aestivum* and *T. compactum*) and *S. cereale* grew. By the beginning of the 2nd millennium AD, rye and bread wheat, together with barley, were the most widespread cereals in the region. The 13th–14th century brought a much broader diversity to macrobotanical assemblages, with the advent of *F. esculentum* and *L. culinaris* cultivation. Moreover, abundant remains of oil and fibre plants, such as *Cannabis sativa, Camelina sativa*, *L. usitatissimum* and *P. somniferum,* were found. At the same time, horticulture products spread and exotic fruits, such as *F. carica* and *V. vinifera,* were imported.

The Eastern Baltic region comprises a territory of more than 150 000 km²; people living in different parts of the region were subject to different cultural and natural influences. It is interesting to note that, in the northern part of the region, i.e. north of the river Daugava, *P. miliaceum* was absent until medieval times. Moreover, during the medieval period, it is considered to have been imported rather than grown locally (Sillasoo and Hiie 2007). Such a strict and clear northern boundary can be explained as a climatic limit; millets are warm-weather plants sensitive to late frosts (Murphy 2007). Moreover, the geographical distribution of the earliest sites where *S. cereale* and *T. aestivum* are the dominant species suggests that it may have spread from south to north.

The data and perceptions presented in this chapter are based on our current knowledge. Future investigations and new data analyses will further clarify these insights and possibly modify some of our propositions; larger datasets are needed to clarify the timing of the earliest farmer communities in the region and the date of the appearance of some crop species. However, in order to proceed, one has to first summarise any past research. We believe that this chapter will be a good basis for future investigations.

Acknowledgements

This research was funded by the European Social Fund under to the activity *Improvement of researchers' qualification by implementing world-class R&D projects* of Measure No. 09.3.3-LMT-K-712. We would like to thank the staff of the Archaeology Department of the National Museum of Lithuania and of Vytautas the Great War Museum, especially Dalia Ostrauskienė, Gediminas Petrauskas, Vytenis Podėnas, Gytis Grižas and Kęstutis Šeškevičius (1970–2018), for their help in finding botanical material from earlier excavations. Moreover, we thank the editors of this volume and two anonymous reviewers for their constructive and very useful comments on our chapter.

References

Aalto M (1997) The cultivated plants of finnish Iron Age. In: Кирпичников АН, Рябинин ЕА, Сакс АИ (eds) Славяне и финно-угры: археология, история, культура. С.-Петербург, pp 47–61

Akeret Ö (2005) Plant remains from a Bell Beaker site in Switzerland, and the beginnings of *Triticum spelta* (spelt) cultivation in Europe. Veg Hist Archaeobot 14:279–286

Antanaitis I, Ogrinc N (2000) Chemical analysis of bone: Stable isotope evidence of the diet of Neolithic and Bronze Age people in Lithuania. Istorija 45:3–12

Antanaitis-Jacobs I, Girininkas A (2002) Periodization and chronology of the Neolithic in Lithuania. Archaeologia Baltica 5:9–39

Antanaitis-Jacobs I, Kisielienė D, Stančikaitė M (2004) Archeobotanika Lietuvoje: Makrobotaniniai ir palinologiniai Tyrinėjimai. Lietuvos Archeologija 26:77–98

Banerjea RY, Badura M, Kalējs U et al. (2017) A multi-proxy, diachronic and spatial perspective on the urban activities within an indigenous community in medieval Riga, Latvia. Quaternary International 460:3–21

Baubonis Z, Zabiela G (2005a) Lietuvos piliakalniai: Atlasas vol 1. Leidykla Briedis, Vilnius

Baubonis Z, Zabiela G (2005b) Lietuvos piliakalniai: Atlasas vol 3. Leidykla Briedis, Vilnius

Behre K-E (1992) The history of rye cultivation in Europe. Veg Hist Archaeobot 1:141–156

Bliujienė A, Stančikaitė M, Kisielienė D et al. (2013) Skomantai hill-fort in western Lithuania: A case study on habitation site and environment. Archaeologia Baltica 17:101–135

Brown DA, Badura M (2017) Plant macrofossil and pollen analysis from the moat and outer bailey at Cēsis castle: Indicators of local environment and plant-use. In: Kalnins G (ed) Cēsu pils raksti, vol 1. Cēsu pils saglabāšanas fonds, pp 101–113

Czeczuga B, Kossacka W (1973) Nasiona roslin uprawnych i chwastow z osady jacwieskiej w Osowej, pow. Suwalki (III–V w.n.e.). Acta Baltico-Slavica 8:151–166

Daugnora L, Girininkas A (2004) Rytų Pabaltijo bendruomenių gyvensena XI–II tūkst. pr. Lietuvos veterinarijos akademija, Kaunas

Daugudis V (1962) Aukštadvario piliakalnio įtvirtinimai ir pastatai. Lietuvos TSR Mokslų Akademijos darbai, serija A 12:43–57

de Klerk P, Couwenberg J, Joosten H (2015) Pollen and macrofossils attributable to *Fagopyrum* in western Eurasia prior to the late medieval: An intercontinental mystery. Palaeogeogr Palaeocl 440:1–21

Gackowski J (2000) On the dating and cultural aspects of the West Baltic Barrow culture lake dwellings. In: Kola A, Sosnowska A (eds) Studies in the lake dwellings of the West Baltic Barrow culture. Uniwersitet Mikolaja Kopernika, Instytut Archeologii i etnologii, Torun, pp 9–63

Grikpėdis M, Motuzaitė Matuzevičiūtė G (2016) The beginnings of rye (*Secale cereale*) cultivation in the East Baltics. Veg Hist Archaeobot 25:601–610

Grikpėdis M, Motuzaite Matuzeviciute G (2017) A review of the earliest evidence of agriculture in Lithuania and the earliest direct AMS date on cereal. Eur J Archaeol 21(2):1–16. https://doi.org/10.1017/eaa.2017.36

Heydeck J (1909) Kultur- und Wohnstätten der Steinzeit in Ostpreussen. Sitzungsberichte der Altertumsgesellschaft Prussia 22:202–206

Jacomet S (2006) Identification of cereal remains from archaeological sites, 2nd edn. Archaeobotany Lab, IPAS, Basel University, Basel

Karczewski M (2012) On the road to the other world: Plants in the burial rites of Bogaczewo culture (Roman period, northeast Poland). Archaeologia Baltica 18:126–146

Karczewski M, Banaszuk P, Kupryjanowicz M, Mueller-Bieniek A (2009) Uwarunkowania środowiskowe mikroregionu osadniczego z okresu wpływów rzymskich nad północnym brzegiem dawnego jeziora Wąż w Krainie Wielkich Jezior Mazurskich. Domańska L, Środowisko—Człowiek—Cywilizacja 2:131–145

Kihno K, Hiie S (2008) Evidence of pollen and plant macro-remains from the sediments of suburban area of medieval Tartu. Est J Archaeol 12:30–50

Konsa M, Kivi K (2012) Village at the foot of the fort: Settlement sites I–III at Linnaaluste. Est J Archaeol 21:63–91

Kriiska A (2009) The beginning of farming in the Eastern Baltic area. In: Dolukhanov PM, Sarson GR, Shukurov AM (eds) The East European plain on the eve of agriculture. BAR Int. Ser., vol. 1964. Archaeopress, Oxford, pp 159–179

Kriiska A, Lavento M, Peets J (2005) New AMS dates of the Neolithic and Bronze Age ceramics in Estonia: Preliminary results and interpretations. Est J Archaeol 9:3–31

Kulikauskas P (1955) Kai kurie archeologiniai duomenys apie seniausiai Lietuvos TSR teritorijoje augintus javus. Lietuvos TSR Mokslų Akademijos darbai, serija A 1:75–85

Kulikauskas P (1958) Nemenčinės piliakalnis. Iš lietuvių kultūros istorijos 1:20-43

Kulikauskas P (1970) Kaukų ir Obelytės piliakalnių tyrinėjimai 1967–1969 metais. In: Tautavičius A, Daniliauskas A (eds) Archeologiniai ir etnografiniai tyrinėjimai Lietuvoje 1968–1969 metais. Lietuvos TSR Mokslų akademija, Vilnius, pp 12–22

Kulikauskas P (1982) Užnemunės piliakalniai I–XIII amžiuje. Mokslas, Vilnius

Kulikauskas P, Luchtanas A (1980) Archeologiniai tyrinėjimai Kernavėje 1979 metais. In: Tautavičius A (ed) Archeologiniai tyrinėjimai lietuvoje 1978–1979 metais. Lietuvos TSR mokslų akademijos Istorijos institutas, Vilnius, pp 35–38

Lang V (1996) Archaeological evidence for human settlement near the Tondi mire, northern Estonia. In: Hackens T, Hicks S, Lang V, Miller U, Saarse L (eds) Coastal Estonia: Recent Advances in Environmental and Cultural History (PACT 51), pp 69–83

Lang V (2007) The Bronze and Early Iron Ages in Estonia. Tartu University Press, Tartu

Lang V (2010) The Early Bronze Age in Estonia: Sites, finds, and the transition to farming. In: Wallin H (ed) Baltic prehistoric interactions and transformations, vol 5. Gotland University Press, Visby, pp 5–22

Latałowa M, Badura M, Jarosińska J, Święta-Musznicka J (2007) Useful plants in medieval and post-medieval archaeobotanical material from the Hanseatic towns of northern Poland (Kołobrzeg, Gdańsk and Elbląg). In: Karg S (ed) Medieval food traditions in northern Europe, vol 12. Publications from the National Museum Studies in Archaeology & History. National Museum of Denmark, Copenhagen

Lavento M (1998) Sisämaan vanhemman metallikauden väestö tutkimusongelmana. Muinaistutkija 4:44–55

Lideikytė-Šopauskienė A (1935) Javai iš Lietuvos piliakalnių. VDU Gamtos-matematikos fakulteto Darbai 9:131–149

Litynska-Zajac M (1997) Roslinnosc i gospodarka rolna w okresie rzymskim: Studium archeobotaniczne. Instytut Archeologii i Etnologii PAN, Krakow

Loze I (1997) Zemkopibas ieviešana Latvijas teritorija (Lubana ezera baseins). Arheologija un etnografija 19:25–41

Luchtanas A (2015) Maistas ir mityba Lietuvos geografinėje aplinkoje nuo ankstyviausių laikų iki XIV a. In: Luchtanas A (ed) Eksperimentinė archeologija, vol 1. Akademinė leidyba, Vilnius, pp 49–65

Matlakówna M (1926) Średniowieczne szczątki roślinne ze Żmudzi oraz niektóre zagadnienia pochodzenia zbóż. Acta Soc Bot Pol 3:196–231

Matlakówna M (1929) Dalsze badania nad zbożem średniowiecznem z Litwy. Acta Soc Bot Pol 6:370–384

Mellin-Wyczolkowska I (2007) Osada kultury bogaczewskiej w Starych Kiełbonkach (Ławnym Lasku), stan. III, gm. Piecki. In: Bitner-Wroblewska A, Brzezinski W (eds) Kultura bogaczewska w 20 lat pozniej. Materialy z konferencji, Warszawa, 26–27 marca 2003. Seminarium baltyjskie, vol 1. Państwowe Muzeum Archeologiczne, Warszawa, pp 437–446

Michelbertas M (1986) Senasis geležies amžius Lietuvoje. Mokslas, Vilnius

Motuzaitė Matuzevičiūtė G, Jonaitis R, Kaplūnaitė I (2017) Archaeobotanical investigations of a burnt-down storage shed in the Civitas Rutenica quarter: An insight into the daily lives of foreigners in Vilnius in the late fourteenth century–the first half of the fifteenth century. Lituanistica 63:219–233

Motuzaitė Matuzevičiūtė G, Rusteikytė A (2017) Paleomitybos tyrimai archeologijoje ir eksperimentinės archeologijos galimybės. In: Luchtanienė D (ed) Eksperimentinė archeologija, vol. 2. Akademinė leidyba, Vilnius, pp 57–80

Motuzaite Matuzeviciute G, Staff RA, Hunt HV, Liu X, Jones MK (2013) The early chronology of broomcorn millet (*Panicum miliaceum*) in Europe. Antiquity 83:1073–1085

Murphy DJ (2007) People, plants and genes: The story of crops and humanity. Oxford University Press, Oxford. https://doi.org/10.1093/acprof:oso/9780199207145.001.0001

Nesbitt M, Samuel D (1996) From staple crop to extinction? The archaeology and history of the hulled wheats. In: Padulosi S, Hammer K, Heller J (eds) Hulled wheats: Proceedings of the First International Workshop on Hulled Wheats, 21–22 July 1995, Castelvecchio Pascoli, Tuscany, Italy. Promoting the conservation and use of underutilized and neglected crops, vol 4. International Plant Genetic Resources Institute, Rome, pp 40–99

Oinonen M, Vasks A, Zarina G, Lavento M (2013) Stones, bones, and hillfort: Radiocarbon dating of Ķivutkalns bronze-working center. Radiocarbon 55:1252–1264

Piličiauskas G (2016) Coastal Lithuania during the Neolithic. In: Zabiela G, Baubonis Z, Marcinkevičiūtė E (eds) A hundred years of archaeological discoveries in Lithuania. Society of the Lithuanian Archaeology, Vilnius, pp 96–109

Piličiauskas G, Kisielienė D, Piličiauskienė G (2017) Deconstructing the concept of Subneolithic farming in the southeastern Baltic. Veg Hist Archaeobot 26:183–193

Podėnas V, Luchtanas A, Čivilytė A (2016) Narkūnų piliakalnių ir papėdės gyvenvietės keramika: Elgsenos atspindžiai. Lietuvos Archeologija 42:191–241

Polcyn M (1995) Pszenica orkisz (*Triticum spelta*) i jęczmien zwyczajny (*Hordeum vulgare*) z osady nawodnej ludnosci kultury kurhanow zachodniobaltyjskich w Moltajnach woj. Olzstyn (Stan. 1). Acta Universitatis Nicolai Copernici, Archeologia 24:53–71

Polcyn M (2000) Archaeobotanical finds from the West Baltic Barrow culture lake dwelling in Pieczarki (Great Masurian Lakeland): An attempt at the reconstruction of plant economy. In: Kola A, Sosnowska A (eds) Studies in lake dwellings of the West Baltic Barrow culture. Uniwersitet Mikolaja Kopernika, Instytut Archeologii i etnologii, Torun, pp 131–190

Pollmann B (2014) Environment and agriculture of the transitional period from the Late Bronze to Early Iron Age in the Eastern Baltic: An archaeobotanical case study of the lakeshore settlement Luokesa 1, Lithuania. Veg Hist Archaeobot 23:403–418

Rasinš A (1958) Manuscript: Report of macrobotanical analysis from Aukštadvaris hillfort. Library of Lithuanian Institute of History, F. 1, file No. 148

Rasiņš A, Tauriņa M (1983) Pārskats par Latvijas PSR arheoloģiskajos izrakumos konstatētajām kultūraugu un nezāļu sēklām. Arheoloģija un etnogrāfija 14:152–175

Reimer PJ, Bard E, Bayliss A et al. (2013) IntCal13 and Marine13 radiocarbon age calibration curves 0–50,000 years cal BP. Radiocarbon 55:1869–1887

Reiner H, Holzner W, Ebermann R (1995) The development of turnip-type and oilseed-type *Brassica rapa* crops from the wild-type in Europe: An overview of botanical, historical and linguistic facts. Rapeseed Today and Tomorrow 4:1066–1069

Rimantienė R (1960) Stoyanki kamenovo i bronzovovo vekov Samantonis (Стоянки каменного и бронзового веков Самантонис). Sovetskaya archeologia (Советская археология) 2:120

Rimantienė R (1992) The Neolithic of the Eastern Baltic. J World Prehist 6:97–143

Sillasoo Ü, Hiie S (2007) An archaeobotanical approach to investigating food of the Hanseatic period in Estonia. In: Karg S (ed) Medieval food traditions in northern Europe, vol 12. Studies in Archaeology & History. National Museums of Denmark, Copenhagen, pp 73–96

Šimkūnaitė E (1957) Manuscript: Report of macrobotanical analysis from Aukštadvaris hillfort. Library of Lithuanian Institute of History, F. 1, file No. 148

Šimkūnaitė E (1981) Manuscript: Report of macrobotanical analysis from Kernavė Mindaugo sosto hillfort. Library of Lithuanian Institute of History, F. 1, file No. 975

Stančikaitė M, Kabailienė M, Ostrauskas T, Guobytė R (2002) Environment and man around Lakes Dūba and Pelesa, SE Lithuania, during the Late Glacial and Holocene. Geol Q 46:391–410

Stančikaitė M, Kisielienė D, Strimaitienė A (2004) Vegetation response to the climatic and human impact changes during the Late Glacial and Holocene: Case study of the marginal area of Baltija Upland, NE Lithuania. Baltica 17:17–33

Stančikaitė M, Kisielienė D, Mažeika J, Blaževičius P (2008) Environmental conditions and human interference during the 6th and 13th–15th centuries AD at Vilnius Lower Castle, east Lithuania. Veg Hist Archaeobot 17:239–250

Stivrins N, Brown A, Reitalu T, Veski S, Heinsalu A, Banerjea RY, Elmi K (2015) Landscape change in central Latvia since the Iron Age: Multi-proxy analysis of the vegetation impact of conflict, colonization and economic expansion during the last 2,000 years. Veg Hist Archaeobot 24:377–391

Swederski W (1926) Chwasty z wykopalisk archeologicznych na Żmudzi i w Małopolsce. Acta Soc Bot Pol 3:242–254

Tvauri A, Vanhanen S (2016) The find of pre-Viking age charred grains from fort-settlement in Tartu. Est J Archaeol 20:33–53

van der Veen M (2007) Formation processes of desiccated and carbonized plant remains—the identification of routine practice. J Archaeol Sci 34:968–990

Vanhanen, S. (in press) Roman Iron Age and Migration period plant cultivation at Salo Isokylä, south-western Finland. In Vanhanen S, Lagerås P (eds) Archaeobotanical studies of past plant cultivation in northern Europe. Barkhuis

Vanhanen S, Koivisto S (2015) Pre-Roman Iron Age settlement continuity and cereal cultivation in coastal Finland as shown by multiproxy evidence at Bäljars 2 site in SW Finland. J Archaeol Sci Reports 1:38–52

Vasks A, Zarina G (2014) Ķivutkalna pilskalns un kapulauks: Jauni dati un jaunas problēmas. Latvijas Vēstures Institūta Žurnāls 92:3–36

Vengalis R (2008) Rytų Lietuvos keramika VIII–XII a. Lietuvos archeologija 33:41–70

Vikkula A, Seppälä S-L, Lempiäinen T (1994) The ancient field of Rapola. Fennoscandia Archaeologica 11:41–59

Volkaitė-Kulikauskienė R (1974) Nauji duomenys apie žemdirbystę ir gyvulininkystę Rytų Lietuvoje XIII–XV amžiais. Lietuvos TSR Mokslų Akademijos darbai, serija A 48:51–65

Volkaitė-Kulikauskienė R (1978) Žemdirbystė, gyvulininkystė ir medžioklė. In: Volkaitė-Kulikauskienė R (ed) Lietuvių materialinė kultūra IX–XIII amžiuje, vol 1. Mokslas, Vilnius, pp 48–72

Vuorela I, Lempiäinen T (1988) Archaeobotany of the site of the oldest cereal grain find in Finland. Annales Botanici Fennici 25:33–45

Wasylikowa K, Cârciumaru M, Hajnalová E, Hartyányi BP, Pashkevich GA, Yanushevich ZV (1991) East-Central Europe. In: Van Zeist W, Wasylikowa K, Behre K-E (eds) Progress in Old World palaeoethnobotany. A.A. Balkema, Rotterdam, pp 207–239

Zohary D, Hopf M, Weiss E (2013) Domestication of plants in the Old World: The origin and spread of domesticated plants in Southwest Asia, Europe, and the Mediterranean Basin. Oxford University Press, Oxford. https://doi.org/10.1093/acprof:oso bl/9780199549061.001.0001

Iron Age emmer and spelt: Where, when and why? A review of archaeobotanical evidence from southern Sweden, c. AD 1–600

Per Lagerås[1] & Mikael Larsson[2]

Author addresses:
[1] The Archaeologists, National Historical Museums, Sweden, per.lageras@arkeologerna.com
[2] Department of Archaeology and Ancient History, Lund University, Sweden,

E-mail: per.lageras@arkeologerna.com, mikael.larsson@ark.lu.se

Abstract

Hulled wheats were common crops in Sweden during the Neolithic and the Bronze Age but seem to have lost their importance from the Late Bronze Age onwards. There are, however, a few macrofossil assemblages from different sites in southern Sweden that indicate temporary growing of hulled wheats, particularly emmer, long after their general decline. The finds are dated to the period c. AD 1–600. In this chapter, we review these finds and their archaeobotanical and archaeological contexts. We also review the distribution of hulled wheats in Europe and discuss possible areas of influence. Two of the richest and youngest assemblages of hulled wheats in southern Sweden, dating to the Migration period or the early Vendel period, are from its eastern part and may indicate influences from the east, for instance the Baltic countries.

Introduction

The general outline of the long-term history of cereal cultivation in southern Sweden is relatively well known from published compilations of archaeobotanical data (Hjelmqvist 1955; Engelmark 1992; Gustafsson 1998; Viklund 1998; Grabowski 2011). These compilations show emmer (*Triticum dicoccum*) and naked barley (*Hordeum vulgare* var. *nudum*) as the principal crops of the Neolithic and the Early Bronze Age, with einkorn (*Triticum monococcum*) and eventually spelt (*Triticum spelta*) being minor crops. Changes to the arable production in the Middle Bronze Age, around 1000 BC (Table 1), included a decline in the cultivation of hulled wheats (emmer, spelt, einkorn) and naked barley, whereas hulled barley (*Hordeum vulgare* var. *vulgare*) became the principal crop. Hulled barley remained the dominant cereal throughout the Iron Age (and later), with bread wheat (*Triticum aestivum*) and oats (*Avena sativa*) as minor crops. Rye (*Secale cereale*) gained economic importance in the Late Iron Age, and from the Viking period and onwards it was a principal crop, together with barley.

According to the above compilations, the cultivation of hulled wheats started to diminish in the Early Bronze Age and more distinctly so in the Late Bronze Age, and by the Pre-Roman Iron Age and onwards, hulled wheats were no longer of economic importance. In this view of a general linear development and replacement of arable crops across time, local variations in cereal cultivation are not considered. The occurrence of grain-rich hulled wheat assemblages in Sweden after their general decline is of particular interest to this study.

Table 1. Archaeological periods mentioned in the text.

Neolithic			4000–1800 BC
Bronze Age	Early Bronze Age		1800–1000 BC
	Late Bronze Age		1000–500 BC
Early Iron Age	Pre-Roman Iron Age		500–1 BC
	Roman Iron Age	Early Roman Iron Age	AD 1–200
		Late Roman Iron Age	AD 200–400
Late Iron Age	Migration Period		AD 400–550
	Vendel Period		AD 550–800
	Viking Period		AD 800–1050

Table 2. Details on radiocarbon dating of hulled wheats mentioned in the text.

Site	Sample	Lab.no	C14 BP	Cal. 2s	Dated material
Lussabacken Norr	PM66109	Ua-44191	1592 ± 33	AD 400–550	*Triticum dicoccum*
Lussabacken Norr	PM32418	Ua-46004	1503 ± 33	AD 430–640	*Triticum dicoccum*
Lussabacken Norr	PM31031	Ua-44189	1441 ± 32	AD 560–655	*Triticum dicoccum*
Örja 9:1	PM231828	Ua-29967	1516 ± 31	AD 430–620	*Triticum dicoccum/spelta*
St Uppåkra 12:110	PM12961	Ua-29587	1836 ± 30	AD 80–250	*Triticum dicoccum*

In this chapter, we present both new and previously published archaeobotanical finds from southern Sweden that, contrary to the general view, indicate that cultivation of hulled wheats took place long after the general decline of these crops. The finds consist of hulled wheat assemblages dated c. AD 1–600, including the Roman Iron Age, the Migration period and the early Vendel period, the latter referring to the time between the Migration period and the Viking period in Sweden (Tables 1 and 2). We present the finds and their archaeological contexts and, based on a brief European overview of published finds of emmer and spelt, we discuss possible far-reaching contacts, trade and migration. We also give a short background on the identification of hulled wheats and on their past and present distribution and ecology.

On the identification of emmer and spelt

The wheat genus (*Triticum*) comprises several species, and these can be divided into hulled and naked wheats. The hulled wheat species identified in Sweden—einkorn (*Triticum monococcum*), emmer (*T. dicoccum*) and spelt (*T. spelta*)—have inedible, tough glumes that tightly enclose the grain and need to be processed after threshing to remove them (Hillman 1981). Threshing breaks the spikes into separate spikelets, and succeeding parching and pounding break up each spikelet into spikelet forks, rachis segments, glume bases, grains and light fragments (glumes, lemmas and paleas). Free-threshing forms of wheat—in Sweden particularly bread wheat (*Triticum aestivum*)—have light glumes and tougher rachis that do not disarticulate at maturity. When threshed, pounding is not required, as the spike immediately breaks up into rachis segments, grains and light chaff.

In charred archaeobotanical assemblages, hulled wheats may easily be distinguished from free-threshing (naked) wheats, but the distinction between the different hulled wheat species is more difficult. Einkorn may be relatively easily identified, but it is rare in Swedish contexts and will not be further discussed in this chapter. The other two species—emmer and spelt—are difficult to distinguish from each other if only the grains (caryopses) are present. Although their average shape may differ, there is a considerable variation and overlap in morphological criteria, which makes distinction based on grains only tentative. If chaff is present in the samples—in particular spikelet forks and glume bases—distinction may be easier (cf. Jacomet 2006).

Ecological characteristics

Emmer and spelt can both be cultivated on poor soils and are resistant to a range of different pests and diseases (Percival 1921; Körber-Grohne 1987: 28; van der Veen 1992: 146). Ecological differences have, however, been noted between the species. Emmer tends to thrive on humus-rich loamy soil and does not grow as well on loose sandy soils or peaty or wet clayey soils (de Hingh 2000). Spelt tolerates a wider range of soil conditions, including heavier clay soils, where it out-yields emmer (Campbell 1997). The ecology of emmer and spelt contrasts to that of naked wheats, which are often more demanding but have the advantage of producing on average a greater yield (Percival 1921).

Grain maturation is an ecological factor that may have been disadvantageous for the growing of hulled wheats, particularly in times of climatic deterioration. The crop cycle, from sowing to harvest, is longer among the hulled wheats compared with barley, bread wheat, rye and oats. Hulled barley, which became the principal crop after the decline of hulled wheats, is considered the most adaptable of all cereals in terms of ecological requirements (Mills 2006: 97–98). It ripens under cool conditions and is characterised by a short crop cycle (Mills 2006: 136–37). In addition, rye—increasingly cultivated in the Late Iron Age—grows in a wide variety of climatic and soil conditions but grows best in light loams or sandy soils in cool, temperate zones (Clark 2007). Like barley, it grows rapidly, and its grain mature early.

A brief European overview

Emmer was the principal wheat crop in much of continental Europe in the Neolithic and the Bronze Age, but major changes in arable production during the 1st millennium BC saw it starting to be replaced by spelt. Around the same time, naked barley was gradually replaced by hulled barley, and several new crops, such as rye and oats, were introduced (Jones 1981).

Although emmer was no longer a crop of major importance after these changes, studies reveal local variation in arable practice and show that emmer was still cultivated in some regions in continental western and north-western Europe in the first half of the 1st millennium AD.

The north-western expansion of the Roman Empire into parts of the Netherlands and Britain in the late 1st century BC and the 1st century AD affected arable production among local agrarian communities (Groot and Kooistra 2009). Before the arrival of the Romans, farmers cultivated a range of mixed cereal crops and focused on self-sufficiency. The presence of the Roman army in these regions triggered an increased demand for agricultural products, which led to surplus production by large farms growing only one or two crops. The Romans seem to have favoured spelt, which together with hulled barley became the dominant crop (Kooistra 1996).

Regional variation occurred, however, in the west and north of England, where several settlements continued to grow emmer well into the Roman period (Jones 1981). The emmer production in this region is regarded as indicative of small-scale subsistence farming, whereas spelt assemblages are believed to represent more large-scale arable production (van der Veen 1992).

In the Dutch river area north of the loess, around the border area of the Roman *limes*, cereal cultivation was intensified during the Roman period, but farmers maintained a mixed farming regime. Hulled barley and emmer were the main cereals, and oats and millet were less dominant (Kooistra 1996: 63–66; Kooistra 2008; Bakels 2009; Groot et al. 2009; Groot and Kooistra 2009). One of the largest Roman villas in the Netherlands, Villa Voerendaal, in the northern loess region, produced emmer, spelt and bread wheat, but later, during the heyday of the Roman period, almost exclusively spelt was grown (Kooistra 1996: 146–176). A Roman ship dated to the end of the 2nd century AD, which sunk with a cargo mainly of emmer, was recovered close to a Roman fortification along the Lower Rhine, which formed the northern *limes* of the Roman Empire. The grain cargo is believed to have originated from the loess area of Belgium and provides an example not only of the importance of emmer still at that time, but also of the cereal as a monocrop (Pals and Hakbijl 1992).

After the fall of the Roman Empire in Britain, free-threshing cereals, such as bread wheat and rye, gained importance. Localised finds of emmer from the latter half of the 1st millennium have, however, been reported from the Thames valley in quantities to indicate emmer as a cereal crop of economic importance (Pelling and Robinson 2000), pointing to a reintroduction of this crop to the region after the end of the Roman period. It has been suggested that groups of Saxons, who colonised the Thames valley in the mid-1st millennium AD, traditionally used emmer and may have brought seed corn with them (Knörzer 1991: 197f).

In the Netherlands after the fall of the Roman Empire, rye became the principal crop, but cultivation also included bread wheat, hulled barley and oats (Bakels 1991). The growing of spelt wheat in the southern Netherlands and in the Rhineland continued, however, though mostly as a minor crop. Spelt no longer occurred in late historic times (Bakels 2009: 211f). Finds of emmer in the second half of the 1st millennium AD in this area are largely restricted to small finds and likely do not represent a true crop plant, but, rather, a crop plant that had turned into a weed (Bakels 2005).

The major changes in wheat production from emmer to spelt seen in Britain and parts of the Netherlands during the Roman period did not occur in northern Germany, Denmark and other areas outside *limes*. In these areas, emmer remained more important than spelt, although barley was more important than wheat (van Zeist 1968, 1981; Mills 2006: 38; Robinson et al. 2009).

In Poland, both emmer and spelt had already lost their prominent role by the Roman period, appearing only as an admixture to other cereals (Wasylikowa et al. 1991; Litynska-Zajac 1997). They were replaced by mostly barley and millet. After the Roman period, barley decreased and rye and bread wheat became the dominant crops.

In the Eastern Baltic region (Lithuania, Latvia and Estonia), hulled wheats continued to be cultivated until the 6th century AD. Among the hulled wheats, emmer seems to have been more common than spelt (Grikpėdis and Motuzaitė Matuzevičiūtė, this volume). Barley was otherwise the principal crop during the Iron Age.

In Norway, emmer declined in the 1st century AD, while cereal production became gradually dominated by hulled barley, with some cultivation of spelt and bread wheat (Prøsch-Danielsen and Soltvedt 2011). In Denmark, cultivation of emmer and spelt declined by the end of the Bronze Age, but some cultivation of emmer took place until the Roman Iron Age (Robinson 2003; Robinson et al. 2009).

Present-day distribution

Despite their general decline long ago, emmer and spelt have survived today as relic crops in small areas. Emmer is grown in mountainous regions in Anatolia, Iran, Caucasia and India, and has been maintained as a traditional crop in central regions of Italy and Spain as well as in the Carpathian Mountains (Perrin et al. 1995; Mills 2006). Spelt is grown in isolated clusters in Germany, Switzerland and northern France and throughout south-eastern Europe, and it is maintained as a traditional crop in northern Spain and in the Alpine areas of Italy and Austria (Rösch et al. 1992;

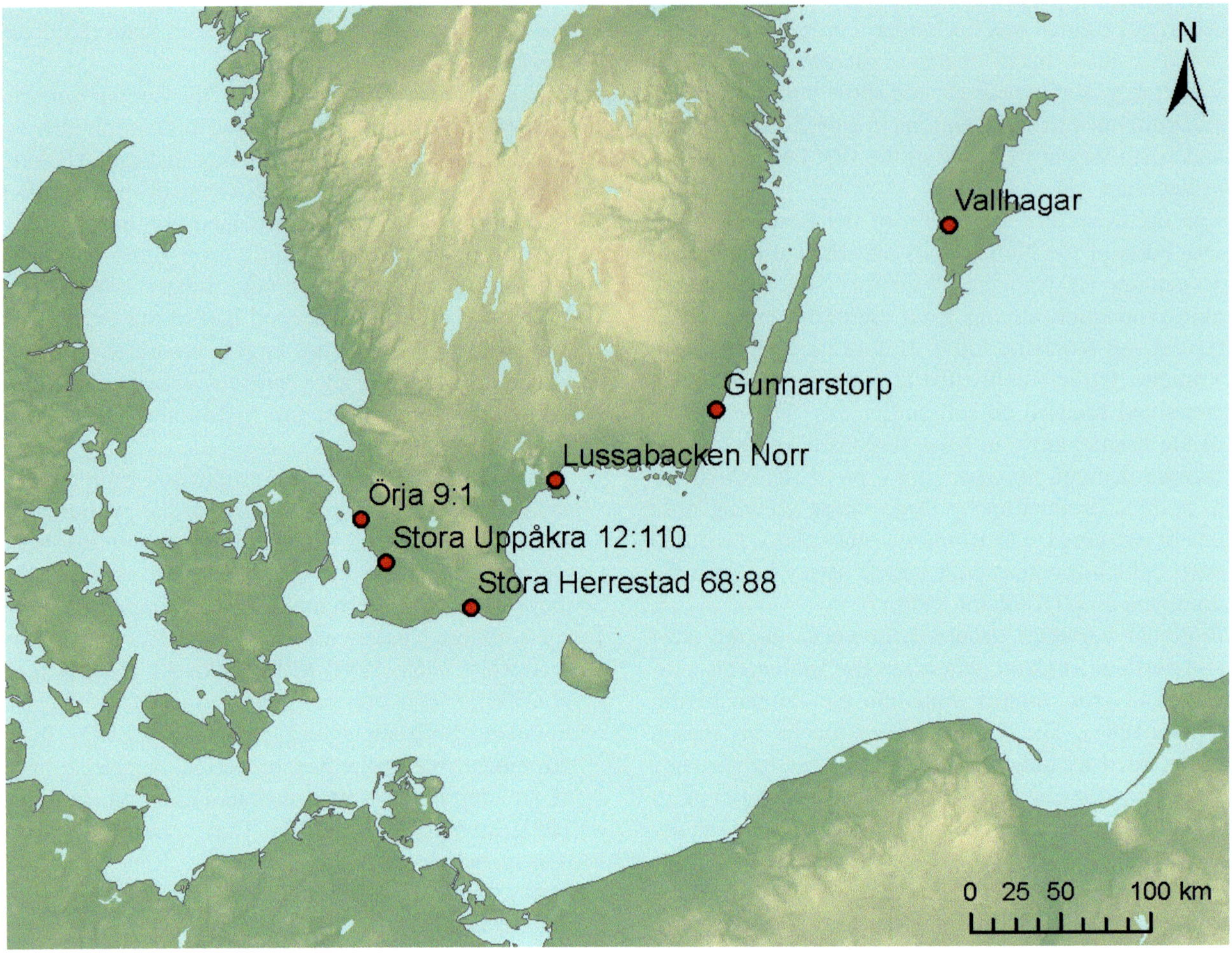

Figure 1. Map of southern Sweden showing sites mentioned in the text.

Perrin 1995; Mills 2006). On the island of Gotland, in Sweden, small-scale cultivation of emmer and spelt is known to have existed as late as the early 20th century (Hjelmqvist 1966).

Although marginal, the hulled wheats have had an upswing recently due to a growing interest in their nutritional content and their suitability for organic farming, and because of increased attention to local traditions and agricultural diversity (Giuliani et al. 2009; Zaharieva et al. 2010).

Iron Age finds of hulled wheats in Sweden

Cultivation of hulled wheats was widespread in southern Sweden during the Neolithic (Table 1) and reached as far north as the Mälaren region of central Sweden (Larsson and Broström 2011; Sørensen and Karg 2014). However, the importance of these crops in relation to other cereals diminished during the Bronze Age, and by the Pre-Roman Iron Age, they were uncommon crops (Engelmark 1992). The reason behind the general decline may have been a climatic change towards cooler conditions (Mayewski et al. 2004) and, possibly, higher precipitation (Mellström et al. 2005).

In Iron Age grain assemblages, hulled wheats are usually absent or represented only by single grains. Such small and scattered occurrences are frequently interpreted as impurities, meaning remains of weeds growing among other crops. However, in some grain assemblages they are quantitatively well represented, both in absolute and relative numbers, which indicates that they were indeed cultivated and harvested by some farmers long after their general decline in arable production.

Below, we present grain-rich assemblages of hulled wheats from six Iron Age settlements and their archaeological contexts (Fig. 1). In the presented cases, the number of hulled wheats ranges from 61 to a reconstructed estimate of c.10 000 grains and relative numbers of 45–98 per cent (Table 3). They are presented here in chronological order.

Stora Herrestad 68:88

Stora Herrestad 68:88, north-east of Ystad, in southernmost Scania, was excavated in 1995 prior to commercial development. Longhouses and other settlement remains dating from the Neolithic to the Roman Iron Age were investigated (Andersson et al. 1999).

Table 3. Summary of the hulled-wheat assemblages. PRIA=Pre-Roman Iron Age, RIA=Roman Iron Age, MP=Migration Period, EVP=Early Vendel Period.

Site	Sample	Context	Absolute numbers	Relative numbers	Chaff present	Dating
Örja 9:1	PM236549, 236550	Posthole	89	45%		MP–EVP
Vallhagar	29/16	Floor layer	c. 10 000	c. 98%	X	MP
Lussabacken Norr	PM66109	Posthole	219	98%	X	MP
Gunnarstorp	A467	Posthole	186	63%		RIA
St Uppåkra 12:110	PM12961	Posthole	75	70%	X	RIA
St Herrestad 68:88	A1750	Oven	61	72%		PRIA–RIA

One sample from a pit interpreted as an oven (A1750) was relatively rich in grains of hulled wheats (Regnell 1999), of which 8 were identified to spelt, 4 to emmer and 49 to emmer/spelt. The sample contained also barley (24), unspecified cereals (73) and some seeds of common weeds (including *Chenopodium album*, *Persicaria maculosa*, *Solanum nigrum* and Poaceae indet.). It is not stated in the report if any chaff was present. Based on three radiocarbon dates, the oven was dated to the period around the transition between the Pre-Roman and the Roman Iron Age.

Stora Uppåkra 12:110

The site of Stora Uppåkra 12:110 was excavated in 2010, prior to commercial development (Becker 2012). It is situated in Hjärup, south of Lund, in south-western Scania, only 1 km west of the Iron Age regional centre of Uppåkra. An Iron Age settlement was investigated, and several phases of two farms, covering the period 300 BC–AD 300, were identified (Becker 2012: 5).

One sample (PM12961) from a posthole (AS8536) in one of the longhouses (House 10) at the site was relatively rich in emmer (Larsson 2015: 53 and Appendix 2, Table 15). The posthole belonged to a roof-bearing post in the easternmost part of the house. The sample contained 67 grains of emmer and 8 of emmer/spelt. It also contained 28 glume bases of emmer. Other cereals in the sample were oats (20), naked barley (8), hulled barley (3), bread wheat (1) and unspecified cereals (78). In addition, there were seeds of flax (*Linum usitatissimum*) (7) and some common weeds (*Fallopia convolvulus*, *Chenopodium*, *Scleranthus annuus*, Poaceae indet.). One of the emmer grains from the sample was ^{14}C dated to AD 80–250, while a grain of hulled barley from another posthole (AS3315) in the same house was ^{14}C dated to AD 230–390 (Becker 2012: 60) (Table 2). Based on these dates, House 10 was dated to c. AD 150–300, which falls within the Roman Iron Age (Becker 2012: 28). The house was interpreted as a dwelling house. No other samples from the house or elsewhere on the site contained more than a few emmer grains.

Gunnarstorp

The site of Gunnarstorp, situated 30 km south of Kalmar, in the province of Småland, was excavated in 1998, prior to commercial development (Gustafsson and Olsson 2001). Three houses and some other remains were dated to the Late Roman Iron Age, and samples were subjected to archaeobotanical analysis (Engelmark and Olofsson 2001).

One sample in particular was rich in emmer/spelt. The sample came from a posthole (A467) in the westernmost part of House 2. The sample contained 186 grains of emmer/spelt, 103 of rye, 5 of barley and 100 of unspecified cereals. In addition, there were 120 seeds of dodder (*Camelina sativa*), interpreted as cultivated, and a few seeds of weeds (*Chenopodium album*, *Persicaria lapathifolia*, *Galium spurium*, *Spergula arvensis*). Emmer/spelt in small numbers was found also in four other postholes of the same house, as well as a few grains in some postholes belonging to the other two houses (Engelmark and Olofsson 2001).

House 2 was interpreted as a burnt-down dwelling house. Based on several ^{14}C dates from charcoal, it was dated to the Late Roman Iron Age (Gustafsson and Olsson 2001: 17, 21).

Lussabacken Norr

The site of Lussabacken Norr, in western Blekinge, was investigated in 2011, prior to commercial development (Björk et al. 2014). Remains from several different prehistoric periods were documented on the site, and among them was an Iron Age settlement with longhouses, pithouses and a boathouse. The buildings were dated to the Migration period and the early Vendel period (Åstrand 2014: 255ff). Several samples from the site were subjected to archaeobotanical analysis (Broström and Lagerås in prep.).

One of the samples (PM66109) from a posthole (A36042) of a roof-bearing post, situated close to a hearth in the eastern part of House 3, contained a large amount of hulled wheats. A total of 219 grains were identified to emmer/spelt. Although distinguishing with certainty between emmer and spelt may not

Figure 2. Charred grains of emmer from Lussabacken Norr, dated to the Migration period (photo: Per Lagerås).

be possible from the grains alone, their shape indicates that the majority (and possibly all of them) belonged to emmer (Fig. 2). There are also 140 fragments of unspecified cereals that probably also belong to the same species. In addition to grains, there are 20 spikelet forks and glume bases of emmer. The sample also contains grain of einkorn (3) and rye (1). Weeds are absent. An emmer grain from the sample was ^{14}C dated to AD 400–550, i.e. the Migration period (Table 1).

Other samples from the settlement contain only a few cereals. Among them are some bread wheat, oat, and single grains of barley, rye and emmer. There was also one seed of millet (*Panicum miliaceum*). Two of the emmer grains from these samples were ^{14}C dated: one to AD 430–640, covering most of the Migration period and the early Vendel period, and one to AD 560–655, covering the early Vendel period (Tables 1 and 2).

Vallhagar

The Iron Age settlement of Vallhagar, on the island of Gotland, was excavated in 1946–1950 (Stenberger 1955a). Twenty-four houses were documented and interpreted as belonging to five or six separate but contemporaneously occupied farms. The first houses on the site were established by the end of the Roman Iron Age, but the main period of occupation was during the Migration period. The settlement burnt down (probably during a raid) and was deserted some time during the Migration period (Stenberger 1955c: 1161ff). Since radiocarbon analysis was not available at the time, the dating was based on the typology of pottery and other archaeological finds (Stenberger 1955b: 1065ff).

The archaeobotanical analysis was restricted to five samples, all taken from concentrations of charred cereal grains from two houses (Helbæk 1955). Three samples were taken from a floor layer in house 16, and one of them (Sample 29), contained a large amount (225 cc) of hulled wheats. The grains were not counted by the original investigator but we estimate that 225 cc would approximately be equivalent to 10 000 grains. Most of the grains were identified as emmer, with only a small inclusion of einkorn and probable spelt grains. According to Helbæk, further identification was not possible (Helbæk 1955: 661). In addition to the grains, 1100 spikelet forks and glume fragments of hulled wheats were also noted in the sample. They were mostly of emmer and einkorn, while 37 were identified as spelt. In addition to these remains from cultivated plants, the samples yielded

240 seeds identified as wild oat (*Avena fatua*) based on their small size, as well as a few seeds of other weeds (*Fallopia convolvulus*, *Bromus secalinus*).

The other two samples from House 16 contained similarly large amounts of grains, identified as barley (1110 cc) and rye (326 cc), respectively. The barley concentration contained some weeds, but the rye sample was almost free of weeds. Hence, there were three concentrations of cereal grains in House 16—one with a mixture of different hulled wheats, one with barley and one with rye. The concentration of barley grain was recovered from a charred but still identifiable wooden box (Lundström 1955: 207).

The assortment of other archaeological finds from House 16 all consisted of everyday objects. Two short iron scythes point, for example, to harvest activities. The house was interpreted to have been an ordinary dwelling house, hastily abandoned during the raid (Lundström 1955: 207).

The other two analysed samples from Vallhagar contained no remains of hulled wheats, only grains of barley and rye.

Örja 9:1

The village of Örja is situated close to the city of Landskrona, in western Scania. An abandoned part of the village—Örja 9:1—was excavated in 2010, prior to commercial development (Schmidt Sabo 2013). The medieval and post-medieval settlement remains were the focus of the excavation. Prehistoric remains were, however, also investigated, including 25 Iron Age longhouses representing the period from the Late Roman Iron Age to the late Vendel period/early Viking period (Carlie 2013: 82ff).

Grains of hulled wheats identified as emmer/spelt were found in two samples (PM236549, PM236550) from an Iron Age house—House 48 (Lagerås 2013). The building itself was small, and it was interpreted as having been part of the biggest farm on the settlement (Carlie 2013: 93). The samples were recovered from two adjacent postholes (A205091, A214644) of roof-bearing posts in the western part of the building. Together, they contained 89 grains of emmer/spelt. Other cereals included barley (103 of hulled or unspecified barley and 2 of naked), oats (5) and unspecified cereal grains (c. 300), as well as ordinary weeds (*Persicaria*, *Fallopia convolvulus*, *Chenopodium*, *Raphanus raphanistrum*, *Vica/Lathyrus*, *Galium*, Poaceae indet.). The samples contained no threshing remains. The high number of unspecified cereal grains is due to their poor preservation and high degree of fragmentation (Lagerås 2013: 220). A barley grain from one of the samples (PM236549) was ^{14}C dated to AD 430–610, covering most of the Migration period and the early Vendel period (Table 1).

Single grains of emmer/spelt were found in two other houses on the site—House 44 and House 50. A grain of emmer/spelt from House 44 was ^{14}C dated to AD 430–620, which corresponds to the dating from House 48. The sample from House 50 was not ^{14}C dated, but based on typology, the building was dated to the Migration period. The rich archaeobotanical material from later periods on the site contained no hulled wheats.

Primary conclusions on the Swedish finds

The primary conclusions from the above examples are as follows (see also Table 3):

The finds rich in hulled wheats presented above cover the period c. AD 1–600. One of the samples is dated to the Pre-Roman/Roman Iron Age, two to the Roman Iron Age, two to the Migration period and one to the Migration/early Vendel period. There are several earlier finds, predating the Pre-Roman Iron Age, but to our knowledge there are no later finds, i.e. from the late Vendel period or later, of rich hulled wheat assemblages from Sweden. The finds are all from sites interpreted as ordinary agrarian settlements. Within the sites, the hulled wheat assemblages belong to household contexts that appear to reflect everyday activities. One sample was recovered from an oven, the others from structural house remains. Of the latter, one derived from a floor layer (Vallhagar) and four from postholes of roof-bearing posts. Together, these contexts indicate handling and storing of the crops indoors. The houses were interpreted as dwelling houses, except for the small building at Örja 9:1, tentatively interpreted as a storage space next to a dwelling house. At Örja 9:1, the rich hulled wheat assemblages were found in two adjacent postholes, and in the other houses they were found in single postholes. This suggests that grain of hulled wheat was stored on the floor, probably in wooden containers. This is true also for Vallhagar, where the hulled wheats were concentrated in one spot in the house.

There are both clean and mixed assemblages. In two contexts, at Vallhagar and Lussabacken Norr, the samples rich in hulled wheats contained almost no other cereals. In the other contexts, the hulled wheats were mixed with other cereals, particularly barley, but to a smaller degree also oats, bread wheat and rye. There were also seeds of flax and gold of pleasure. Regarding weeds, Vallhagar and Lussbacken Norr are also noteworthy. Their grain assemblages were almost clean, whereas the grain assemblages from the other sites were relatively rich in seeds of ordinary arable weeds and ruderals.

Emmer appears to have been more common than spelt. Because of the difficulties in distinguishing grain of emmer from that of spelt, it is not possible to be specific about the proportional relationship between the two species. The hulled wheats were identified only to emmer/spelt at Gunnarstorp and Örja 9:1, and at Stora Herrestad 68:88, the majority was referred to as emmer/spelt. At Stora Uppåkra 12:110, all hulled wheats were identified to emmer. At Lussabacken Norr, they were tentatively referred to as emmer/spelt, but it was noted that they probably belong to emmer. Most of the hulled wheats at Vallhagar were identified as emmer. Hence, the overall pattern of the grain from all the assemblages is that emmer was more common than spelt. Only a few grains from two of the sites (Stora Herrestad 68:88 and Vallhagar) were specifically referred to as spelt. The chaff, which is a more distinctive characteristic for species identification, further supports this proportional composition of cereal types. Chaff (spikelet forks and glume bases) was reported from three of the sites. At Vallhagar some of it was identified to spelt but the clear majority to emmer; at Stora Uppåkra 12:110 and Lussbacken Norr all of it was identified to emmer.

Discussion

This review shows that hulled wheats, particularly emmer, have been handled and probably cultivated at some farms in southern Sweden during the Iron Age, long after they lost their importance in arable production. They are exceptions to the general range of crops cultivated during this time, which consisted of hulled barley as the principal cereal, with bread wheat, oats and, later, rye being secondary crops (e.g. Engelmark 1992; Grabowski 2011). Although few in number, the hulled wheat assemblages indicate that Iron Age agriculture may have been more diverse than appears from regional compilations and overviews. In fact, their rare occurrence makes them particularly interesting, and raises question about possible far-reaching contacts and influences.

It may be difficult from archaeobotanical data to distinguish cereals that have been intentionally cultivated from those that may only have been contaminations among other crops. In all the examples presented in this review, however, the representation of hulled wheats is sufficient, both in relative and absolute numbers, to indicate harvested crops rather than contamination. This interpretation is supported by the archaeological contexts. All the finds are from household contexts on agrarian settlements, and the grains have been stored indoors, commonly in dwelling houses. Hence, the charred grain of hulled wheats appears in similar ways on archaeological sites as that of other cereals routinely interpreted as harvested crops.

The hulled wheat assemblages may reflect trade or locally harvested crops or both. Trading of grain has been discussed regarding finds of rye from some Iron Age sites, such as Vallhagar, in Sweden (Helbæk 1955), and Fyrkat, in Denmark (Helbæk 1977), but also regarding the introduction of crops to Britain and grain trade by the Romans (Helbæk 1964). Non-local weed species found among grain assemblages have been considered an indication of long-distance grain trade. In Sweden, a more local redistribution of crops has been suggested between the Iron Age regional centre of Uppåkra and surrounding settlements (Larsson 2015: 64–66). It is important to note that in these cases grain appears to have been transported from agricultural production areas to non-rural sites, such as military campsites or centres of religion, trade and artisanship. Similarly, in Continental Europe, the trading of grain is primarily expected to have supported cities, hillforts, *oppida* and other non-rural sites (Dreslerová et al. 2017: 522). Agriculture-based settlements, on the other hand, are believed to have relied mainly on self-sufficiency (Pedersen and Widgren 2011), and cereal remains from such sites are for good reasons routinely interpreted as locally produced. The most straightforward interpretation, therefore, is that the hulled wheat assemblages from agrarian sites in southern Sweden reflect local cultivation of these crops. This can be further supported by the absence of non-local weed species in the hulled wheat assemblages. The possibility of imported grain cannot, however, be entirely ruled out.

We conclude that some farms in southern Sweden were growing hulled wheats (particularly emmer) long after these crops had lost their importance as a staple crop. This small-scale cultivation may reflect a continuous tradition that had been maintained since the Neolithic and the Bronze Age. For the province of Scania, however, this is less likely. Scania has been subjected to rather comprehensive archaeobotanical research for decades, and still has produced only three of the six assemblages presented above. A continuous tradition of producing hulled wheats would most probably have resulted in more finds. Hence, the infrequent occurrence of hulled wheats in Iron Age Scania points to occasional cultivation by single farms over limited periods of time.

The other three assemblages are from the eastern parts of southern Sweden, namely, the province of Blekinge, the easternmost part of Småland and the island of Gotland (Fig. 1). From these areas, contrary to Scania, the total amount of archaeobotanical data is small. The fact that three of the six examples come from the eastern part of southern Sweden, despite the low archaeobotanical activity in that area, suggests a geographical variation, whereby hulled wheats were more common in the east than in Scania.

The question of continuity versus periodic cultivation of hulled wheats is important, because if these crops were only temporarily grown, access to grain for sowing may have been a problem. The first sowing of a new crop by any farm would have depended on traded seed corn, and in regions dominated by other crops, it would have had to be imported from far away. If emmer and possibly other hulled wheats were locally grown in southern Sweden during the Iron Age, they therefore seem to imply some far-reaching contacts and influences. All the sites are situated within relatively short distances from the coast, which may have facilitated such contacts (Fig. 1). Movement of people may also have been involved. The re-introduction of emmer to some areas of southern England after the fall of the Roman Empire was connected to the arrival and settling of new people (in this case Saxons) with other farming traditions (Pelling and Robinson 2000). It seems plausible that emmer-growing people may have settled in southern Sweden, but such an interpretation would need to be supported by archaeological data.

The scattered distribution of emmer cultivation in Europe at the time makes it difficult to pinpoint from which areas such possible migration or influences would have come. There seem not to have been any large regions dominated by emmer; instead, small and scattered areas of emmer traditions were maintained in many parts of Europe. One region where emmer was still cultivated after the Roman Iron Age is the Eastern Baltic region (Grikpėdis and Motuzaitė Matuzevičiūtė, this volume). It is interesting to note that the Swedish emmer finds seem to be relatively more common in the eastern part of southern Sweden, the area closest to the Eastern Baltic region.

Two finds in south-eastern Sweden are particularly worth mentioning, Vallhagar, on the island of Gotland, and Lussabacken Norr, in Blekinge. They are the two largest emmer assemblages (although Vallhagar is by far the largest), they are both dated to the Migration period (AD 400–550), and they are both situated near the coast of the Baltic. For these two sites, and possibly for some of the others, the emmer assemblages may indicate contacts with emmer-growing communities on the eastern side of the Baltic. However, this interpretation also needs to be supported by archaeological data. It is interesting to note that small-scale cultivation of the hulled wheats emmer, spelt and einkorn was practiced on the island of Gotland in the early 20th century (Hjelmqvist 1966; Leino 2017: 133). But whether it was the final phase of a long, continuous tradition or a late re-introduction is not known.

References

Andersson T (1999) Boplatslämningar från stenålder–äldre järnålder (Arkeologisk slutundersökning). Skåne, St Herrestads socken, Herrestad 68:88 m.fl., RAÄ 60, 1995. Riksantikvarieämbetet UV Syd Rapport 1999:8

Åstrand J (2014) Järnåldersboplatsen. In: Björk T, Henriksson M, Larsson F, Persson C, Rudebeck E, Åstrand J (eds) Lussabacken norr: Boplatslämningar från mesolitikum, neolitikum, bronsålder och järnålder (Särskild arkeologisk undersökning 2011). Ysane socken, Sölvesborgs kommun i Blekinge län. Blekinge Museum Rapport 2014:9

Bakels CC (1991) Western continental Europe. In: van Zeist W, Wasylikowa K, Behre K-E (eds) Progress in Old World palaeoethnobotany. AA Balkema, Rotterdam

Bakels CC (2005) Crop produced in the southern Netherland and northern France during the early medieval period: A comparison. Veg Hist Archaeobot 14:394–399

Becker N (2012) Stora Uppåkra 12:110 (tidigare 12:2): Äldre järnåldersbebyggelse i centralplatsens omland (Riksantikvarieämbetet UV rapport) 2012:51

Björk T, Henriksson M, Larsson F, Persson C, Rudebeck E, Åstrand J (2014) Lussabacken norr: Boplatslämningar från mesolitikum, neolitikum, bronsålder och järnålder (Särskild arkeologisk undersökning 2011). Ysane socken, Sölvesborgs kommun i Blekinge län. Blekinge Museum Rapport 2014:9

Campbell KG (1997) Spelt: Agronomy, genetics, and breeding. Plant Breeding Review 115:187–213

Carlie A (2010) Järnåldersby och urnegravfält. In: Schmidt Sabo K (ed) *Örja* 1:9. Riksantikvarieämbetet UV Rapport 2013:68

Clark A (ed) (2007) Managing cover crops profitably, 3rd edn. Sustainable Agriculture Research and Education (SARE) Handbook Series Book 9. Sustainable Agriculture Network, Beltsville

de Hingh AE (2000) Food production and food procurement in the Bronze Age and Early Iron Age (2000–500 BC). Archaeological Studies 7. Faculty of Archaeology, Leiden University, Leiden

Dreslerová D, Kocár P, Chuman T, Pokorná A (2017) Cultivation with deliberation: Cereals and their growing conditions in prehistory. Veg Hist Archaeobot 26: 513–526

Engelmark R (1992) A review of the farming economy in south Scania based on botanical evidence. In: Larsson L, Callmer J, Stjernquist B (eds) The archaeology of the cultural landscape: Fieldwork and research in a south Swedish rural region. Acta Archaeologica Lundensia Ser in 4°, 19:369–375

Engelmark R, Olofsson J (2001) Miljöarkeologisk undersökning av delsträcka 1, delområde 8, RAÄ 298, Söderåkra sn, Kalmar län (Rapport). Miljöarkeologiska laboratoriet, Institutionen för arkeologi och samiska studier, Umeå University, Umeå

Giuliani A, Karagöz A, Zencirci N (2009) Emmer (*Triticum dicoccon*) production and market potential in marginal mountainous areas of Turkey. Mt Res Dev 29:220–229

Grabowski R (2011) Changes in cereal cultivation during the Iron Age in southern Sweden: A compilation and interpretation of the archaeobotanical material. Veg Hist Archaeobot 20:479–494

Groot M, Heeren S, Kooistra LI, Vos WK (2009) Surplus production for the market? The agrarian economy in the non-villa landscapes of Germania Inferior. J Roman Archaeol 22:231–252

Groot M, Kooistra LI (2009) Land use and the agrarian economy in the Roman Dutch River area. Internet Archaeol 27

Gustafsson M, Olsson R (2001) Gunnarstorp: En boplats från äldre järnålder. Fornlämning 298, Söderåkra socken, Småland. E22-projektet. Kalmar Läns Museum, Rapport 2001:1. Kalmar Läns Museum, Kalmar

Gustafsson S (1998) The farming economy in south and central Sweden during the Bronze Age: A study based on carbonised botanical evidence. Current Swedish Archaeol 6:63–71

Helbæk H (1955) The botany of the Vallhagar Iron Age field. In: Stenberger M (ed) Vallhagar: A Migration period settlement on Gotland/Sweden, Part II. Munksgaard, Copenhagen

Helbæk H (1964) The Isca grain: A Roman plant introduction in Britain. New Phytol 63:158–164

Helbæk H (1977) The Fyrkat grain: A geographical and chronological study of rye. Nordiske Fortidsminder Serie B 3:1–41

Hillman GC (1981) Reconstructing crop husbandry practices from charred remains of crops. In: Mercer R (ed) Farming practices in British prehistory. Edinburgh University Press, Edinburgh, pp 123–162

Hjelmqvist H (1955) Die *älteste* Geschichte der Kulturpflanzen in Schweden. Opera Botanica 1:3

Hjelmqvist H (1966) Some notes on the old wheat species of Gotland. Hereditas 56:382–394

Jacomet S (2006) Identification of cereal remains from archaeological sites, 2nd edn. Archaeobotany Lab, IPAS, Basel University, Basel

Jones MK (1981) The development of crop husbandry. In: Jones M, Dimbleby G (eds) The environment of man: The Iron Age to the Anglo-Saxon period. BAR British Ser. 87. Archaeopress, Oxford, pp 95–127

Knörzer KH (1991) Deutschland nördlich der Donau. In: van Zeist W, Wasylikowa K, Behre KE (eds) Progress in Old World Palaeoethnobotany. AA Bakema, Rotterdam, pp 189–206

Kooistra LI (1996) Borderland farming: Possibilities and limitations of farming in the Roman period and the Early Middle Ages between Rhine and Meuse. Rijksdienst voor het Oudheidkundig Bodemonderzoek, Amersfoort

Kooistra LI (2008) Vegetation history and agriculture in the cover-sand area west of Breda (province of Noord-Brabant, The Netherlands). Veg Hist Archaeobot 17:113–125

Körber-Grohne U (1987) Nutzpflanzen in Deutschland. Theiss Verlag, Stuttgart

Lagerås P (2013). Den agrara ekonomin och växtutnyttjandet i Örja: Analyser av växtmakrosfossil och pollen. In: Schmidt Sabo K (ed) *Örja* 1:9. Riksantikvarieämbetet UV Rapport 2013:68.

Larsson L, Broström S-G (2011) Meeting for transformation: A locality for ritual activities during the Early Neolithic Funnel Beaker culture in central Sweden. Current Swedish Archaeol 19:183–201

Larsson M (2015) Agrarian plant economy at Uppåkra and the surrounding area: Archaeobotanical studies of an Iron Age regional center. Acta Archaeologica Lundensia, Series in 4° 33: 1–88 –

Leino MW (2017) Spannmål: Svenska lantsorter. Nordiska Museets Förlag, Stockholm

Lundström P (1955) The Vallhagar buildings: Building 16. In: Stenberger M (ed) Vallhagar: A Migration period settlement on Gotland/Sweden, Part I. Munksgaard, Copenhagen

Litynska-Zajac M (1997) Roslinnose i gospodarka rolna w okresie rzymskim. Instytut Archeologii Etnologii PAN: 258–262

Mayewski PA, Rohling EE, Stager JC, Karlén W, Maasch KA, Meeker LD, Meyerson EA, Gasse F, van Kreveld S, Holmgren K, Lee-Thorp J, Rosqvist G, Rack F, Staubwasser M, Schneider RR, Steig EJ (2004) Holocene climate variability. Quaternary Res 62:243–255

Mellström A, van der Putten N, Muscheler R, de Jong R, Björck S (2015) A shift towards wetter and windier conditions in southern Sweden around the prominent solar minimum 2750 cal BP. J Quaternary Sci 30:235–244

Mills T (2006) A study of European cereal frequency change during the Iron Age and Roman periods, Volume 1. Dissertation, University of Sheffield

Pals JP, Hakbijl T (1992) Weed and insect infestation of a grain cargo in a ship at the Roman fort of Laurium in Woerden (Province of Zuid-Holland). Rev Palaeobot Palyno 73:287–300

Pedersen EA, Widgren M (2011) Agriculture in Sweden 800 BC–AD 1000. In: Myrdal J, Morell M (eds) The agrarian history of Sweden 4000 BC–AD 2000. Nordic Academic Press, Stockholm, pp 46–71

Pelling R, Robinson M (2000) Saxon emmer wheat from the upper and middle Thames valley, England. Environ Archaeol 5:117–119

Percival I (1921) The wheat plant. Duckworth, London

Perrin P, Laghetti G, D'Antuono LF, Al Ajlouni M, Kanbertay M, Szabó AT, Hammer K (1995) Ecogeographical distribution of hulled wheat species. In: Padulosi S, Hammer K, Heller J (eds) Hulled wheats (Proceedings of the First International Workshop on Hulled Wheats, 21–22 July 1995) International Plant Genetic Resources Institute, Rome, pp 100–127

Prøsch-Danielsen L, Soltvedt EC (2011) From saddle to rotary hand querns in south-western Norway and the corresponding crop plant assemblages. Acta Archaeologica 82:129–162

Regnell M (1999) Appendix. In: Andersson T (ed) Boplatslämningar från stenålder–äldre järnålder: Arkeologisk slutundersökning. Skåne, St Herrestads socken, Herrestad 68:88 m.fl., RAÄ 60, 1995. Riksantikvarieämbetet UV Syd Rapport 1999:8. Riksantikvarieämbetet, Stockholm

Regnell M (2002) Skånska järnåldersskördar: Växtmakrofossilanalyser och odlingshistoriska iakttagelser från tolv boplatser. In: Carlie A (ed) Skånska regioner: Tusen år av kultur och samhälle i förändring. Arkeologiska Undersökningar, Skrifter 40. Riksantikvarieämbetet, Stockholm

Robinson DE (2003) Neolithic and Bronze Age agriculture in southern Scandinavia: Recent archaeobotanical evidence from Denmark. Environ Archaeol 8:145–165

Robinson DE, Hambro Mikkelsen P, Malmros C (2009) Agerbrug, driftsformer og planteressourcer i jernalder og vikingetid (500 f.Kr.–1100 e.Kr.). In: Odgaard B, Rydén Römer J (eds) Danske landbrugslandskaber gennem 2000 år: Fra digevoldinger til støtteordninger. Aarhus Universitetsforlag, Aarhus, pp 117–142

Rösch M, Jacomet S, Karg S (1992) The history of cereals in the region of the former Duchy of Swabia (Herzogtum Schwaben) from the Roman to the post-medieval period: Results of archaeobotanical research. Veg Hist Archaeobot 1:193–231

Schmidt Sabo K (ed) (2013) *Örja* 1:9. Riksantikvarieämbetet UV Rapport 2013:68. Riksantikvarieämbetet, Stockholm

Stenberger M (ed) (1955a). Vallhagar: A Migration period settlement on Gotland/Sweden, Parts I–II. Munksgaard, Copenhagen

Stenberger M (1955b) The finds and the dating of the Vallhagar settlement. In: Stenberger M (ed) Vallhagar: A Migration period settlement on Gotland/Sweden, Part II. Munksgaard, Copenhagen

Stenberger M (1955c) The reasons for the abandonment of Vallhagar. In: Stenberger M (ed) Vallhagar: A Migration period settlement on Gotland/Sweden, Part II. Munksgaard, Copenhagen

Sørensen L, Karg S (2014) The expansion of agrarian societies towards the north: New evidence for agriculture during the Mesolithic/Neolithic transition in southern Scandinavia. J Archaeol Sci 51:98–114 van der Veen M (1992) Crop husbandry regimes: An archaeobotanical study of farming in northern England 1000 BC–AD 500. Sheffield Archaeological Monographs 3. JR Collis Publications, Sheffield

Viklund K (1996) Makrofossilanalys, Raä 35, Landa sn, Halland. Miljöarkeologiska Laboratoriet, Arkeologiska Institutionen, Umeå University, Umeå

Viklund K (1997) Elestorp, Tjärby sn, Halland: Arkeobotanisk och markkemisk analys av jordprover från stolphål. Miljöarkeologiska Laboratoriet, Arkeologiska Institutionen, Umeå University, Umeå

Viklund K (1998) Cereals, weeds and crop processing in Iron Age Sweden. Archaeol and Environment 14. Umeå University, Umeå

Wasylikowa K, Carciumaru M, Hajnalova E, Hartyanyi BP, Pashkevich A, Yanushevich ZV (1991) East-Central Europe. In: van Zeist W, Wasylikowa K, Behre KE (eds) Progress in Old World palaeoethnobotany. AA Bakema, Rotterdam, pp 207–240

Zaharieva M, Ayana NG, Al Hakimi A, Misra SC, Monneveux P (2010) Cultivated emmer wheat (*Triticum dicoccon* Schrank), an old crop with a promising future: A review. Genet Resour Crop Ev 57:927–962

van Zeist W (1968) Prehistoric and early historic food plants in the Netherlands. Palaeohistoria 14:41–173

van Zeist W (1981) Plant remains from the Iron Age Noordbarge, Province of Drenthe, The Netherlands. Palaeohistoria 23:169–193